T0137109

CISM International Centre for Mechanical Sciences

Courses and Lectures

Volume 595

For more than 40 years the book series edited by CISM, "International Centre for Mechanical Sciences: Courses and Lectures", has presented groundbreaking developments in mechanics and computational engineering methods. It covers such fields as solid and fluid mechanics, mechanics of materials, micro- and nanomechanics, biomechanics, and mechatronics. The papers are written by international authorities in the field. The books are at graduate level but may include some introductory material.

More information about this series at http://www.springer.com/series/76

Michael Le Bars · Daniel Lecoanet
Editors

Fluid Mechanics of Planets and Stars

Editors
Michael Le Bars
IRPHE (UMR 7342)
CNRS, Aix Marseille University
Centrale Marseille
Marseille, France

Daniel Lecoanet
Department of Astrophysical Sciences
Princeton University
Princeton, NJ, USA

ISSN 0254-1971 ISSN 2309-3706 (electronic)
CISM International Centre for Mechanical Sciences
ISBN 978-3-030-22076-1 ISBN 978-3-030-22074-7 (eBook)
https://doi.org/10.1007/978-3-030-22074-7

This Springer imprint is published by the registered company Springer Nature Switzerland AG
The registered company address is: Gewerbestrasse 11, 6330 Cham, Switzerland

Preface

The course “Fluid mechanics of planets and stars” was held at the International Centre for Mechanical Sciences in Udine, Italy, from April 16 to 20, 2018. It was part of the research project FLUDYCO, supported by the European Research Council (ERC) under the European Union’s Horizon 2020 research and innovation program.

The scientific focus of this course was the dynamics of planetary and stellar fluid layers, including atmospheres, oceans, iron cores, convective and radiative zones in stars, etc. Our first motivation for organizing this school came from the following ascertainment: this scientific domain is by its essence interdisciplinary and multi-method. But while much effort has been devoted to solving open questions within the various communities of Mechanics, Applied Mathematics, Engineering, Physics, Planetary and Earth Sciences, Astrophysics, and while much progress has been made within each enclosed domain using theoretical, numerical, and experimental approaches, cross-fertilizations have remained marginal. The objective of this CISM School was to go beyond this state, by providing participants with a global introduction and an up-to-date overview of relevant studies, fully addressing the wide range of involved disciplines and methods.

44 participants attended the 35 lectures given by 6 lecturers, chosen so as to cover the widest possible range of skills and knowledge in fundamental mechanics as well as geo- and astrophysical applications. Professor Gordon Ogilvie from University of Cambridge (UK) was in charge of the theme “Waves in fluids and in stellar interiors”. Dr. Daniel Lecoanet from Princeton University (USA) focused on the dynamics and interactions of convective and radiative zones in stars. Professor Bruce Sutherland from University of Alberta (Canada) provided an overview of instabilities in atmospheres and oceans. Dr. Michael Le Bars from CNRS (France) reviewed numerous instabilities in planetary interiors. Dr. Renaud Deguen from University Claude Bernard (France) discussed various aspects of the fluid mechanics of planetary cores. And finally, Dr. Benjamin Favier from CNRS (France) offered a large overview of various aspects of turbulences. All lectures were stimulating, of the top scientific level, and entertaining, while simultaneously highlighting the many

connections between different fields and communities. The six chapters of this book summarize this intense, but scientifically enlightening week.

Before starting, let us thank all the people from the International Centre for Mechanical Sciences, and especially its highly qualified and sympathetic secretariat, who allowed us to focus on science and made this week highly enjoyable for all participants.

Marseille, France Michael Le Bars
Princeton, USA Daniel Lecoanet

Contents

Chapter 1
Internal Waves and Tides in Stars and Giant Planets

Gordon I. Ogilvie

Abstract Internal waves play an important role in tidal dissipation in stars and giant planets. This chapter provides a pedagogical introduction to the study of astrophysical tides, with an emphasis on the contributions of inertial waves and internal gravity waves.

Introduction to Internal Waves

Internal waves are those restored by Coriolis or buoyancy forces in rotating or stably stratified fluids. In stars and giant planets, internal waves can propagate at frequencies that are much lower than those of acoustic or surface gravity waves and are usually more suitable for excitation by tidal forcing when the body has a close orbital companion. I begin this chapter with an exploration of some of the basic properties of internal waves, using the simplest possible models.

Plane Inertial Waves

Consider an unbounded, inviscid, incompressible fluid that is rotating uniformly with angular velocity $\boldsymbol{\Omega}$. In the rotating frame, arbitrary velocity perturbations $\boldsymbol{u}$ to this basic state satisfy the equation of motion and incompressibility condition

$$\frac{D\boldsymbol{u}}{Dt} + 2\boldsymbol{\Omega} \times \boldsymbol{u} = -\nabla q, \qquad \nabla \cdot \boldsymbol{u} = 0,$$

where $D/Dt = \partial/\partial t + \boldsymbol{u} \cdot \nabla$ is the Lagrangian time derivative and q is the pressure perturbation divided by the density. Initially neglecting the nonlinear term $\boldsymbol{u} \cdot \nabla \boldsymbol{u}$,

G. I. Ogilvie (✉)
DAMTP, Cambridge, UK
e-mail: gio10@cam.ac.uk

M. Le Bars and D. Lecoanet (eds.), *Fluid Mechanics of Planets and Stars*,
CISM International Centre for Mechanical Sciences 595,
https://doi.org/10.1007/978-3-030-22074-7_1

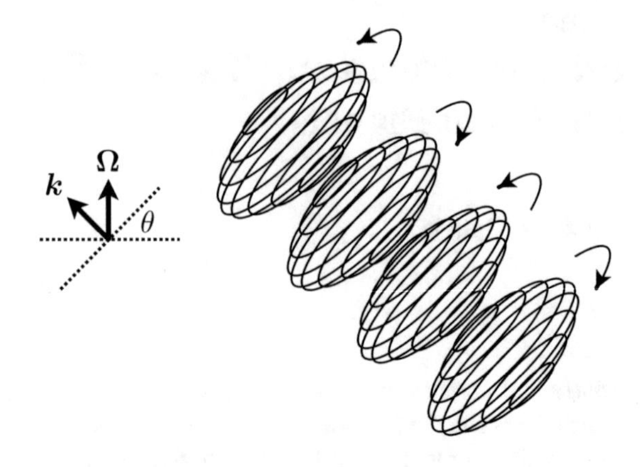

Fig. 1.1 Illustration of the fluid motion in a plane inertial wave

let us seek plane-wave solutions with wavevector $\boldsymbol{k}$ and angular frequency ω:

$$\boldsymbol{u} = \mathrm{Re}\left[\tilde{\boldsymbol{u}}\, e^{i(\boldsymbol{k}\cdot\boldsymbol{x}-\omega t)}\right], \qquad q = \mathrm{Re}\left[\tilde{q}\, e^{i(\boldsymbol{k}\cdot\boldsymbol{x}-\omega t)}\right].$$

Choosing axes such that $\boldsymbol{\Omega} = \Omega\,\boldsymbol{e}_z$, we obtain the algebraic equations

$$\begin{aligned} -i\omega\tilde{u}_x - 2\Omega\tilde{u}_y &= -ik_x\tilde{q}, \\ -i\omega\tilde{u}_y + 2\Omega\tilde{u}_x &= -ik_y\tilde{q}, \\ -i\omega\tilde{u}_z &= -ik_z\tilde{q}, \\ i\boldsymbol{k}\cdot\tilde{\boldsymbol{u}} &= 0, \end{aligned}$$

for which a non-zero solution exists if the dispersion relation

$$\omega^2 = 4\Omega^2\left(\frac{k_z^2}{k_x^2+k_y^2+k_z^2}\right)$$

is satisfied. This can be written in the form $\omega = \pm 2\Omega\cos\theta$, where θ is the angle between $\boldsymbol{k}$ and $\boldsymbol{\Omega}$.

These *plane inertial waves* (Fig. 1.1) are in fact exact solutions of the nonlinear equations (although a superposition of plane waves is not); the $\boldsymbol{u}\cdot\nabla\boldsymbol{u}$ term vanishes because $\boldsymbol{u} \perp \boldsymbol{k}$.

Plane Internal Gravity Waves

Now consider a stratified atmosphere in a uniform gravitational field $-g\,\boldsymbol{e}_z$. In the Boussinesq approximation (which is valid for highly subsonic motions on scales that are small compared to the scale height of the atmosphere), an ideal fluid satisfies the equations

$$\frac{D\boldsymbol{u}}{Dt} = -\nabla Q + B\,\boldsymbol{e}_z, \qquad \frac{DB}{Dt} = 0, \qquad \nabla\cdot\boldsymbol{u} = 0,$$

where Q is a modified pressure and

$$B = g\left(\frac{\rho_0 - \rho}{\rho_0}\right)$$

is a buoyancy variable, proportional to the difference between a constant reference density ρ_0 and the actual density ρ of the fluid.

An equilibrium atmosphere is a solution depending only on z, in which $\boldsymbol{u} = \boldsymbol{0}$, $Q = Q(z)$ and $B = B(z)$, with $dQ/dz = B$. Arbitrary perturbations $\boldsymbol{u}$, q and b to this basic state satisfy the nonlinear equations

$$\frac{D\boldsymbol{u}}{Dt} = -\nabla q + b\,\boldsymbol{e}_z, \qquad \frac{Db}{Dt} + N^2 u_z = 0, \qquad \nabla\cdot\boldsymbol{u} = 0,$$

where $N^2 = dB/dz$ is the square of the buoyancy frequency.

In the case of uniform stable stratification, N^2 is a positive constant. Initially neglecting the nonlinear terms $\boldsymbol{u}\cdot\nabla\boldsymbol{u}$ and $\boldsymbol{u}\cdot\nabla b$, let us seek plane-wave solutions

$$\boldsymbol{u} = \mathrm{Re}\left[\tilde{\boldsymbol{u}}\, e^{i(\boldsymbol{k}\cdot\boldsymbol{x} - \omega t)}\right],$$

etc., leading to the algebraic equations

$$-i\omega\tilde{u}_x = -ik_x\tilde{q}, \qquad -i\omega\tilde{u}_y = -ik_y\tilde{q}, \qquad -i\omega\tilde{u}_z = -ik_z\tilde{q} + \tilde{b},$$
$$-i\omega\tilde{b} + N^2\tilde{u}_z = 0, \qquad i\boldsymbol{k}\cdot\tilde{\boldsymbol{u}} = 0.$$

A non-zero solution exists if the dispersion relation

$$\omega^2 = N^2\left(\frac{k_x^2 + k_y^2}{k_x^2 + k_y^2 + k_z^2}\right)$$

is satisfied. This can be written in the form $\omega = \pm N\sin\theta$, where θ is the angle between $\boldsymbol{k}$ and $\boldsymbol{g}$.

These *plane internal gravity waves* are also exact solutions of the nonlinear equations; the $\boldsymbol{u}\cdot\nabla\boldsymbol{u}$ and $\boldsymbol{u}\cdot\nabla b$ terms vanish because $\boldsymbol{u}\perp\boldsymbol{k}$.

Properties of Internal Waves

There is a close similarity between the dispersion relations of inertial and internal gravity waves. These internal waves have properties that are opposite to those of acoustic or electromagnetic waves, being strongly anisotropic and dispersive. Their frequency is independent of the magnitude $k = |\boldsymbol{k}|$ of the wavevector and depends only on its direction $\hat{\boldsymbol{k}} = \boldsymbol{k}/k$. This means that the group velocity $\partial\omega/\partial\boldsymbol{k}$ is perpendicular to $\boldsymbol{k}$ and is proportional to the wavelength $2\pi/k$. The frequency is also bounded (by 2Ω or N, respectively), resulting in a dense or continuous spectrum when suitable boundary conditions are imposed.

Linear inertial waves satisfy the differential equation

$$\frac{\partial^2}{\partial t^2}\nabla^2 q + 4\Omega^2 \frac{\partial^2 q}{\partial z^2} = 0.$$

If the time-dependence $e^{-i\omega t}$ is assumed, this reduces to *Poincaré's equation*,

$$\omega^2 \nabla^2 q = 4\Omega^2 \frac{\partial^2 q}{\partial z^2}. \tag{1.1}$$

In frequency range $-2\Omega < \omega < 2\Omega$, this equation is hyperbolic in the *spatial* coordinates. Its characteristic curves or surfaces are inclined at a constant angle θ to the plane perpendicular to the rotation axis, where $\omega = \pm 2\Omega \cos\theta$, and coincide with the rays determined from the dispersion relation and group velocity. However, when we seek modal solutions in a contained fluid, the boundary conditions are specified on a closed surface, which is generally unsuitable for a hyperbolic equation because of the way information is propagated along the characteristics from one part on the boundary to another. The problem is generally ill-posed and smooth modal solutions may not exist; similar considerations apply to internal gravity waves.

Internal Wave Beams

An internal wave beam can be formed from a superposition of waves with the same ω and $\hat{\boldsymbol{k}}$, but different k. Consider inertial waves:

$$\frac{D\boldsymbol{u}}{Dt} + 2\boldsymbol{\Omega} \times \boldsymbol{u} = -\nabla q, \qquad \nabla \cdot \boldsymbol{u} = 0.$$

For waves of frequency $\omega = 2\Omega \cos\theta$, the beam is inclined at an angle θ to the horizontal. Introduce coordinates parallel (ξ) and perpendicular (η) to the beam in the xz plane, with $\boldsymbol{\Omega} = \Omega\, \boldsymbol{e}_z$ (Fig. 1.2):

$$\xi = x\cos\theta + z\sin\theta, \qquad \eta = x\sin\theta - z\cos\theta.$$

In these coordinates, we can find solutions with $u_\eta = 0$ that are independent of ξ and y, oscillating at frequency ω and satisfying

$$\frac{\partial u_\xi}{\partial t} - \omega u_y = 0, \qquad \frac{\partial u_y}{\partial t} + \omega u_\xi = 0, \qquad -\omega u_y \tan\theta = -\frac{\partial q}{\partial \eta}.$$

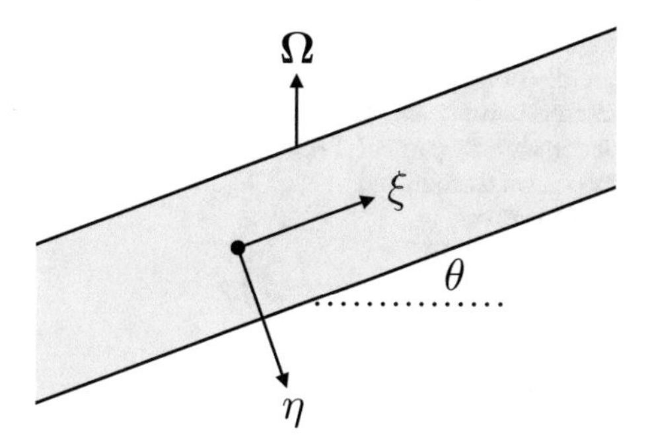

Fig. 1.2 Coordinates parallel and perpendicular to an internal wave beam

Any complex beam profile $u_\xi = \mathrm{Re}\left[U(\eta)\, e^{-i\omega t}\right]$ allows an exact nonlinear solution.

When a small viscosity ν is included, a monochromatic beam spreads as it propagates, and is accompanied by a small transverse velocity u_η. It is described approximately by

$$(u_\xi, u_\eta, u_y, q) = \mathrm{Re}\left[\left(\frac{\partial\Psi}{\partial\eta}, -\frac{\partial\Psi}{\partial\xi}, -i\frac{\partial\Psi}{\partial\eta}, -i\omega\Psi\tan\theta\right)e^{-i\omega t}\right],$$

where $\Psi(\xi, \eta)$ is a streamfunction that varies more rapidly with η than with ξ. Viscous spreading of the beam along its length is described by the equation

$$\frac{\partial\Psi}{\partial\xi} = i\lambda\frac{\partial^3\Psi}{\partial\eta^3}, \qquad \lambda = \frac{\nu}{\omega\tan\theta},$$

which can be derived by an asymptotic expansion of the solution in the limit of small viscosity. If $\lambda > 0$, then waves propagating in the $+\xi$ direction have negative transverse wavenumbers $k_\eta < 0$ and are attenuated in the $+\xi$ direction. As $\xi \to +\infty$, a generic beam tends towards a similarity solution proportional to

$$\int_{-\infty}^{0} e^{\lambda k^3\xi} e^{ik\eta}\, dk = (\lambda\xi)^{-1/3} f(\tilde{\eta}),$$

where

$$f(\tilde{\eta}) = \int_0^\infty e^{-\tilde{k}^3} e^{-i\tilde{k}\tilde{\eta}}\, d\tilde{k}, \qquad \tilde{k} = -(\lambda\xi)^{1/3}k, \qquad \tilde{\eta} = \eta(\lambda\xi)^{-1/3}$$

is a complex function (Moore and Saffman 1969) describing the transverse structure of the spreading beam in a dimensionless similarity variable (Fig. 1.3). The width of the beam is proportional to $\xi^{1/3}$, where ξ is measured along the beam from its (virtual) source. This type of structure is commonly seen in problems in which internal waves are generated by periodic forcing, as described below in the section ‘Forced Internal Waves’.

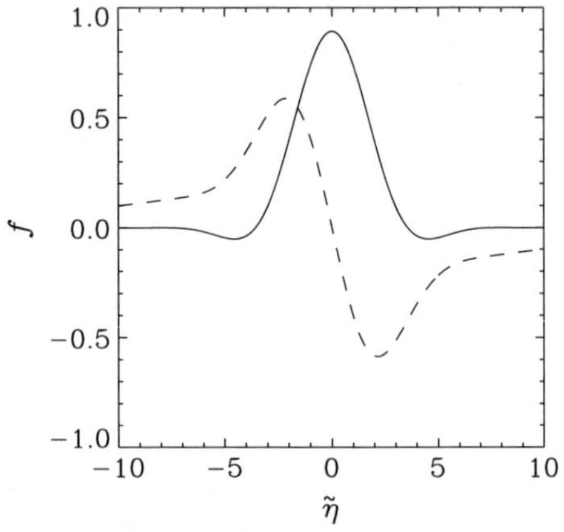

Fig. 1.3 Complex wave profile across a spreading internal wave beam. The real and imaginary parts of $f(\tilde{\eta})$ are shown as solid and dashed curves

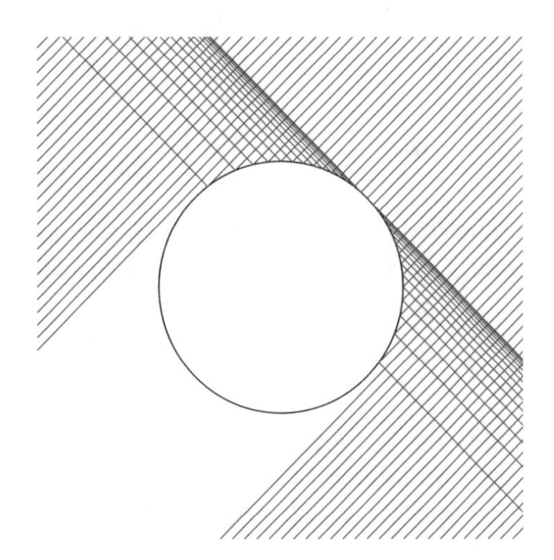

Fig. 1.4 Formation of a singularity at the critical latitude through reflection of inertial waves from a sphere. Rays incident from the top right are blue and rays reflected from the sphere are red

Reflections and Singularities

When a beam of internal waves reflects from a boundary, preservation of the frequency and therefore of the angle between $\boldsymbol{k}$ and $\boldsymbol{\Omega}$ (or $\boldsymbol{g}$) means that the reflection is generally non-specular and leads to focusing or defocusing of the beam. Reflection of inertial waves from a sphere creates a singularity at the *critical latitude* (at which the rays are tangent to the boundary) through this focusing effect (Fig. 1.4).

In a closed container in which the boundaries are not all parallel or perpendicular to $\boldsymbol{\Omega}$ (or $\boldsymbol{g}$), internal waves are generically focused into stable limit cycles known as *wave attractors* (Maas and Lam 1995). A simple example is a square container that is tilted with respect to the axis of rotation (or gravity). In the left panel of Fig. 1.5, for the purposes of illustration, the tilt angle is $\arctan(1/3)$ and the wave frequency has been chosen to be $1/\sqrt{2}$ of the maximum frequency so that the rays propagate at

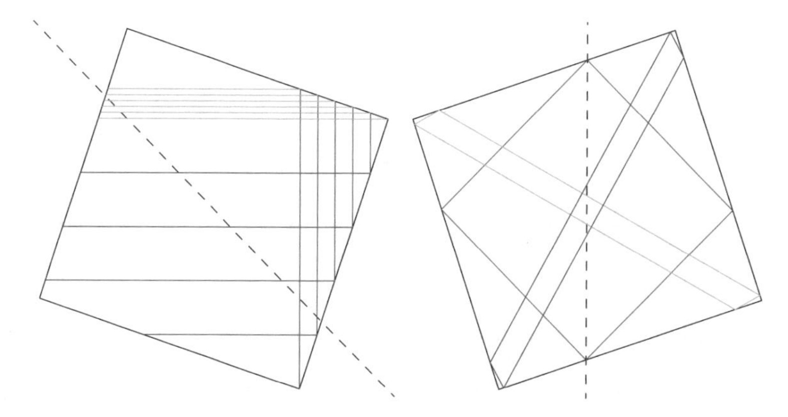

Fig. 1.5 Left: Formation of a wave attractor through focusing reflections. The dashed line indicates the axis of rotation (or gravity). The figure is oriented so that the rays propagate either horizontally or vertically. Right: Variation of the attractor with wave frequency

45° with respect to the axis. For ease of visualization, the figure is oriented so that the rays propagate either horizontally or vertically. Consider the blue rays propagating towards the right; at each reflection (producing the red rays, then the green . . .), the width of the beam is halved and the rays are focused towards a square attractor. As the wave frequency is varied in the range $1/\sqrt{5} < \omega/\omega_{\max} < 2/\sqrt{5}$ (right panel, where the axis is drawn vertically), the attractor maintains a continuous existence, transforming through a family of parallelograms. The central member of the family is the square attractor, which has a total focusing power of 16 (this being the factor by which the width of the beam is reduced after a complete circuit). The outer extremes of this family are the two diagonals of the box, each of which has a total focusing power of 49.

The propagation of inertial waves in a uniformly rotating spherical shell involves both critical-latitude singularities and wave attractors (Rieutord et al. 2001). As an example of the complexity of this behaviour, Fig. 1.6 shows the frequency dependence of a measure of the focusing power of the strongest attractor, in a spherical shell with a radius ratio of $1/2$. The bandwidth of each attractor is relatively small because of the sensitivity of the trajectories of the waves to the angle of propagation and therefore to the wave frequency.

Inertial Waves in a Sphere

A problem that can be solved analytically is to find inertial wave modes in a uniformly rotating, homogeneous, incompressible fluid in a spherical container. Despite the ill-posedness of the eigenvalue problem, there does exist a complete set of modes (Ivers et al. 2015), whose frequencies are dense in the interval $-2\Omega < \omega < 2\Omega$ and

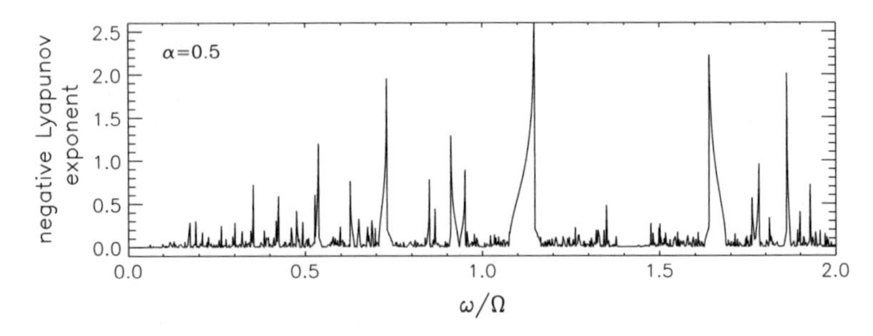

Fig. 1.6 Focusing power of the strongest inertial wave attractors in a spherical shell, as a function of wave frequency. The quantity plotted is the natural logarithm of the focusing power, divided by the number of reflections on the outer sphere; this picks out the strongest and shortest attractors. The radius ratio of the shell is $1/2$

whose eigenfunctions are orthogonal polynomial vector fields. Similar results hold for an ellipsoidal container (Backus and Rieutord 2017), and appear plausible for a Maclaurin spheroid, which is a self-gravitating, homogeneous, incompressible fluid body with a free surface (Bryan 1889). Qualitatively similar modes have been found analytically in certain non-uniform spheres (Wu 2005) and computed in polytropes (Lockitch and Friedman 1999), both in a low-frequency approximation that excludes acoustic and surface gravity waves.

There are numerous methods to describe these special solutions. One is to transform Poincaré's equation (1.1), for $-2\Omega < \omega < 2\Omega$, into Laplace's equation by rescaling the z coordinate by an imaginary factor: $(\tilde{x}, \tilde{y}, \tilde{z}) = (x, y, i\alpha z)$. The homogeneous solutions of this equation that are regular at the origin are of the form $q \propto \tilde{r}^l Y_l^m(\tilde{\theta}, \tilde{\phi})$, where $\tilde{r} = \sqrt{\tilde{x}^2 + \tilde{y}^2 + \tilde{z}^2} = \sqrt{x^2 + y^2 - \alpha^2 z^2}$ is the new radial coordinate and Y_l^m is a spherical harmonic. These solutions are polynomial in $\tilde{\boldsymbol{x}}$ and therefore in $\boldsymbol{x}$. (The homogeneous solutions that are regular at infinity are of the form $q \propto \tilde{r}^{-(l+1)} Y_l^m(\tilde{\theta}, \tilde{\phi})$, but these involve negative powers of $\tilde{r}$ and therefore have singularities on the cones $\tilde{r} = 0$; as argued by Goodman and Lackner (2009), this property explains why non-singular modes cannot be found in a spherical shell.)

It is also possible to find the modal solutions by directly calculating the radial velocity in spherical polar coordinates (Ogilvie 2009). For example, in the case $m = 2$ one separable solution has

$$u_r \propto r Y_2^2$$

and another has

$$u_r \propto r^3 \left[Y_2^2 + \left(\frac{7\omega^2 + 7\Omega\omega - 2\Omega^2}{6\sqrt{3}\,\Omega^2} \right) Y_4^2 \right].$$

For free modes in a full sphere of radius R with a rigid outer boundary, u_r must vanish at $r = R$. This is possible using a linear combination of these solutions:

$$u_r \propto r(R^2 - r^2)Y_2^2,$$

when ω is one of the two roots of

$$7\omega^2 + 7\Omega\omega - 2\Omega^2 = 0, \qquad \text{i.e.} \quad \omega = \left(-1 \pm \sqrt{\frac{15}{7}}\right)\frac{\Omega}{2}.$$

These are two of the lowest order inertial modes of a full sphere.

Instabilities of Internal Waves

If internal waves exceed a critical amplitude, they break and are strongly dissipated. In the case of a stably stratified atmosphere with a plane internal gravity wave, the total buoyancy is

$$B = N^2 z + \mathrm{Re}\left[\tilde{b}\, e^{i(\boldsymbol{k}\cdot\boldsymbol{x}-\omega t)}\right] + \text{constant}.$$

The vertical gradient

$$\frac{\partial B}{\partial z} = N^2 + \mathrm{Re}\left[ik_z\tilde{b}\, e^{i(\boldsymbol{k}\cdot\boldsymbol{x}-\omega t)}\right]$$

becomes inverted at some phase if $|k_z\tilde{b}| > N^2$ (Fig. 1.7). This leads to a local convective instability that generates small-scale motions and causes the wave to break. The breaking criterion is equivalent to $|k_z\tilde{\xi}_z| > 1$, where $\boldsymbol{\xi}$ is the displacement, related to the velocity perturbation by $\boldsymbol{u} = \partial\boldsymbol{\xi}/\partial t$.

Similarly, for a plane inertial wave, the vertical component of the absolute vorticity is

$$2\Omega + \mathrm{Re}\left[i(k_x\tilde{u}_y - k_y\tilde{u}_x)\, e^{i(\boldsymbol{k}\cdot\boldsymbol{x}-\omega t)}\right]$$

and becomes inverted at some phase if $|k_x\tilde{u}_y - k_y\tilde{u}_x| > 2\Omega$, leading to a local inertial instability. The breaking criterion is again equivalent to $|k_z\tilde{\xi}_z| > 1$. Since $\nabla\cdot\boldsymbol{\xi} = 0$ for internal waves, $|k_z\tilde{\xi}_z|$ is equal to $|\boldsymbol{k}_\mathrm{h}\cdot\tilde{\boldsymbol{\xi}}_\mathrm{h}|$, where 'h' denotes the horizontal components.

Internal wave beams undergo similar breaking instabilities when the amplitude exceeds a critical value (Jouve and Ogilvie 2014; Dauxois et al. 2018). This is particularly relevant when the amplitude is increased by a focusing reflection.

Plane internal waves in an unbounded or periodic domain are in fact unstable at *any* non-zero amplitude in the absence of dissipation (Phillips 1981, and references therein). The *parametric subharmonic instability* involves the destabilization of a pair of secondary plane waves through their coupling with the primary wave. Parametric

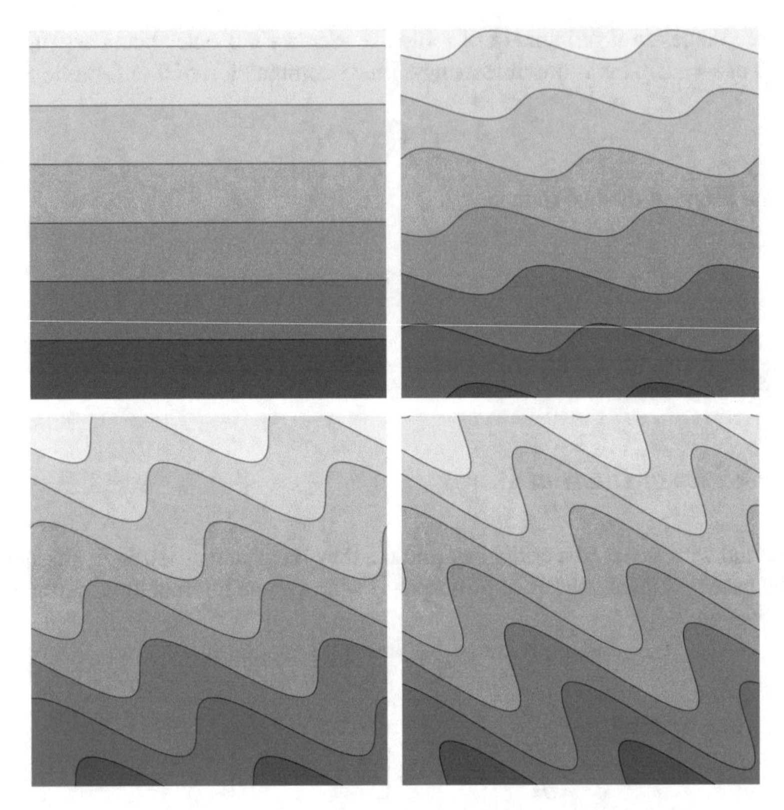

Fig. 1.7 Contours of $z + A\cos(x + z)$ for $A = 0.0$ (top left), 0.5, 1.0 and 1.5 (bottom right). The last case illustrates the overturning of stable stratification by a plane internal wave of sufficient amplitude

resonance occurs if the wavevectors of the three waves sum to zero and similarly for their frequencies, within some tolerance that depends on the amplitude of the primary wave. Owing to the denseness of the spectrum, this condition can always be achieved. However, this type of instability relies on the spatial periodicity of the waves and does not apply to single beams in the same way.

Forced Internal Waves

We have seen that internal waves fill a restricted range of the spectrum of oscillation frequencies in a rotating or stably stratified fluid system, and that the waves may involve singularities. What happens when a system with a dense or continuous spectrum of internal waves is forced at a frequency within that range? For the application to astrophysical tides, we are interested in the total dissipation rate and

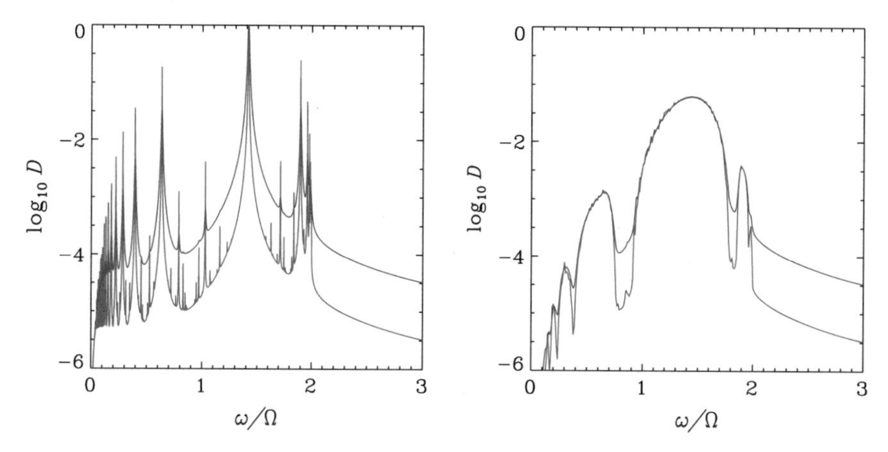

Fig. 1.8 Variation of the dissipation rate (on a logarithmic scale) with forcing frequency, when a rotating fluid in a square domain is subjected to a large-scale body force in the linear regime. The blue and red curves are for frictional damping coefficients $\gamma = 0.01\Omega$ and 0.001Ω, respectively. Left: untilted case, showing classical resonances with separable normal modes in the frequency range of inertial waves. Right: tilted case, showing robust dissipation in wave attractors of finite bandwidth

its dependence on the forcing frequency, especially in the limit that the dissipation coefficients of the fluid are very small.

Figure 1.8 shows the linear response of a rotating fluid in a square container to periodic forcing. The body force is harmonic in time and has a uniform curl. A scale-independent frictional damping is applied to the fluid motion; similar results can be expected for scale-dependent viscous damping. The response curves show the total dissipation rate in a steady state versus the forcing frequency, for two values of the damping coefficient γ.

In the left panel, the rotation axis is aligned with the container and the response is dominated by resonances with normal modes, which have a rectangular structure and can be obtained by separation of variables. Each mode contributes a Lorentzian peak to the dissipation rate, with a height $\propto \gamma^{-1}$ and a width $\propto \gamma$, like a damped harmonic oscillator. In the limit of small γ, the response shows a forest of very narrow peaks, although the frequency-averaged dissipation rate is independent of γ. As expected, the resonances occur at frequencies less than 2Ω.

In the right panel, the rotation axis is tilted by $\arctan(1/3)$ as considered in the section ‘Reflections and Singularities’. Now the dissipation rate shows a sequence of smooth, broad ridges. Each is associated with a wave attractor, occupying a certain band of frequency, the broadest of which is the one shown in Fig. 1.5. Within each band, the dissipation rate has a smooth dependence on frequency and becomes independent of γ in the limit $\gamma \to 0$. This behaviour was noted and explained by Ogilvie (2005) using an asymptotic analysis that separates the essentially inviscid large-scale dynamics from the dissipative behaviour on small scales close to the attractor. Large-scale forcing generates waves that are focused towards the attractor and carry a certain energy flux towards it. Provided that there is a mechanism to

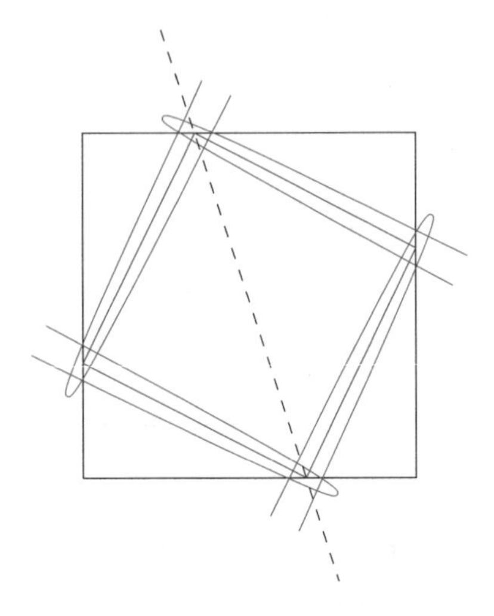

Fig. 1.9 A beam of internal waves (red curves) that achieves a balance between focusing by a wave attractor (blue lines) and viscous spreading

dissipate wave energy on small scales, the attractor absorbs this flux rather like a black hole, without feedback on the large scales. If the dissipation coefficients are lowered, then the waves must undergo more focusing reflections to reach the scale on which they can be dissipated.

Numerical simulations of the tilted square with a viscous fluid (Jouve and Ogilvie 2014) confirm this behaviour in the linear regime. The physics of the forced wave attractor is illustrated in Fig. 1.9. A spreading beam of the type discussed in the section 'Internal Wave Beams', with a virtual source outside the container, is focused at each reflection in a way that compensates exactly for the viscous spreading. The simulations can also explore the nonlinear regime in which the inertial waves become unstable, and it is found that approximately the same total dissipation rate is obtained in the nonlinear regime as if the instability did not occur. As the waves are focused towards the attractor, they reach a length scale on which they become unstable before they can be dissipated directly by viscosity; the instability merely provides an alternative channel of dissipation, by diverting energy into secondary waves of smaller scale that are dissipated more easily by viscosity.

Interiors of Stars and Giant Planets

Interior Models

The interior structure of stars, and their evolution as a result of nuclear reactions, is fairly well understood and described by spherically symmetric, hydrostatic models that neglect rotation and magnetic fields. The structure of the Sun inferred from such models is largely confirmed by helioseismology. It consists of a stably stratified core in which energy is transported by radiation, surrounded by an unstably strat-

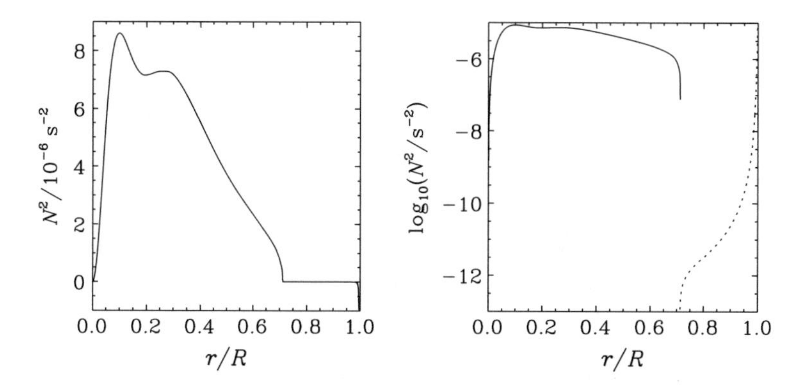

Fig. 1.10 Variation of the squared buoyancy frequency with fractional radius in a standard solar model (Christensen-Dalsgaard et al. 1996). In the logarithmic plot (right), the dashed curve indicates the negative values estimated in the convective zone

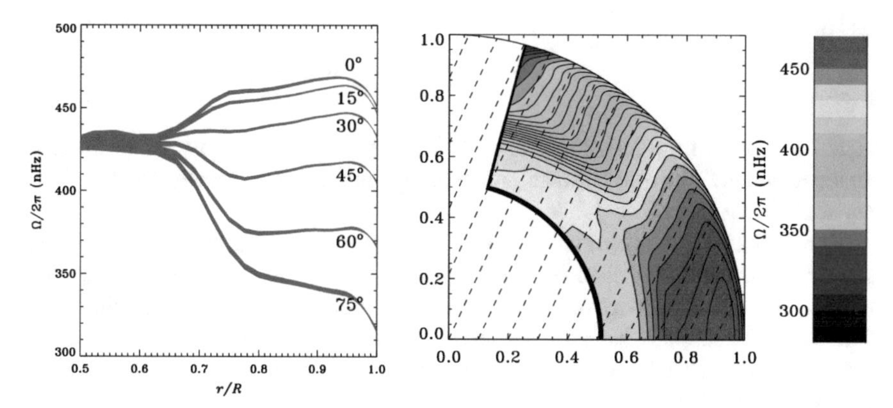

Fig. 1.11 Variation of the solar rotational frequency $\Omega/2\pi$ with radius and latitude, as deduced from helioseismology. Credit: NSO/GONG

ified envelope in which convection dominates. The radial profile of the buoyancy frequency is shown in Fig. 1.10. The distribution of radiative and convective zones varies considerably with the mass and age of the star.

The interior rotation of the Sun has also been deduced from helioseismology, using the rotational splitting of acoustic mode frequencies (Thompson et al. 1996). The latitudinal differential rotation seen at the surface through the motion of sunspots continues nearly to the base of the convective zone (Fig. 1.11).

The interior structure of giant planets, even within the solar system, is much less well understood. Not only are there very few seismological constraints but the interior physics is also more complicated and less certain. Nevertheless, considerable progress has been made recently in modelling the interiors of Jupiter and Saturn (e.g. Helled 2019), and Saturn's rings have been used as a seismometer for the planet (French et al. 2019, and references therein).

Waves and Instabilities in a Stratified, Rotating Body

Basic equations and basic state. The basic equations for an ideal, compressible fluid in cylindrical polar coordinates (r, ϕ, z) are

$$\frac{Du_r}{Dt} - \frac{u_\phi^2}{r} = -\frac{\partial \Phi}{\partial r} - v\frac{\partial p}{\partial r},$$

$$\frac{Du_\phi}{Dt} + \frac{u_r u_\phi}{r} = -\frac{1}{r}\frac{\partial \Phi}{\partial \phi} - \frac{v}{r}\frac{\partial p}{\partial \phi},$$

$$\frac{Du_z}{Dt} = -\frac{\partial \Phi}{\partial z} - v\frac{\partial p}{\partial z},$$

$$\frac{D\ln v}{Dt} = -\frac{1}{\gamma}\frac{D\ln p}{Dt} = \Delta = \frac{1}{r}\frac{\partial (ru_r)}{\partial r} + \frac{1}{r}\frac{\partial u_\phi}{\partial \phi} + \frac{\partial u_z}{\partial z},$$

where

$$\frac{D}{Dt} = \frac{\partial}{\partial t} + u_r\frac{\partial}{\partial r} + \frac{u_\phi}{r}\frac{\partial}{\partial \phi} + u_z\frac{\partial}{\partial z}$$

is the Lagrangian time derivative, $v = 1/\rho$ is the specific volume and γ is the first adiabatic exponent. The gravitational potential Φ is related to the density via Poisson's equation, $\nabla^2\Phi = 4\pi G\rho$.

Consider a steady, axisymmetric basic state representing a (differentially) rotating star or giant plant, described by $v(r, z)$, $p(r, z)$, $\Phi(r, z)$ and $u_\phi = r\Omega(r, z)$, while $u_r = u_z = 0$. This description neglects magnetic fields, convection, meridional circulation, diffusion, etc. Nevertheless, the linear theory of this basic state contains a rich theory of internal waves that is relevant to both free and forced oscillations of stars and giant planets.

The basic equations are satisfied if

$$-r\Omega^2 = -\frac{\partial \Phi}{\partial r} - v\frac{\partial p}{\partial r},$$

$$0 = -\frac{\partial \Phi}{\partial z} - v\frac{\partial p}{\partial z}.$$

The *thermal wind equation* is obtained by eliminating Φ by cross differentiating:

$$-r\frac{\partial \Omega^2}{\partial z} = \frac{\partial v}{\partial r}\frac{\partial p}{\partial z} - \frac{\partial v}{\partial z}\frac{\partial p}{\partial r}.$$

The basic state is called *barotropic* if the right-hand side

$$\frac{\partial (v, p)}{\partial (r, z)} = \boldsymbol{e}_\phi \cdot (\boldsymbol{\nabla} p \times \boldsymbol{\nabla} v)$$

vanishes; otherwise, it is *baroclinic*. According to the fundamental thermodynamic identity $de = T\,ds - p\,dv$ (where e is specific internal energy, T is temperature and s is specific entropy),

$$\frac{\partial(v,p)}{\partial(r,z)} = \frac{\partial(s,T)}{\partial(r,z)}, \qquad \nabla p \times \nabla v = \nabla T \times \nabla s.$$

Linearized equations. Now consider small perturbations, representing waves or instabilities on this background. The linearized equations in the *Cowling approximation* (in which perturbations of the gravitational potential are neglected) are

$$\frac{Du_r'}{Dt} - 2\Omega u_\phi' = -v'\frac{\partial p}{\partial r} - v\frac{\partial p'}{\partial r},$$

$$\frac{Du_\phi'}{Dt} + u_r'\frac{\partial(r\Omega)}{\partial r} + u_z'\frac{\partial(r\Omega)}{\partial z} + \Omega u_r' = -\frac{v}{r}\frac{\partial p'}{\partial \phi},$$

$$\frac{Du_z'}{Dt} = -v'\frac{\partial p}{\partial z} - v\frac{\partial p'}{\partial z},$$

$$\frac{Dv'}{Dt} + u_r'\frac{\partial v}{\partial r} + u_z'\frac{\partial v}{\partial z} = v\Delta',$$

$$\frac{Dp'}{Dt} + u_r'\frac{\partial p}{\partial r} + u_z'\frac{\partial p}{\partial z} = -\gamma p\Delta',$$

with

$$\frac{D}{Dt} = \frac{\partial}{\partial t} + \Omega\frac{\partial}{\partial \phi}, \qquad \Delta' = \frac{1}{r}\frac{\partial(ru_r')}{\partial r} + \frac{1}{r}\frac{\partial u_\phi'}{\partial \phi} + \frac{\partial u_z'}{\partial z}.$$

Let us introduce the *Lagrangian displacement* $\boldsymbol{\xi}$, which is the difference in position of a fluid element in the perturbed and unperturbed flows. It is related to the Eulerian velocity perturbation $\boldsymbol{u}'$ by

$$\frac{D\boldsymbol{\xi}}{Dt} = \boldsymbol{u}' + \boldsymbol{\xi}\cdot\nabla\boldsymbol{u},$$

or, in components,

$$u_r' = \frac{D\xi_r}{Dt}, \qquad u_\phi' = \frac{D\xi_\phi}{Dt} - r\boldsymbol{\xi}\cdot\nabla\Omega, \qquad u_z' = \frac{D\xi_z}{Dt},$$

and their divergences are related by

$$\Delta' = \nabla\cdot\boldsymbol{u}' = \frac{D}{Dt}(\nabla\cdot\boldsymbol{\xi}).$$

Rewritten in terms of the Lagrangian displacement, the linearized equation of motion has components

$$\frac{D^2\xi_r}{Dt^2} - 2\Omega\frac{D\xi_\phi}{Dt} + 2r\Omega\,\boldsymbol{\xi}\cdot\nabla\Omega = -v'\frac{\partial p}{\partial r} - v\frac{\partial p'}{\partial r},$$
$$\frac{D^2\xi_\phi}{Dt^2} + 2\Omega\frac{D\xi_r}{Dt} = -\frac{v}{r}\frac{\partial p'}{\partial \phi},$$
$$\frac{D^2\xi_z}{Dt^2} = -v'\frac{\partial p}{\partial z} - v\frac{\partial p'}{\partial z},$$

and we find by integration that

$$v' = v\left[\frac{1}{r}\frac{\partial(r\xi_r)}{\partial r} + \frac{1}{r}\frac{\partial \xi_\phi}{\partial \phi} + \frac{\partial \xi_z}{\partial z}\right] - \xi_r\frac{\partial v}{\partial r} - \xi_z\frac{\partial v}{\partial z},$$
$$p' = -\gamma p\left[\frac{1}{r}\frac{\partial(r\xi_r)}{\partial r} + \frac{1}{r}\frac{\partial \xi_\phi}{\partial \phi} + \frac{\partial \xi_z}{\partial z}\right] - \xi_r\frac{\partial p}{\partial r} - \xi_z\frac{\partial p}{\partial z}.$$

Harmonic disturbances. Let us now consider free or forced harmonic disturbances of the form

$$\xi_r = Re\left[\tilde{\xi}_r(r,z)\exp(-i\omega t + im\phi)\right],$$

etc., where ω is the wave frequency and m (an integer) is the azimuthal wavenumber. Then, the Lagrangian derivative reduces to multiplication by $-i\hat{\omega}$, where $\hat{\omega} = \omega - m\Omega$ is the *intrinsic wave frequency*: the wave frequency seen in the fluid frame. Dropping the tildes and eliminating ξ_ϕ and v' algebraically, we obtain

$$(-\hat{\omega}^2 + A)\xi_r + B\xi_z = -v\frac{\partial p'}{\partial r} + \frac{vp'}{\gamma p}\frac{\partial p}{\partial r} + \frac{2\Omega}{\hat{\omega}}\frac{v}{r}mp',$$
$$C\xi_r + (-\hat{\omega}^2 + D)\xi_z = -v\frac{\partial p'}{\partial z} + \frac{vp'}{\gamma p}\frac{\partial p}{\partial z},$$
$$\left(1 - \frac{m^2 v_s^2}{r^2\hat{\omega}^2}\right)p' = -\gamma p\left[\frac{1}{r}\frac{\partial(r\xi_r)}{\partial r} + \frac{m}{r}\frac{2\Omega}{\hat{\omega}}\xi_r + \frac{\partial \xi_z}{\partial z}\right] - \xi_r\frac{\partial p}{\partial r} - \xi_z\frac{\partial p}{\partial z},$$

where $v_s = \sqrt{\gamma p/\rho}$ is the adiabatic sound speed and the four coefficients are given by

$$A = 4\Omega^2 + r\frac{\partial \Omega^2}{\partial r} - \left(\frac{\partial v}{\partial r} + \frac{v}{\gamma p}\frac{\partial p}{\partial r}\right)\frac{\partial p}{\partial r},$$
$$B = r\frac{\partial \Omega^2}{\partial z} - \left(\frac{\partial v}{\partial z} + \frac{v}{\gamma p}\frac{\partial p}{\partial z}\right)\frac{\partial p}{\partial r},$$
$$C = -\left(\frac{\partial v}{\partial r} + \frac{v}{\gamma p}\frac{\partial p}{\partial r}\right)\frac{\partial p}{\partial z},$$
$$D = -\left(\frac{\partial v}{\partial z} + \frac{v}{\gamma p}\frac{\partial p}{\partial z}\right)\frac{\partial p}{\partial z}.$$

According to the thermal wind equation, $B = C$ and the coefficients form a symmetric matrix

$$\mathbf{M} = \begin{pmatrix} A & B \\ C & D \end{pmatrix} = \begin{pmatrix} A & B \\ B & D \end{pmatrix}.$$

These can be written in terms of gradients of the specific angular momentum $\ell = r^2\Omega$ and the specific entropy s (if the composition is uniform), as well as the effective gravity $\boldsymbol{g} = -\boldsymbol{\nabla}\Phi + r\Omega^2\,\boldsymbol{e}_r = v\boldsymbol{\nabla}p$:

$$\begin{aligned} A &= \frac{1}{r^3}\frac{\partial \ell^2}{\partial r} - \delta\,\frac{g_r}{c_p}\frac{\partial s}{\partial r}, \\ B &= \frac{1}{r^3}\frac{\partial \ell^2}{\partial z} - \delta\,\frac{g_r}{c_p}\frac{\partial s}{\partial z} = -\delta\,\frac{g_z}{c_p}\frac{\partial s}{\partial r}, \\ D &= -\delta\,\frac{g_z}{c_p}\frac{\partial s}{\partial z}, \end{aligned}$$

where we have used

$$\left(\frac{\partial \ln v}{\partial s}\right)_p = \left(\frac{\partial \ln v}{\partial \ln T}\right)_p \left(\frac{\partial \ln T}{\partial s}\right)_p = \delta\,\frac{1}{c_p}$$

and note that $\delta = 1$ for an ideal gas.

Our equations are reducible to a second-order partial differential equation (PDE) for p'. Without writing this out in full, we can note that the structure of the second-derivative terms is

$$(\hat{\omega}^2 - D)\frac{\partial^2 p'}{\partial r^2} + 2B\frac{\partial^2 p'}{\partial r\,\partial z} + (\hat{\omega}^2 - A)\frac{\partial^2 p'}{\partial z^2} + \cdots = 0,$$

implying that the PDE is hyperbolic when

$$B^2 > (\hat{\omega}^2 - A)(\hat{\omega}^2 - D).$$

This is true for squared frequencies in the range

$$\frac{(A+D) - \sqrt{(A-D)^2 + 4B^2}}{2} < \hat{\omega}^2 < \frac{(A+D) + \sqrt{(A-D)^2 + 4B^2}}{2}, \quad (1.2)$$

corresponding to *inertia-gravity waves*—internal waves that are restored by a combination of inertial and buoyancy forces and generalize the waves discussed in the sections 'Plane Inertial Waves' and 'Plane Internal Gravity Waves'. The characteristics of the hyperbolic PDE are generally curved, unlike those of Poincaré's equation for pure inertial waves in a uniformly rotating fluid.

Axisymmetric perturbations. In the special case of axisymmetric perturbations ($m = 0$), our equations can be written in the form

$$\omega^2 \boldsymbol{\xi} = \mathbf{M} \cdot \boldsymbol{\xi} + v \left(\nabla p' - \frac{p'}{\gamma p} \nabla p \right), \tag{1.3}$$

$$p' = -\gamma p \nabla \cdot \boldsymbol{\xi} - \boldsymbol{\xi} \cdot \nabla p,$$

where only the meridional (r, z) components of $\boldsymbol{\xi}$ are included. Taking the scalar product of Eq. (1.3) with $\rho \boldsymbol{\xi}^*$ and integrating over the volume of the star with $p = 0$ and $\rho = 0$ at the surface, we obtain

$$\omega^2 = \left[\int \rho Q(\boldsymbol{\xi}) \, dV + \int \frac{|p'|^2}{\gamma p} \, dV \right] \bigg/ \int \rho |\boldsymbol{\xi}|^2 \, dV,$$

where $Q(\boldsymbol{\xi}) = \boldsymbol{\xi}^* \cdot \mathbf{M} \cdot \boldsymbol{\xi}$ is the real Hermitian form associated with the matrix $\mathbf{M}$. This expression for ω^2 has two contributions: the first involves inertial and buoyancy forces and is associated with internal waves, while the second involves pressure perturbations and is associated with acoustic waves. It has the usual variational property associated with self-adjoint eigenvalue problems: the eigenvalues ω^2 are the stationary values of this expression among trial displacements that are regular at the surface. The star is unstable (with respect to axisymmetric perturbations) if and only if the expression can be made negative by such a trial displacement. Clearly, a necessary condition for instability is that $\mathbf{M}$ have a negative eigenvalue in some region because, otherwise, $Q \geq 0$. This condition can also be shown to be sufficient for instability; if it is satisfied, we can construct a trial displacement $\boldsymbol{\xi}$ that is localized in this region, achieves $Q < 0$ by being nearly parallel to the eigenvector $\boldsymbol{e}$ with the negative eigenvalue, and avoids any positive acoustic contribution by having $p' = 0$. (This can be arranged by giving $\boldsymbol{\xi}$ a short wavelength in the direction perpendicular to $\boldsymbol{e}$ and setting the subdominant component of $\boldsymbol{\xi}$ perpendicular to $\boldsymbol{e}$ so as to make p' vanish.)

Noting that $\mathbf{M}$ has non-negative eigenvalues if $\operatorname{tr} \mathbf{M} = A + D \geq 0$ and $\det \mathbf{M} = AD - B^2 \geq 0$, we obtain the *Høiland stability criteria*: for stability we require

$$\frac{1}{r^3} \frac{\partial \ell^2}{\partial r} - \frac{\delta}{c_p} \boldsymbol{g} \cdot \nabla s \geq 0 \quad \text{and} \quad -g_z \left(\frac{\partial \ell^2}{\partial r} \frac{\partial s}{\partial z} - \frac{\partial \ell^2}{\partial z} \frac{\partial s}{\partial r} \right) \geq 0.$$

Two special (barotropic) cases can be understood more easily. In the non-rotating case ($\ell = 0$), the first criterion reduces to the *Schwarzschild criterion* for convective stability, $-\boldsymbol{g} \cdot \nabla s \geq 0$ (i.e. $N^2 \geq 0$). In the homentropic case ($s =$ constant), it reduces instead to the *Rayleigh criterion* for inertial stability, $d\ell^2/dr \geq 0$.

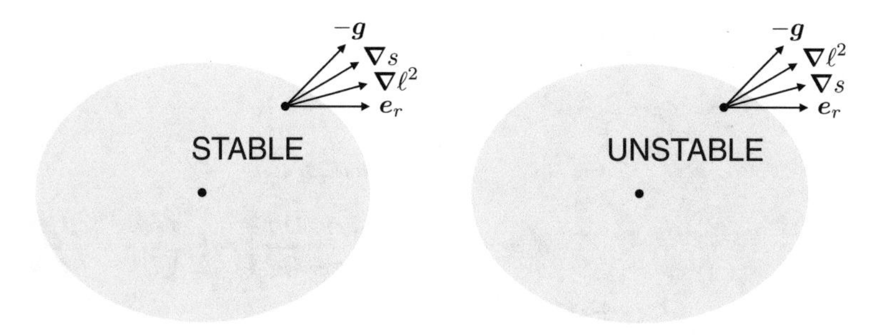

Fig. 1.12 Stable and unstable baroclinic configurations of a rotating star

In baroclinic equilibria, the second Høiland criterion needs to be considered:

$$(\boldsymbol{e}_r \times (-\boldsymbol{g})) \cdot (\nabla \ell^2 \times \nabla s) \geq 0.$$

For stability, the specific angular momentum should not decrease with r on each surface of constant entropy (Fig. 1.12).

Short-wavelength disturbances. Let us now consider short-wavelength (and generally non-axisymmetric) disturbances of the form

$$\xi_r \sim Re\left[\tilde{\xi}_r(r, z)\, \exp(-i\omega t + im\phi + i\varphi)\right],$$

etc., where $\varphi(r, z)$ is a rapidly varying phase variable. Then, when the gradient operator acts on perturbations, the result at leading order is multiplication by $i\boldsymbol{k}$, where $\boldsymbol{k} = \nabla\varphi$ is the local wavevector.

Two different ordering schemes are possible in the short-wavelength limit $|\boldsymbol{k}| \to \infty$. In the *acoustic-wave ordering*, $\hat{\omega} = O(|\boldsymbol{k}|)$ also diverges and we obtain the dominant balances

$$-\hat{\omega}^2\boldsymbol{\xi} \sim -v\, i\boldsymbol{k}\, p', \qquad p' \sim -\gamma p\, i\boldsymbol{k} \cdot \boldsymbol{\xi},$$

leading to the local dispersion relation for acoustic waves,

$$\hat{\omega}^2 \sim v_{\mathrm{s}}^2|\boldsymbol{k}|^2.$$

In this scheme, m/r can be $O(|\boldsymbol{k}|)$ in general, forming the ϕ component of $\boldsymbol{k}$.

In the *internal wave ordering*, $\hat{\omega} = O(1)$ remains finite as $|\boldsymbol{k}| \to \infty$. In this scheme, m is at most $O(1)$, so $\boldsymbol{k}$ is essentially meridional. We require p' to be an order smaller than $\boldsymbol{\xi}$; otherwise, the $\nabla p'$ terms are unbalanced. The dominant balances are

$$
\begin{aligned}
-\hat{\omega}^2\xi_r + 2i\Omega\hat{\omega}\xi_\phi + 2r\Omega\boldsymbol{\xi}\cdot\nabla\Omega &\sim -v'\frac{\partial p}{\partial r} - v\,ik_r p',\\
-\hat{\omega}^2\xi_\phi - 2i\Omega\hat{\omega}\xi_r &\sim 0,\\
-\hat{\omega}^2\xi_z &\sim -v'\frac{\partial p}{\partial z} - v\,ik_z p',\\
v' &\sim -\xi_r\left(\frac{\partial v}{\partial r} + \frac{v}{\gamma p}\frac{\partial p}{\partial r}\right) - \xi_z\left(\frac{\partial v}{\partial z} + \frac{v}{\gamma p}\frac{\partial p}{\partial z}\right),\\
ik_r\xi_r + ik_z\xi_z &\sim 0,
\end{aligned}
$$

while p' becomes algebraically independent of $\boldsymbol{\xi}$. Eliminating ξ_ϕ, we obtain (in terms of the meridional components of the displacement only)

$$
-\hat{\omega}^2\boldsymbol{\xi} + \mathbf{M}\cdot\boldsymbol{\xi} \sim -v\,i\boldsymbol{k}\,p', \qquad i\boldsymbol{k}\cdot\boldsymbol{\xi} \sim 0,
$$

leading to the local dispersion relation for internal (inertia-gravity) waves,

$$
\hat{\omega}^2 \sim Q(\hat{\boldsymbol{\xi}}),
$$

with $\hat{\boldsymbol{\xi}} = \boldsymbol{\xi}/|\boldsymbol{\xi}| \perp \boldsymbol{k}$. The range of $\hat{\omega}^2$ for internal waves is bounded by the two eigenvalues of $\mathbf{M}$ (Eq. 1.2), leading to a dense or continuous spectrum, and $\hat{\omega}^2$ is positive if the equilibrium is Høiland stable.

Approaches to Internal Waves in Stars and Giant Planets

A variety of approaches have been taken to understanding and calculating internal waves in stars and giant planets. Global linear computations of internal waves on idealized axisymmetric basic states have been carried out for both free oscillations (e.g. Dintrans and Rieutord 2000; Mirouh et al. 2016) and tidally forced disturbances (e.g. Ogilvie and Lin 2004). Nonlinear numerical simulations have also been used, particularly to study the internal waves generated by convection (e.g. Alvan et al. 2014). More analytical approaches have also been taken: these include the analysis of characteristics, wave attractors, etc. (e.g. Rieutord et al. 2001), ray tracing (e.g. Prat et al. 2016) and extensions of the WKB method (e.g. Ivanov and Papaloizou 2010). These methods are in fact closely related and it would be valuable to unify them.

Tidal Interaction of Two Bodies

Introduction

It is well known that the motion of two gravitationally bound point masses in Newtonian dynamics is described by a Keplerian elliptical orbit. If the bodies are extended, however, then each is deformed from a spherical shape by tidal forces, as well as by rotation, and this perturbs the orbital motion. We are particularly interested in dissipative aspects of the tidal interaction that lead to an irreversible evolution of the orbit and the spins of the two bodies.

Gravitational Interaction of Particulate Systems

Consider first two bodies composed of particles that interact only through gravity. The particles of body 1 have masses m_i and position vectors $\boldsymbol{r}_i(t)$, so that the total mass M and centre of mass $\boldsymbol{R}(t)$ of body 1 are given by

$$M = \sum_i m_i, \qquad M\boldsymbol{R} = \sum_i m_i \boldsymbol{r}_i.$$

The positions $\boldsymbol{x}_i = \boldsymbol{r}_i - \boldsymbol{R}$ of its particles relative to the centre of mass satisfy

$$\sum_i m_i \boldsymbol{x}_i = \boldsymbol{0}. \tag{1.4}$$

The total linear momentum of body 1 is

$$\sum_i m_i \dot{\boldsymbol{r}}_i = M\dot{\boldsymbol{R}},$$

its total angular momentum about the origin is

$$\sum_i m_i \boldsymbol{r}_i \times \dot{\boldsymbol{r}}_i = M\boldsymbol{R} \times \dot{\boldsymbol{R}} + \sum_i m_i \boldsymbol{x}_i \times \dot{\boldsymbol{x}}_i$$

and its total kinetic energy is

$$\sum_i \frac{1}{2} m_i \, |\dot{\boldsymbol{r}}_i|^2 = \frac{1}{2} M |\dot{\boldsymbol{R}}|^2 + \sum_i \frac{1}{2} m_i \, |\dot{\boldsymbol{x}}_i|^2 .$$

We denote the equivalent quantities for body 2 with primes: m_i', $\boldsymbol{r}_i'$, M', $\boldsymbol{R}'$, $\boldsymbol{x}_i'$.

The total kinetic energy of the system is

$$T = \frac{1}{2} M |\dot{\boldsymbol{R}}|^2 + \sum_i \frac{1}{2} m_i \, |\dot{\boldsymbol{x}}_i|^2 + \frac{1}{2} M' |\dot{\boldsymbol{R}}'|^2 + \sum_i \frac{1}{2} m_i' \, |\dot{\boldsymbol{x}}_i'|^2$$

and its total potential energy is

$$V = -\frac{1}{2}\sum_i \sum_{j \neq i} \frac{G m_i m_j}{|\boldsymbol{r}_j - \boldsymbol{r}_i|} - \frac{1}{2}\sum_i \sum_{j \neq i} \frac{G m'_i m'_j}{|\boldsymbol{r}'_j - \boldsymbol{r}'_i|} - \sum_i \sum_j \frac{G m_i m'_j}{|\boldsymbol{r}'_j - \boldsymbol{r}_i|}.$$

The last double sum represents the gravitational interaction between the two bodies. Let us expand this expression in powers of the ratio of the size of the bodies to the separation $\boldsymbol{d} = \boldsymbol{R}' - \boldsymbol{R}$ of their centres of mass, by writing $\boldsymbol{r}'_j - \boldsymbol{r}_i = \boldsymbol{d} + \boldsymbol{x}'_j - \boldsymbol{x}_i$ and using the expansion

$$\frac{1}{|\boldsymbol{d} + \boldsymbol{x}|} = \frac{1}{|\boldsymbol{d}|} - \frac{\boldsymbol{d} \cdot \boldsymbol{x}}{|\boldsymbol{d}|^3} + \frac{3(\boldsymbol{d} \cdot \boldsymbol{x})^2 - |\boldsymbol{d}|^2 |\boldsymbol{x}|^2}{2|\boldsymbol{d}|^5} + \cdots$$

Terms that are linear in $\boldsymbol{x}_i$ (or $\boldsymbol{x}'_j$) do not contribute to V because of the property (1.4). The interaction energy to this order becomes

$$-\frac{G M M'}{|\boldsymbol{d}|} - \frac{G M' (\boldsymbol{d}^\top \mathbf{Q} \boldsymbol{d})}{2|\boldsymbol{d}|^5} - \frac{G M (\boldsymbol{d}^\top \mathbf{Q}' \boldsymbol{d})}{2|\boldsymbol{d}|^5},$$

where the gravitational quadrupole moment of body 1 (e.g.) is the tensor

$$\mathbf{Q} = \sum_i m_i \left(3\boldsymbol{x}_i \boldsymbol{x}_i - |\boldsymbol{x}_i|^2\, \mathbf{I}\right)$$

($\mathbf{I}$ being the unit tensor). The leading-order tidal interaction is therefore between the monopole of one body and the quadrupole of the other.

Lagrange's equations of motion for the system are

$$\frac{d}{dt}\frac{\partial L}{\partial \dot{q}} = \frac{\partial L}{\partial q},$$

where $L = T - V$ is the Lagrangian function and q is any of the generalized coordinates. These can be regarded as $\boldsymbol{R}$ and $\boldsymbol{x}_i$, subject to the linear dependence (1.4), together with the equivalent primed quantities.

In particular, the motion of the centres of mass of the two bodies is given by

$$M\ddot{\boldsymbol{R}} = -M'\ddot{\boldsymbol{R}}' = -\boldsymbol{F},$$

with mutual force

$$\boldsymbol{F} = -\frac{\partial V}{\partial \boldsymbol{d}} = -\frac{G M M' \boldsymbol{d}}{|\boldsymbol{d}|^3} + \frac{G M' (\mathbf{Q}\boldsymbol{d})}{|\boldsymbol{d}|^5} - \frac{5 G M' (\boldsymbol{d}^\top \mathbf{Q} \boldsymbol{d}) \boldsymbol{d}}{2|\boldsymbol{d}|^7} + \frac{G M (\mathbf{Q}'\boldsymbol{d})}{|\boldsymbol{d}|^5} - \frac{5 G M (\boldsymbol{d}^\top \mathbf{Q}' \boldsymbol{d}) \boldsymbol{d}}{2|\boldsymbol{d}|^7}.$$

The first term on the right-hand side is the force between point masses, while the remaining terms represent the net tidal force between the bodies, involving the interaction of monopoles (M, M') and quadrupoles ($\mathbf{Q}$, $\mathbf{Q}'$).

The relative motion of particles in body 1 (e.g.) is given by

$$
\begin{aligned}
m_i\ddot{\boldsymbol{x}}_i &= -\frac{\partial V}{\partial \boldsymbol{x}_i} \\
&= \sum_{j\neq i} \frac{G m_i m_j \left(\boldsymbol{x}_j - \boldsymbol{x}_i\right)}{\left|\boldsymbol{x}_j - \boldsymbol{x}_i\right|^3} + \frac{G M' m_i}{|\boldsymbol{d}|^5}\left[3\left(\boldsymbol{d}\cdot\boldsymbol{x}_i\right)\boldsymbol{d} - |\boldsymbol{d}|^2\boldsymbol{x}_i\right],
\end{aligned}
$$

satisfying $\sum_i m_i\ddot{\boldsymbol{x}}_i = \mathbf{0}$, as required. The first term on the right-hand side is the sum of the internal gravitational forces within body 1, while the second term is tidal force due to body 2. This force varies linearly with the relative position $\boldsymbol{x}_i$ and depends only on the total mass M' of body 2, which can be regarded as a point mass.

The interaction of the bodies conserves the total energy $T + V$ and the total angular momentum $\boldsymbol{L}$ of the system. If we decompose $\boldsymbol{L} = \boldsymbol{L}_\mathrm{o} + \boldsymbol{L}_\mathrm{s} = \text{constant}$ into orbital and spin angular momenta (or, more generally, external and internal angular momenta),

$$
\begin{aligned}
\boldsymbol{L}_\mathrm{o} &= M\boldsymbol{R}\times\dot{\boldsymbol{R}} + M'\boldsymbol{R}'\times\dot{\boldsymbol{R}}', \\
\boldsymbol{L}_\mathrm{s} &= \sum_i m_i\boldsymbol{x}_i\times\dot{\boldsymbol{x}}_i + \sum_i m_i'\boldsymbol{x}_i'\times\dot{\boldsymbol{x}}_i',
\end{aligned}
$$

then the rate of exchange is given by the tidal torque

$$
\dot{\boldsymbol{L}}_\mathrm{o} = -\dot{\boldsymbol{L}}_\mathrm{s} = \boldsymbol{d}\times\boldsymbol{F} = \frac{GM'\boldsymbol{d}\times(\mathbf{Q}\boldsymbol{d})}{|\boldsymbol{d}|^5} + \frac{GM\boldsymbol{d}\times(\mathbf{Q}'\boldsymbol{d})}{|\boldsymbol{d}|^5}.
$$

This represents a spin–orbit (or internal–external) coupling through the gravitational interaction of monopoles and quadrupoles.

Gravitational Interaction of Fluid Bodies

Similar ideas can be applied to continuous media such as fluid bodies in which dissipation can occur, e.g. by viscous forces. In a non-rotating, but accelerated, frame of reference with origin at the centre of mass of body 1, the equation of motion of the fluid is

$$
\rho\frac{D\boldsymbol{u}}{Dt} = \nabla\cdot\mathbf{T} - \rho\nabla\Phi,
$$

where the stress tensor $\mathbf{T}$ includes collective effects such as pressure, viscosity, magnetic fields, etc., and

$$\Phi = -\int_1 \frac{G\,dm'}{|\boldsymbol{x}' - \boldsymbol{x}|} - \frac{GM'}{2|\boldsymbol{d}|^5}\left[3(\boldsymbol{d}\cdot\boldsymbol{x})^2 - |\boldsymbol{d}|^2|\boldsymbol{x}|^2\right]$$

consists of the self-gravitational potential of body 1 together with the tidal potential due to body 2. The total mechanical energy (of both bodies) is no longer conserved:

$$\frac{d}{dt}(T + V) = -\int_{1+2} \mathbf{T} : \nabla\boldsymbol{u}\,dV.$$

A viscous stress acting on the tidal deformation contributes to a loss of mechanical energy and an irreversible evolution of the system.

Incompressible Fluid Ellipsoids

A particular example, which can be solved semi-analytically, is provided by homogeneous, incompressible ellipsoids (Sridhar and Tremaine 1992). In the centre-of-mass frame of body 1, its free surface is the ellipsoid $S_{ij}x_ix_j = 1$, where $S_{ij}(t)$ is a positive-definite symmetric tensor. The velocity field is $u_i = A_{ij}x_j$, where $A_{ij}(t)$ is a traceless tensor, and this deforms the shape according to the kinematic equation

$$\dot{\mathbf{S}} + \mathbf{SA} + (\mathbf{SA})^{\top} = \mathbf{0}. \tag{1.5}$$

If the pressure (vanishing at the free surface) is $p = \rho P(1 - S_{ij}x_ix_j)$, where $P(t)$ is a positive scalar, the dynamic viscosity is $\mu = kp$ for some positive constant k, and the gravitational potential is $\Phi = B_{ij}x_ix_j$, where $B_{ij}(t)$ is a symmetric tensor, then the equation of motion is satisfied provided that

$$\dot{\mathbf{A}} + \mathbf{A}^2 = -2\mathbf{B} + 2P\left[\mathbf{S} - k\left(\mathbf{A} + \mathbf{A}^{\top}\right)\mathbf{S}\right]. \tag{1.6}$$

The trace of this equation determines P in terms of $\mathbf{S}$ and $\mathbf{A}$.

The potential tensor can be decomposed into contributions from self-gravity and tides: $\mathbf{B} = \mathbf{B}_{\text{self}} + \mathbf{B}_{\text{tide}}$. A classical result of potential theory is that the self-gravitational potential within a homogeneous ellipsoid is of the quadratic form assumed here; $\mathbf{B}_{\text{self}}$ is aligned with the shape tensor $\mathbf{S}$ and their eigenvalues are related nonlinearly through elliptic integrals. For small departures from a sphere, we may write $\mathbf{S} = S(\mathbf{I} + \mathbf{a})$ with a small traceless anisotropy ($\|\mathbf{a}\| \ll 1$), in which case

$$\mathbf{B}_{\text{self}} = \frac{2}{3}\pi G\rho\left[\mathbf{I} + \frac{3}{5}\mathbf{a} + O\left(\|\mathbf{a}\|^2\right)\right].$$

The tidal potential in the approximation described above is also of this quadratic form, with

$$\mathbf{B}_{\text{tide}} = -\frac{GM'}{2|\boldsymbol{d}|^5}\left(3\boldsymbol{d}\boldsymbol{d} - |\boldsymbol{d}|^2\mathbf{I}\right).$$

The quadrupole moment of the ellipsoid can be shown to be

$$\mathbf{Q} = \frac{M}{5}\left[3\,\mathbf{S}^{-1} - \left(\operatorname{tr}\mathbf{S}^{-1}\right)\mathbf{I}\right].$$

In the absence of tidal forces, Eqs. (1.5) and (1.6) admit the classical rotating equilibria (Maclaurin spheroids and Jacobi ellipsoids; Chandrasekhar 1969) and a variety of oscillation modes, damped by viscosity. For example, a non-rotating spherical equilibrium is described by $\mathbf{S} = S\,\mathbf{I}$, $\mathbf{A} = \mathbf{0}$ and $PS = \frac{2}{3}\pi G\rho$. Small departures from this equilibrium satisfy the linearized equations $\dot{\mathbf{a}} + \mathbf{A} + \mathbf{A}^\top = \mathbf{0}$ and $\dot{\mathbf{A}} = -\frac{4}{5}\pi G\rho\,\mathbf{a} + \frac{4}{3}\pi G\rho\left[\mathbf{a} - k\left(\mathbf{A} + \mathbf{A}^\top\right)\right]$, which combine to give the equation of a damped harmonic oscillator,

$$\ddot{\mathbf{a}} + \frac{8}{3}\pi G\rho k\,\dot{\mathbf{a}} + \frac{16}{15}\pi G\rho\,\mathbf{a} = \mathbf{0},$$

describing the quadrupolar fundamental (surface gravity) modes, subject to viscous damping.

In order to evaluate the tidal tensor $\mathbf{B}_{\text{tide}}$, we need to know how the separation $\boldsymbol{d}$ evolves in time. Since the tidal force between the bodies is supposed to be small compared to the gravitational force between point masses, the dominant motion of $\boldsymbol{d}$ is a Keplerian orbit. One approach is to specify a particular Keplerian orbit and solve equations (1.5) and (1.6) for $\mathbf{S}$ and $\mathbf{A}$ to determine the time-dependent tidal deformation.

Some relevant results obtained in the limit $M_1 \ll M_2$ (Sridhar and Tremaine 1992) are as follows. On a circular orbit of radius a, the body settles to a tidally distorted, synchronously rotating 'Roche ellipsoid' if

$$a > 1.52\left(\frac{M_2}{\rho}\right)^{1/3},$$

otherwise it is tidally disrupted. Within this model, the 'disruption' takes the form of a rapid growth of the major axis of the ellipsoid as the body is deformed into a needle-like configuration (eventually invalidating the approximations made in deriving the model). In the opposite extreme of a parabolic orbit of minimum distance q, the body is disrupted if

$$q < 1.05\left(\frac{M_2}{\rho}\right)^{1/3},$$

or less if the viscosity is significant.

More generally, the orbit can be evolved self-consistently, applying the tidal forces appropriate to the monopole–quadrupole interactions assumed here. This model is dynamically rich and contains much of the basic physics of astrophysical tidal interactions. However, internal waves are missing because the bodies are unstratified and the velocity fields are too simple to capture inertial waves.

Internal Waves and Astrophysical Tides

In this final section, I attempt to summarize the role of internal waves in tidal dissipation in stars and giant planets. Further details can be found in the review by Ogilvie (2014) and the references cited therein.

Equilibrium and Dynamical Tides

When a star or giant planet is subject to periodic forcing from a nearby orbital companion, at a frequency that is low compared to that of its surface gravity or acoustic modes, the response of the fluid body can be thought of as consisting of two parts, as emphasized in the work of J.-P. Zahn.

The *equilibrium* (or *non-wavelike*) tide is a quasi-hydrostatic adjustment to the instantaneous tidal potential. It consists of a spheroidal tidal bulge pointing towards and away from the companion. Unless the orbit is circular and the spin is synchronized and aligned with the orbit, the bulge is time dependent in the frame rotating with body, and is therefore accompanied by a large-scale flow. Tidal dissipation occurs when this flow is resisted either by linear damping (in particular, the effective viscosity associated with turbulent motion in convective zones) or nonlinear processes resulting from instabilities. An important example of the latter is the *elliptical instability*, in which the oscillatory tidal flow in a rotating fluid is unstable through the parametric resonance of inertial waves (see the chapter by Michael Le Bars).

The *dynamical* (or *wavelike*) tide is an additional component of the response in the form of internal waves. These waves are excited linearly by residual, vortical forcing that occurs because the equilibrium tide does not exactly satisfy the equation of motion. The dynamical tide can make a significant, or even dominant, contribution to tidal dissipation if the waves are amplified through resonance, or if they are effectively damped as a result of developing very short length scales or becoming nonlinear.

Both inertial waves and internal gravity waves have suitable properties for this purpose. Indeed, the tidal forcing frequencies in many applications naturally fall within the frequency range of inertial waves. (An important exception is when a body rotates slowly compared to the orbital motion, as in the case of solar-type stars with very short-period planets, but in this case internal gravity waves may be important.)

Tidal Dissipation via Inertial Waves

In convective regions of stars and giant planets, there is no stable stratification to provide the restoring force required for internal gravity waves. Although it is likely to be an oversimplification, convective regions are often modelled as neutrally stratified, barotropic fluids (in which $p = p(\rho)$) that lack buoyancy forces altogether; the convective flows are also neglected, or are assumed to provide an effective viscosity.

When the linear response to harmonic tidal forcing is decomposed into wavelike and non-wavelike parts, the wavelike tide can be calculated using two-dimensional numerical computations (e.g. Ogilvie and Lin 2004). The basic state is usually considered to be a spherically symmetric, hydrostatic equilibrium in which centrifugal effects are neglected, although the full Coriolis force is taken into account in the linearized equations for the tidal perturbations. If the core of the body is radiative, then the calculation of inertial waves may be restricted to the convective spherical shell, with boundary conditions that allow the waves to be reflected by the strongly stably stratified core. High resolution is required to examine the low-viscosity limit of interest, because of the presence of wave singularities associated with the critical latitude and wave attractors.

The results of many calculations of this type (e.g. Rieutord and Valdettaro 2010; Ogilvie 2013) can be summarized as follows. If the inner radius is zero, then the response curve is dominated by a few Lorentzian peaks corresponding to classical resonances with large-scale normal modes of the inertial waves in a full sphere. These were mentioned in the section 'Inertial Waves in a Sphere'; however, in the case of $l = m = 2$ forcing of a homogeneous sphere, these resonances are absent because the forcing has no overlap with the normal modes—indeed, the response in this case consists purely of the special flows described in the section 'Incompressible Fluid Ellipsoids'. For small values of the inner radius, the response remains similar because the large-scale modes are barely affected by the presence of a small core, even though exact inviscid modal solutions may cease to exist. For larger values of the inner radius, the response is complicated by the ray dynamics and wave singularities of the spherical shell and exhibits an erratic dependence on frequency that is exacerbated at lower viscosities.

Despite this complexity, it has been found that a certain frequency average of the response curve is exactly independent of viscosity and depends only on the large-scale properties of the basic state (Ogilvie 2013). This quantity can be determined from an impulse calculation (which is related physically to what happens in a single tidal encounter in a parabolic or hyperbolic orbit) by finding the smooth solution to a relatively simple set of ordinary differential equations. An explicit analytical example of this is provided by $l = m = 2$ forcing in a homogeneous body in which the core (of fractional radius α) is rigid; the dimensionless measure of the frequency-averaged dissipation is then

$$\frac{100\pi}{63}\left(\frac{\alpha^5}{1-\alpha^5}\right)\frac{R^3\Omega^2}{GM}.$$

This expression exemplifies the more widely applicable properties that tidal dissipation via inertial waves is more efficient if the body is more rapidly rotating or if it has a larger core that excludes the inertial waves. The dependence on α in a homogeneous body is very steep and diverges in the limit of a thin shell, $\alpha \to 1$; this dependence is moderated in more realistic models with density stratification, where some dissipation through normal modes remains in the limit $\alpha \to 0$.

The frequency-averaged dissipation is a crude measure and must be viewed with caution; what matters in practice is the time-averaged dissipation, which may differ significantly from the frequency-averaged quantity, because in actual applications the system may either lock into, or stall in between, the peaks of the response curve. It has been found, however, that the presence of a magnetic field provides some smoothing (i.e. small-scale frequency averaging) of the response curve (Lin and Ogilvie 2018) and it is likely that other physical complications such as nonlinearity, convection, imperfect reflections, etc., have a similar effect.

Tidal Dissipation via Internal Gravity Waves

In radiative regions of stars and giant planets, the stable stratification dominates the physics of internal waves. Internal gravity waves are typically excited near a convective–radiative interface where the buoyancy frequency rises steeply from zero and locally matches the tidal frequency. The waves propagate into the radiative region and may reflect to form global, standing 'g modes', modified to some extent by rotation. Linear waves are typically only very weakly damped by radiative thermal diffusion, so the g modes form sharp resonant peaks in the response curve, with negligible tidal dissipation in between.

However, if the waves propagate into a region in which they locally exceed the critical amplitude for instability (as discussed in the section 'Instabilities of Internal Waves'), they will break and be absorbed, instead of forming global modes. In particular, Barker and Ogilvie (2010) found that internal gravity waves excited in a solar-type star by a close planetary companion would break near the centre of the star if

$$0.28\left(\frac{C}{C_\odot}\right)^{5/2}\left(\frac{M_\mathrm{p}}{M_\mathrm{J}}\right)\left(\frac{M_*}{M_\odot}\right)^{-1}\left(\frac{P}{\mathrm{day}}\right)^{1/10}\gtrsim 1,$$

where C is a measure of the stable stratification near the centre, $C_\odot$ is its estimated value in the present Sun, $M_\mathrm{p}/M_\mathrm{J}$ is the planetary mass in Jovian units, $M_*/M_\odot$ is the stellar mass in solar units and P is the orbital period. Because C increases with stellar mass and also with time, as nuclear reactions cause heavier elements to build up, this mechanism could turn on at a critical age. The planet would then be consumed within a few million years as its orbital angular momentum is transferred to the stellar core through gravity waves.

Indeed, observations show that short period and massive hot Jupiters are generally absent around stars in which breaking would be expected to occur. A possible exception is WASP-12 b, for which transit-timing variations have revealed that the orbit is shrinking on a time scale of a few million years. If the star is interpreted as a subgiant with a stably stratified core (Weinberg et al. 2017), then internal wave breaking would be expected in this system and may explain the rapid orbital decay that is observed.

Future Perspectives

Much further work is needed to improve on the idealized calculations that have been made so far. For example, the nonlinear interplay between inertial waves and zonal flows (e.g. Favier et al. 2014), their interaction with convection, magnetic fields and meridional circulation, and their instabilities and breaking all require deeper investigation. Another rich subject worthy of further study is the weakly nonlinear dynamics of large collections of internal waves (e.g. Weinberg et al. 2012). And, given the complexity of the linear response curves, the ability of evolving systems to lock into tidal resonances with internal waves (e.g. Witte and Savonije 1999; Fuller et al. 2016) ought to be investigated in a wide variety of circumstances.

References

Alvan, L., Brun, A. S., & Mathis, S. (2014). Theoretical seismology in 3D: Nonlinear simulations of internal gravity waves in solar-like stars. *Astronomy & Astrophysics*, *565*, A42.

Backus, G., & Rieutord, M. (2017). Completeness of inertial modes of an incompressible inviscid fluid in a corotating ellipsoid. *Physical Review E*, *95*, 053116.

Barker, A. J., & Ogilvie, G. I. (2010). On internal wave breaking and tidal dissipation near the centre of a solar-type star. *Monthly Notices of the Royal Astronomical Society*, *404*, 1849–1868.

Bryan, G. H. (1889). The waves on a rotating liquid spheroid of finite ellipticity. *Philosophical Transactions of the Royal Society of London Series A*, *180*, 187–219.

Chandrasekhar, S. (1969). *Ellipsoidal figures of equilibrium*. New Haven: Yale University Press.

Christensen-Dalsgaard, J., Dappen, W., Ajukov, S. V., et al. (1996). The current state of solar modeling. *Science*, *272*, 1286–1292.

Dauxois, T., Joubaud, S., Odier, P., & Venaille, A. (2018). Instabilities of internal gravity wave beams. *Annual Review of Fluid Mechanics*, *50*, 131–156.

Dintrans, B., & Rieutord, M. (2000). Oscillations of a rotating star: A non-perturbative theory. *Astronomy & Astrophysics*, *354*, 86–98.

Favier, B., Barker, A. J., Baruteau, C., & Ogilvie, G. I. (2014). Non-linear evolution of tidally forced inertial waves in rotating fluid bodies. *Monthly Notices of the Royal Astronomical Society*, *439*, 845–860.

French, R. G., McGhee-French, C. A., Nicholson, P. D., & Hedman, M. M. (2019). Kronoseismology III: Waves in Saturn's inner C ring. *Icarus*, *319*, 599–626.

Fuller, J., Luan, J., & Quataert, E. (2016). Resonance locking as the source of rapid tidal migration in the Jupiter and Saturn moon systems. *Monthly Notices of the Royal Astronomical Society*, *458*, 3867–3879.

Goodman, J., & Lackner, C. (2009). Dynamical tides in rotating planets and stars. *Astrophysical Journal*, *696*, 2054–2067.

Helled, R. (2019). The interiors of Jupiter and Saturn. In Read P., et al. (Eds.), *Oxford research encyclopedia of planetary science*. Oxford: Oxford University Press.

Ivanov, P. B., & Papaloizou, J. C. B. (2010). Inertial waves in rotating bodies: A WKBJ formalism for inertial modes and a comparison with numerical results. *Monthly Notices of the Royal Astronomical Society*, *407*, 1609–1630.

Ivers, D. J., Jackson, A., & Winch, D. (2015). Enumeration, orthogonality and completeness of the incompressible Coriolis modes in a sphere. *Journal of Fluid Mechanics*, *766*, 468–498.

Jouve, L., & Ogilvie, G. I. (2014). Direct numerical simulations of an inertial wave attractor in linear and nonlinear regimes. *Journal of Fluid Mechanics*, *745*, 223–250.

Lin, Y., & Ogilvie, G. I. (2018). Tidal dissipation in rotating fluid bodies: The presence of a magnetic field. *Monthly Notices of the Royal Astronomical Society*, *474*, 1644–1656.

Lockitch, K. H., & Friedman, J. L. (1999). Where are the r-modes of isentropic stars? *Astrophysical Journal*, *521*, 764–788.

Maas, L. R. M., & Lam, F. P. A. (1995). Geometric focusing of internal waves. *Journal of Fluid Mechanics*, *300*, 1–41.

Mirouh, G. M., Baruteau, C., Rieutord, M., & Ballot, J. (2016). Gravito-inertial waves in a differentially rotating spherical shell. *Journal of Fluid Mechanics*, *800*, 213–247.

Moore, D. W., & Saffman, P. G. (1969). The structure of free vertical shear layers in a rotating fluid and the motion produced by a slowly rising body. *Philosophical Transactions of the Royal Society of London Series A*, *264*, 597–634.

Ogilvie, G. I. (2005). Wave attractors and the asymptotic dissipation rate of tidal disturbances. *Journal of Fluid Mechanics*, *543*, 19–44.

Ogilvie, G. I. (2009). Tidal dissipation in rotating fluid bodies: A simplified model. *Monthly Notices of the Royal Astronomical Society*, *396*, 794–806.

Ogilvie, G. I. (2013). Tides in rotating barotropic fluid bodies: The contribution of inertial waves and the role of internal structure. *Monthly Notices of the Royal Astronomical Society*, *429*, 613–632.

Ogilvie, G. I. (2014). Tidal dissipation in stars and giant planets. *Annual Review of Astronomy and Astrophysics*, *52*, 171–210.

Ogilvie, G. I., & Lin, D. N. C. (2004). Tidal dissipation in rotating giant planets. *Astrophysical Journal*, *610*, 477–509.

Phillips, O. M. (1981). Wave interactions—The evolution of an idea. *Journal of Fluid Mechanics*, *106*, 215–227.

Prat, V., Lignières, F., & Ballot, J. (2016). Asymptotic theory of gravity modes in rotating stars I. Ray dynamics. *Astronomy & Astrophysics*, *587*, A110.

Rieutord, M., & Valdettaro, L. (2010). Viscous dissipation by tidally forced inertial modes in a rotating spherical shell. *Journal of Fluid Mechanics*, *643*, 363–394.

Rieutord, M., Georgeot, B., & Valdettaro, L. (2001). Inertial waves in a rotating spherical shell: attractors and asymptotic spectrum. *Journal of Fluid Mechanics*, *435*, 103–144.

Sridhar, S., & Tremaine, S. (1992). Tidal disruption of viscous bodies. *Icarus*, *95*, 86–99.

Thompson, M. J., Toomre, J., Anderson, E. R., et al. (1996). Differential rotation and dynamics of the solar interior. *Science*, *272*, 1300–1305.

Weinberg, N. N., Arras, P., Quataert, E., & Burkart, J. (2012). Nonlinear tides in close binary systems. *Astrophysical Journal*, *751*, 136.

Weinberg, N. N., Sun, M., Arras, P., & Essick, R. (2017). Tidal dissipation in WASP-12. *Astrophysical Journal*, *849*, L11.

Witte, M. G., & Savonije, G. J. (1999). Tidal evolution of eccentric orbits in massive binary systems. A study of resonance locking. *Astronomy & Astrophysics*, *350*, 129–147.

Wu, Y. (2005). Origin of tidal dissipation in Jupiter. I. Properties of inertial modes. *Astrophysical Journal*, *635*, 674–687.

Chapter 2
Waves and Convection in Stellar Astrophysics

Daniel Lecoanet

Abstract This chapter begins with the principles determining a star's structure: hydrostatic and thermal balance, and energy generation and transport. These imply that some stars have stably stratified cores and convective envelopes, whereas other stars have convective cores and stably stratified envelopes. The convection in stars is predominantly low Mach number, but the density at the top of a convection zone can be orders of magnitude smaller than the density at the bottom. We derive the anelastic equations which can model efficient, low Mach number convection. The properties of stars can be inferred by studying the waves at their surface. Here we describe sound and internal gravity waves, both of which have been observed in the Sun or other stars. The second half of this chapter discusses two phenomena at the interface between the convective and stably stratified layers of stars. First we consider convective overshoot, the convective motions which can extend into an adjacent stably stratified fluid. This can lead to substantial mixing in the stably stratified part of stars. Then, we discuss internal gravity wave generation by convection, which can lead to wave-induced energy or momentum transport. These illustrate some important fluid dynamical problems in stellar astrophysics.

Stellar Structure and Evolution

Stars are some of the most important objects in astrophysics. Stars impact the evolution and structure of galaxies and the interstellar medium; form black holes and/or neutron stars which, in binary systems, can radiate gravitational waves prior to

D. Lecoanet (✉)
Department of Astrophysical Sciences, Princeton Center for Theoretical Science, Princeton University, Princeton, NJ, USA
e-mail: lecoanet@princeton.edu

M. Le Bars and D. Lecoanet (eds.), *Fluid Mechanics of Planets and Stars*,
CISM International Centre for Mechanical Sciences 595,
https://doi.org/10.1007/978-3-030-22074-7_2

merger; and are the hosts to planetary systems, including our own. The structure and evolution of stars dictate the impact they can have on their environments.

Stars are fluid bodies held together by gravity, and supported by the pressure of their hot interiors, which remain hot due to nuclear burning in their cores. They are characterized by several important timescales:

$$\tau_d = \sqrt{\frac{R^3}{GM}}, \qquad \tau_{KH} = \frac{GM}{RL}, \qquad \tau_N = \frac{MQ}{L}. \tag{2.1}$$

The *dynamical* time, τ_d is defined in terms of the star's radius R, mass M, and the gravitational constant G. It estimates the time for gravitational disturbances and/or sound waves to propagate across the star. The *Kelvin–Helmholtz* time, τ_{KH} depends on the luminosity of the star, L. It estimates the time for energy generated by nuclear reactions within the star to leave the star. Finally, the *nuclear burning* time, τ_N, is an estimate of the time to burn the star's fuel via nuclear fusion. Here, Q denotes the energy generation per unit mass by the fusion reaction occurring in the star (e.g., hydrogen fusing to helium).

As a concrete example, consider the Sun. For the Sun, we have $M_\odot \sim 2 \times 10^{30}$ kg, $R_\odot \sim 7 \times 10^8$ m, $L_\odot \sim 4 \times 10^{26}$ W, and $Q \sim 4 \times 10^{14}$ J/kg. Combining these gives $\tau_d \sim 30$ min, $\tau_{KH} \sim 3 \times 10^7$ yr, and $\tau_N \sim 10^{10}$ yr. In general, stars typically satisfy

$$\tau_d \ll \tau_{KH} \ll \tau_N. \tag{2.2}$$

Stellar structure and evolution are concerned with stars on the long, nuclear timescale. On this long timescale, the star is in approximate hydrostatic balance ($\tau_d \ll \tau_N$), and thermal equilibrium ($\tau_{KH} \ll \tau_N$). If one neglects rotation, the problem is spherically symmetric, and one can study the radial structure of the star. This gives the base state from which we will study waves, convection, and magnetic fields. These phenomena cannot be fully studied in the context of a one-dimensional stellar model; thus their back-reaction on the stellar structure can only be parameterized.

Lane–Emden Equation

We will now derive stellar structure equations. As described above, on long timescales ($\sim\tau_N$), the star is in hydrostatic balance. If we define $m(r)$ to be the enclosed mass, then hydrostatic balance implies

$$\frac{dm}{dr} = 4\pi r^2 \rho, \tag{2.3}$$

$$\frac{dp}{dr} = -\frac{Gm\rho}{r}. \tag{2.4}$$

Equation 2.3 is the definition of the enclosed mass: $m(r)$ is the volume integral of the density between the center of the star and r. Equation 2.4 is hydrostatic balance: the gravitational force (right hand side of the equation) is balanced by the pressure gradient force (left hand side). To solve for a stellar structure, one must use an *equation of state*, which is a relation between the pressure and other state variables, e.g., $p = p(\rho, T, c_H, c_{He}, \ldots)$, where c_i represents the volume fraction of element i. Thus, Eqs. 2.3 and 2.4, in general, are supplemented with additional equations for T and the chemical composition. However, in this section, we will consider the simplest case of a *polytropic* equation of state where pressure depends only on the density, such that Eqs. 2.3 and 2.4 form a closed system.

It is convenient to write the polytropic equation of state as

$$p = K\rho^{1+1/n}, \tag{2.5}$$

where K is a constant of proportionality, and n is the *polytropic index*. Then, p and ρ can both be represented in terms of a single variable Θ which satisfies

$$\rho = \rho_0 \Theta^n, \qquad p = p_0 \Theta^{n+1}, \tag{2.6}$$

where ρ_0, p_0 are the density and pressure at the center of the star, where $\Theta = 1$. Then, Θ satisfies a second-order differential equation in r

$$\frac{p_0(n+1)}{4\pi G\rho_0^2} \frac{1}{r^2} \frac{d}{dr} r^2 \frac{d\Theta}{dr} = -\Theta^n, \tag{2.7}$$

or in non-dimensional form,

$$\eta^{-2} \frac{d}{d\eta} \left(\eta^2 \frac{d\Theta}{d\eta} \right) = -\Theta^n, \tag{2.8}$$

after defining $r = \alpha\eta$ with $\alpha^2 = p_0(n+1)/(4\pi G\rho_0^2)$.

Equation 2.8 is known as the Lane–Emden equation. The typical boundary conditions are $\Theta(0) = 1$ and $\Theta'(0) = 0$. The first is a normalization condition, and the second is a regularity condition associated with spherical geometry. The boundary of the star is at η_R where $\Theta(\eta_R) = 0$. The only free parameter in the problem is the polytropic index n. One can solve this nonlinear boundary value problem using a variety of techniques. Boyd (2011) describes how to solve the problem using Chebyshev spectral methods, which can be easily implemented in the Dedalus[1] code (Burns et al. 2019).

In Fig. 2.1, we plot the solution to the Lane–Emden equation for several values of n. As n increases, the density and pressure become more centrally peaked, but the radius of the star increases. Although this is a very simple model for the structure of a star, it is one of the few analytic results in stellar structure. Furthermore, the

[1]More information at https://dedalus-project.org.

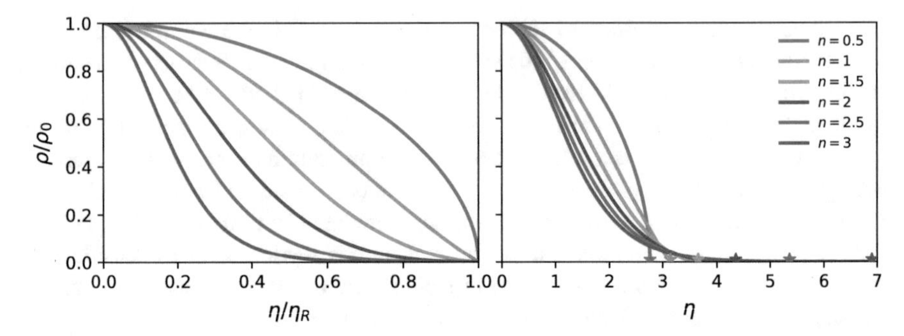

Fig. 2.1 Solutions to the Lane–Emden equation for different n. In the left plot, the normalized density is plotted as a function of the fractional radius. In the right plot, it is plotted as a function of $\eta = r/\alpha$. The outer boundary of the star (where $\rho = 0$) is denoted by a star

polytropic equation of state is a good approximation in regions of stars with efficient convection. As we will describe below, in these regions, $p \sim \rho^{\gamma}$, where γ is the ratio of specific heats, which is $5/3$ for a monatomic ideal gas. For this reason, the $n = 3/2$ polytrope is an important special case.

Stellar Energy Transport

Typically, a star's pressure depends on both its density and temperature, e.g., the ideal gas equation of state, $p = \mathcal{R}\rho T/\mu$, where $\mathcal{R}$ is the ideal gas constant, and μ is the mean molecular weight. If a star is in thermal equilibrium, the energy generated in a region matches the net flux through that region's boundary. Then defining the local luminosity $L(r) = 4\pi r^2 F(r)$, where F is the flux, we have

$$\frac{dL}{dr} = 4\pi r^2 \rho \epsilon, \tag{2.9}$$

where ϵ the energy generation rate. This energy equation states that energy generated in the core by nuclear reactions must propagate out of the star.

But what carries this energy? There are two main transport mechanisms: photon radiation and convection. The energy transport mechanism relates the luminosity to the temperature gradient, allowing us to solve for the temperature profile of the star. This then closes the stellar structure equations.

Radiative Energy Transport Energy can be transported by the random diffusion of photons. The photons emitted by a black body carry an energy density proportional to the fourth power of temperature, aT^4, where $a = 4\sigma/c$, σ is the Stefan–Boltzmann constant and c is the speed of light. Since different parts of a star are at different temperatures, the photons on average carry an energy flux from regions of high temperature to regions of low temperature. This is proportional to the differential energy density, $4aT^3 dT/dr$, times the photon diffusivity, $c\ell$. The mean-free path, ℓ, is often written as $\ell = (\rho\kappa)^{-1}$, where κ is the photon opacity. The *radiative luminosity* is then

$$L_{\rm rad} = 4\pi r^2 F_{\rm rad} = -\frac{64\pi r^2 \sigma T^3}{3\rho\kappa}\frac{dT}{dr}. \quad (2.10)$$

This differs from our estimate by 1/3, a geometric factor which emerges from a more careful analysis (e.g., Sect. 5.1.2 of Kippenhahn and Weigert 1994). Thus, in the radiative zone of a star, where the energy is carried by radiation, the temperature gradient is given by

$$\frac{dT}{dr} = \left(\frac{dT}{dr}\right)_{\rm rad} = -\frac{3\rho\kappa L}{64\pi r^2 \sigma T^3}. \quad (2.11)$$

Convective Energy Transport Energy can alternatively be transported by bulk fluid motions, known as convection. Convection occurs when the fluid in a star is, in some sense, top-heavy. Under the Boussinesq approximation, which assumes density variations are small in comparison to the background density, this occurs when the density gradient is in the opposite sense as gravity (i.e., high density above low density when gravity points downward). The criterion for convection in a star is slightly different because there are large density contrasts and the Boussinesq approximation is not valid. In this chapter, we will sometimes use the Boussinesq approximation to simplify analysis; other times, non-Boussinesq effects are important, so we will include them.

As above, we can gain insight into stellar convection by comparing different timescales. Here we will focus on *efficient convection*, which occurs in the interiors of stars—convection very close to a star's surface can be *inefficient*, and some of our approximations will break down. Efficient convection occurs at low Mach numbers, $\mathrm{Ma}_c = u_c/c_s$, where u_c is a typical convective velocity, and c_s is the sound speed. We will consider the convective time which is a representative length scale of the convection (i.e., scaleheight or depth of the convecting layer), over a typical convective velocity. Thus, the convective time $\tau_c \gg \tau_d$, the dynamical time. This means convective motions stay in pressure equilibrium. Simultaneously, the convective time is typically much shorter than the thermal time, $\tau_{KH} \gg \tau_c$. This means convective motions are nearly adiabatic. For example, in the Sun, the convective time is about 1 month.

Now consider a fluid element with density $\rho_i = \rho(s_i, p_i)$, where s_i and p_i are the initial entropy and pressure. If the fluid element is displaced radially outwards, its pressure will equilibrate to the ambient pressure p_a quickly. Thus, its new density will be $\rho_n = \rho(s_i, p_a)$. This must be compared to the ambient density, $\rho_a = \rho(s_a, p_a)$. If the $\rho_a > \rho_n$, then the fluid element is buoyant and will continue to rise. This occurs when $ds/dr < 0$, because $(d\rho/ds)_p < 0$. On the other hand, if $\rho_a < \rho_n$, the fluid element is denser than its surrounds and will fall down to its original position. Then the fluid is stably stratified, which occurs when $ds/dr > 0$. If the convection is efficient, it will rearrange the entropy of the fluid until the fluid is marginally stable to convection, i.e., $ds/dr = 0$.

Using hydrostatic balance and the equation of state, constant entropy corresponds to a specific temperature gradient, known as the *adiabatic temperature gradient*, $(dT/dr)_{\mathrm{ad}}$. As an example, consider an ideal gas with ratio of specific heats γ. We have $s \sim \log(p^{1/\gamma}/\rho) \sim \log(Tp^{(1-\gamma)/\gamma})$. Taking the derivative of this expression with respect to r, and using that $ds/dr = 0$ for an adiabatic process, and $dp/dr = -\rho g$, we find

$$\left(\frac{dT}{dr}\right)_{\mathrm{ad}} = -\frac{g}{C_p}, \tag{2.12}$$

where C_p is the specific heat at constant pressure. Furthermore, at constant entropy, $p \sim \rho^\gamma$, so the mean structure follows a polytrope solution with $\gamma = 1 + n^{-1}$. For a monotonic ideal gas, $\gamma = 5/3$, which corresponds to polytropic index $n = 3/2$. Thus, convective regions of stars can be modeled using Lane–Emden solutions.

Stellar Evolution Equations

The equations described above can be combined into the *stellar structure equations*

$$\frac{dm}{dr} = 4\pi r^2 \rho, \tag{2.13a}$$

$$\frac{dp}{dr} = -\frac{Gm\rho}{r}, \tag{2.13b}$$

$$\frac{dL}{dr} = 4\pi r^2 \rho \epsilon, \tag{2.13c}$$

$$\frac{dT}{dr} = -\min\left[\frac{3\rho\kappa L}{64\pi r^2 \sigma T^3}, -\left(\frac{dT}{dr}\right)_{\mathrm{ad}}\right]. \tag{2.13d}$$

These equations, respectively, represent continuity, hydrostatic balance, nuclear energy generation, and energy transport. Equation 2.13d says the energy is transported by radiation (the left term in the minimum), unless the radiative temperature gradient is greater in magnitude than the convective temperature gradient (the right term in the minimum). In that case, energy is also transported by convection.

However, these equations are not yet closed. They must be supplemented with *microphysics* equations

$$\rho = \rho(p, T, c_i), \tag{2.14a}$$

$$\epsilon = \epsilon(p, T, c_i), \tag{2.14b}$$

$$\kappa = \kappa(p, T, c_i). \tag{2.14c}$$

These equations are the equation of state, the nuclear reaction rates, and the opacity. Because these depend on the chemical composition (especially the nuclear reaction rate), we must also evolve the c_i according to

$$\frac{\partial c_i}{\partial t} = \sum_k \omega_k \, \Delta c_i^k, \tag{2.15}$$

and the sum is over different nuclear reactions labeled with k. The reaction rate is denoted ω_k, and Δc_i^k is the change in the concentration of chemical i every time reaction k occurs.

Finally, we need to specify four boundary conditions to solve Eqs. 2.13. At the center, we have that

$$L(0) = 0, \qquad m(0) = 0. \tag{2.16}$$

The other two boundary conditions are at the surface of the star. The luminosity of a star can be written in terms of an effective temperature, $T_{\rm eff}$, which satisfies $L(R) = 4\pi r^2 \sigma T_{\rm eff}^4$. The effective temperature is the temperature the star would be at if it was a black body. The surface of a star, called the *photosphere*, is defined to be where the temperature equals the effective temperature, i.e., $T(R) = T_{\rm eff}$. This relation between the temperature and the luminosity is the third boundary condition. The last boundary condition sets the pressure at the photosphere. This comes from integrating hydrostatic balance over the stellar atmosphere. A simple model for this is $P(R) = (2/3)GM/(R^2\kappa)$, where κ is the average opacity in the atmosphere (e.g., Sect. 10.2 of Kippenhahn and Weigert 1994). More realistic boundary conditions can be calculated from 3D simulations of stellar surfaces (e.g., Mosumgaard et al. 2018).

To solve for a stellar structure, one solves Eqs. 2.13 together with a microphysics scheme (Eqs. 2.14). This requires knowing the chemical composition so one can calculate the nuclear reaction rates. However, these nuclear reactions change the chemical composition according to Eq. 2.15, which in turn changes the energy generation rate and the stellar structure. Thus, these equations can be coupled to study how the stellar structure changes with time.

Stars begin their lives mostly composed of hydrogen. They burn hydrogen into helium in their cores. This part of a star's life is known as the *main sequence*. The structure of a star does not change much on the main sequence because the nuclear reaction rates do not depend much on the exact quantity of hydrogen in the core.

The structure of several representative main-sequence stars is depicted in Fig. 2.2. Almost all stars have regions where the energy is transported by radiation ("radiative zone"), and regions where the energy is transported by convection ("convective zone"). Convection occurs when the radiative temperature gradient (Eq. 2.11) is large in magnitude. Typically, this occurs because either L is large or κ is large. Massive main-sequence stars have core convection zones because L is large. Their convection zone ends because the r^{-2} geometry factor decreases the magnitude of the radiative temperature gradient. However, smaller main-sequence stars also have convection

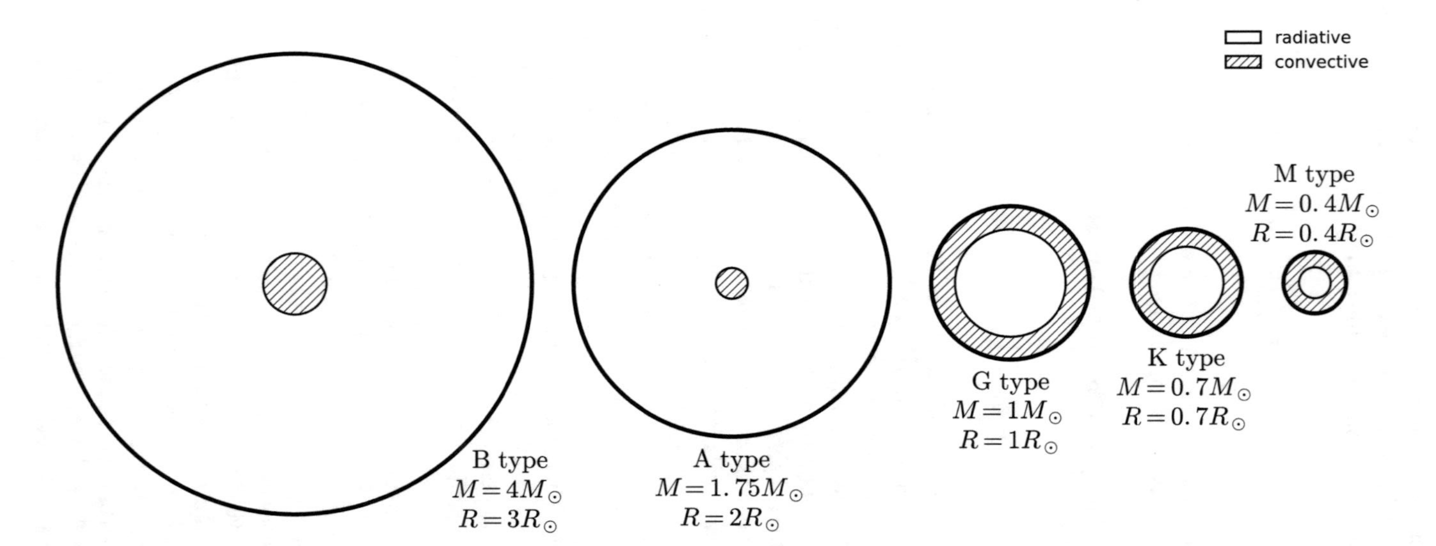

Fig. 2.2 Schematic diagram of several main-sequence stars, labeled by their spectral type, showing the regions of the star that are radiative and convective. The diagrams are based on stellar structure models calculated using MESA (Paxton et al. 2011, 2013, 2015, 2018), and provided by Matteo Cantiello (private communications 2019). $M_\odot$ and $R_\odot$ denote the mass and radius of the Sun, respectively

zones—but now in the outer part of the star. This is because the opacity becomes large due to increasing bound-free opacity as the temperature decreases toward the surface (e.g., Sect. 4.6 of Hansen et al. 2004). Thus, for solar-type stars, the transition between radiation and convection zones is mediated by the temperature dependence of the opacity. For very small M-type stars (e.g., $M \lesssim 0.25 M_{\odot}$), the star becomes completely convective.

After a star runs out of hydrogen in its core, it begins to contract and heat up. If the envelope becomes hot enough, it can start to burn hydrogen itself, causing the outer part of the star to puff out, leading to a red giant star. Stars which are sufficiently massive will burn helium into carbon in their cores. The structure and evolution of these massive stars can become extremely complicated, with many shells comprised of various elements, each having different nuclear reactions, some of which are convective, others radiative (e.g., Chap. 33 of Kippenhahn and Weigert 1994). The details of massive star evolution, including the effects of rotation, are important for determining the final fate of these stars (Langer 2012; Sukhbold and Woosley 2014): Do they collapse to neutron stars? Form black holes? Explode as supernova?

There are many uncertainties in stellar evolution. One uncertainty comes from the microphysics schemes (Eqs. 2.14). Making different assumptions about opacities, equations of state, reaction rates, size of reaction networks, etc., can lead to different stellar properties (Fields et al. 2018). Massive stars exhibit a range of complicated phenomena, such as opacity-driven waves or convection (Townsend et al. 2018; Cantiello and Braithwaite 2011), flames (Denissenkov et al. 2013), and outbursts (Jiang et al. 2018). Mass loss, probably due to launching magnetic winds, is key to understanding the evolution of massive stars (Smith 2014).

Another important uncertainty in stellar evolution is convection. Convection occurs in almost every star, and is a multidimensional phenomenon which must be parameterized in one-dimensional stellar evolution calculations.

Modeling Stellar Convection

The most studied convection problem in the fluid dynamics literature is Rayleigh–Bénard convection. A major difference between Rayleigh–Bénard convection and convection in stars is the importance of compressibility. There can be large density contrasts across the convection zones of many stars, e.g., about 10^6 in the case of the solar convection zone (Hotta et al. 2014). Standard simplifying assumptions are that the entropy gradient is small (i.e., the convection is efficient), and the Mach number is also small.

Although stars are close to spherical, it is sometimes convenient to study convection in stars in cartesian geometry, as we will do in this section. This simplifies the analysis, and can provide a decent model for some phenomena in stars where the convective layer depth is much smaller than the radial location of the layer. Curvature effects play a secondary role in such thin shells.

To model strongly stratified convection, consider the (ideal) Euler equations including compressibility

$$\rho \frac{D\boldsymbol{u}}{Dt} = -\boldsymbol{\nabla} p + \rho \boldsymbol{g}, \tag{2.17a}$$
$$\frac{D\rho}{Dt} = -\rho \boldsymbol{\nabla} \cdot \boldsymbol{u}, \tag{2.17b}$$
$$\frac{Ds}{Dt} = 0, \tag{2.17c}$$

where $D/Dt = \partial_t + \boldsymbol{u} \cdot \boldsymbol{\nabla}$ is the advective derivative. First we must quantify the effects of compressibility. One way is to consider the *scaleheight*, which gives an estimate of the lengthscale on which the density and pressure vary. If g and the sound speed squared $c_s^2 = p/\rho$ are assumed to be constant, then a solution to hydrostatic balance in cartesian geometry, $\partial_z p = -g\rho$, is

$$p \sim \rho \sim \exp(-zg/c_s^2) \equiv \exp(-z/H). \tag{2.18}$$

Even when c_s^2 is not constant, $c_s^2/g \equiv H$ gives an estimate of the scaleheight.

We non-dimensionalize these equations using

$$\begin{aligned} \boldsymbol{u} &\rightarrow \tilde{u}\boldsymbol{u}, & \nabla &\rightarrow \tilde{L}^{-1}\nabla, & \partial_t &\rightarrow \tilde{u}\tilde{L}^{-1}\partial_t, \\ \rho &\rightarrow \tilde{\rho}\rho, & p &\rightarrow \tilde{\rho}\tilde{c}_s^2 p, & g &\rightarrow \tilde{c}_s^2/\tilde{H}, \end{aligned}$$

where after the substitution, quantities with a tilde $(\tilde{\cdot})$ are dimensional, and quantities without a tilde are dimensionless.

The form of the continuity and entropy equations does not change upon non-dimensionalization, but the momentum equations become

$$\rho \frac{D\boldsymbol{u}}{Dt} = -\mathrm{Ma}^{-2} \boldsymbol{\nabla} p - \frac{\tilde{L}}{\tilde{H}} \mathrm{Ma}^{-2} \rho \boldsymbol{e}_z, \tag{2.19}$$

where $\mathrm{Ma} = \tilde{u}/\tilde{c}_s$ is the Mach number, and gravity is assumed to point in the $-\boldsymbol{e}_z$ direction.

If one assumes $\tilde{L}/\tilde{H} \sim \mathcal{O}(\mathrm{Ma})$, then one can derive the Boussinesq equations. Thus, the Boussinesq equations are appropriate for dynamics on lengthscales much smaller than the scaleheight. However, for stellar convection, one is interested in lengthscales comparable to the scaleheight, so we instead assume $\tilde{L}/\tilde{H} \sim \mathcal{O}(\mathrm{Ma}^0)$. In this case, we can write an asymptotic series in the small parameter $\epsilon = \mathrm{Ma}^2$, i.e., $\boldsymbol{u} = \boldsymbol{u}_0 + \epsilon^1 \boldsymbol{u}_1 + \cdots$.[2] Then the lowest order momentum equation is

$$\boldsymbol{\nabla} p_0 = -\rho_0 \boldsymbol{e}_z. \tag{2.20}$$

[2] ϵ should not be confused with the nuclear reaction rate above.

Next, we make a crucial assumption. We assume the entropy gradient is small because the convection is efficient. This means

$$\nabla s_0 = 0. \tag{2.21}$$

Together with an equation of state, we can now derive a background state for the convection problem, which consists of $p_0(z)$, $\rho_0(z)$, and s_0 constant. These are adiabatic atmospheres, and are assumed to be fixed in time. The lowest order continuity equation becomes

$$-\partial_t \rho_0 = \boldsymbol{u}_0 \cdot \nabla \rho_0 + \rho_0 \nabla \cdot \boldsymbol{u}_0 = \nabla \cdot (\rho_0 \boldsymbol{u}_0) = 0. \tag{2.22}$$

This is often referred to as the *anelastic constraint*.

The leading order entropy equation is $\partial_t s_0 = 0$ because $\nabla s_0 = 0$. The next to leading order equation is

$$\partial_t s_1 + \boldsymbol{u}_0 \cdot \nabla s_1 = 0. \tag{2.23}$$

Note that $\boldsymbol{u}_1 \cdot \nabla s_0 = 0$ because $\nabla s_0 = 0$. We cannot derive an equation for $\boldsymbol{u}_1$, so it is crucial that we assumed that $\nabla s_0 = 0$, otherwise the system of equations would not be closed.

The next to leading order momentum equation is

$$\rho_0 \frac{D\boldsymbol{u}_0}{Dt} = -\nabla p_1 - \rho_1 \boldsymbol{e}_z. \tag{2.24}$$

This is an equation for $\boldsymbol{u}_0$ in terms of p_1 and ρ_1. We can solve for p_1 by using Eq. 2.22 and inverting an elliptic operator. To find ρ_1 we must use an equation of state, i.e., $\rho_1 = \rho_1(p_1, s_1)$. It is common to use an ideal gas equation of state.

Put together, we have

$$\rho_0 \frac{D\boldsymbol{u}_0}{Dt} = -\nabla p_1 - \rho_1 \boldsymbol{e}_z, \tag{2.25a}$$

$$\nabla \cdot (\rho_0 \boldsymbol{u}_0) = 0, \tag{2.25b}$$

$$\partial_t s_1 + \boldsymbol{u}_0 \cdot \nabla s_1 = 0, \tag{2.25c}$$

$$\frac{s_1}{c_p} = \frac{p_1}{\gamma p_0} - \frac{\rho_1}{\rho_0}, \tag{2.25d}$$

where we have used the linearized ideal gas equation of state for the last equation. These equations are the *anelastic equations*, and are used to model convection in astrophysical and geophysics fluids.

A crucial advantage of the anelastic equations is that they do not admit sound waves. When numerically integrating the full compressible equations in a low Mach number flow, the largest stable timestep size is typically related to the sound crossing

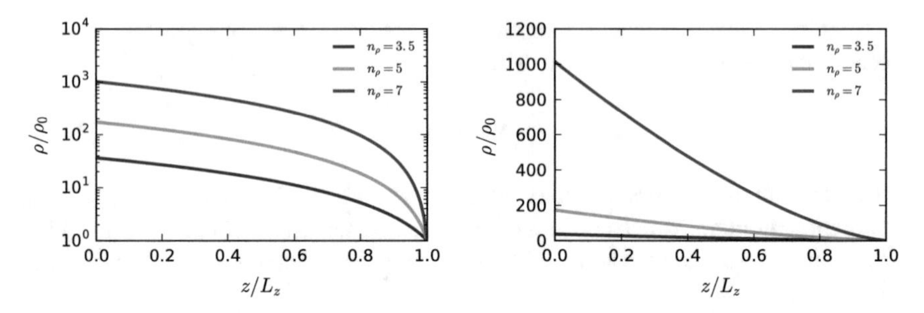

Fig. 2.3 Density as a function of height for adiabatic atmospheres with different density contrasts in cartesian geometry, $n_\rho = \log(\rho(L_z)/\rho_0)$. We assume $\gamma = 5/3$. In the left panel, the density is shown on a log scale, in the right panel, the density is shown on a linear scale. In all cases, most of the fractional density variation occurs near the top of the domain

time across grid cells. This is shorter than the flow crossing time across grid cells, by a factor of the Mach number. The anelastic (and Boussinesq) equations can take timesteps close to the flow crossing time because they do not admit sound waves. Note however, that implicit timestepping can be used to take large timesteps even when solving the full compressible equations (e.g., Viallet et al. 2013; Lecoanet et al. 2014).

The anelastic equations can also be applied to the atmosphere (Ogura and Phillips 1962), and in planetary cores (e.g., Jones et al. 2011, and references within). They are asymptotically valid in the limit of small Mach number, and small entropy gradient. The anelastic equations have been employed extensively in astrophysics to study different types of stars using the Anelastic Spherical Harmonic (ASH) code (e.g., Clune et al. 1999; Browning et al. 2004; Brown et al. 2011; Alvan et al. 2014; Augustson et al. 2016), as well as the Sun, especially using the open-source Rayleigh code[3] (e.g., Featherstone and Hindman 2016a, b).

There are two main differences between the Boussinesq and anelastic equations. First, the anelastic equations have a new length scale—the local scaleheight, e.g., $H = -(d \log \rho/dz)^{-1}$. In Fig. 2.3, we plot the density profile for different cartesian, adiabatic backgrounds with different density contrasts. In each case, the bottom part of the domain only contains a couple of density scale heights. Most of the density change occurs in a thin layer near the very top of the domain.

This is related to the second main difference. The Boussinesq equations have a vertical mirror symmetry, whereas the anelastic equations do not. A falling plume will narrow due to the anelastic constraint (Eq. 2.22), whereas a rising plume expands. There are suggestions that very narrow, falling plumes may stay coherent over many scaleheights, and may play an important role in stellar convection (Brandenburg 2016).

The effects of compressibility play an important role in *mixing length theory*, which is the parameterization of convection used in stellar evolution. The star's

[3]See https://geodynamics.org/cig/software/rayleigh/ for more information.

luminosity must be carried by some combination of radiation and convection. Thus, in a convection zone, the luminosity carried by convection is

$$L_{\text{conv}} = L_{\text{tot}} - L_{\text{rad}}. \tag{2.26}$$

Then the convective luminosity can be estimated as

$$L_{\text{conv}} = 4\pi r^2 \rho u_c^3, \tag{2.27}$$

where u_c is assumed to be the free-fall velocity over a lengthscale ℓ_{ML},

$$u_c = \ell_{\text{ML}} \sqrt{\frac{g}{c_p} \frac{ds}{dr}}. \tag{2.28}$$

ℓ_{ML} is referred to as the *mixing length*, and is normally assumed to be proportional to the scaleheight, $\ell_{\text{ML}} = \alpha H$, where α is the *mixing length parameter*. The interpretation is that the convection can be decomposed into flow structures whose dominant size is similar to ℓ_{ML}.

If ℓ_{ML} is specified, then one can solve for ds/dr in the convection zone given L_{conv}. Also, one can model the mixing of chemical species within the convection zone by assuming a turbulent convective diffusivity $D_t = \ell u_c$, and modifying the chemical evolution equation,

$$\frac{\partial c_i}{\partial t} = \sum_k \omega_k \Delta c_i^k + D_t \nabla^2 c_i. \tag{2.29}$$

These simple relations represent the state-of-the-art modeling of convection in stellar evolution.

Sound and Internal Gravity Waves in Stars

Waves play important diagnostic and dynamical roles in stars. Here we will neglect rotation, magnetic fields, and self-gravity, so there remain only two types of waves: sound waves and internal gravity waves. Waves are linear perturbations to a background state, so in this section, we will consider different simple background states and solve for the linear oscillations. To simplify the analysis we first consider waves in cartesian geometry, i.e., we neglect geometric factors. We then generalize to spherical geometry to discuss observations of waves in stars. Because sound waves play an important diagnostic role in stars, we will study the full compressible equations, rather than the Boussinesq or anelastic equations.

Uniform Background

The simplest background which supports waves is a uniform background with density ρ_0 and pressure p_0. The linearized perturbation equations are

$$\rho_0 \partial_t \boldsymbol{u}' + \boldsymbol{\nabla} p' = 0, \tag{2.30}$$

$$\partial_t p' + \gamma p_0 \boldsymbol{\nabla} \cdot \boldsymbol{u}' = 0, \tag{2.31}$$

where p' and $\boldsymbol{u}'$ represent the pressure and velocity perturbations. One can take the divergence of the momentum equation and substitute it into the pressure equation to get

$$\rho_0 \partial_t^2 \boldsymbol{\nabla} \cdot \boldsymbol{u} - \gamma p_0 \nabla^2 \boldsymbol{\nabla} \cdot \boldsymbol{u} = 0. \tag{2.32}$$

If we now assume that the perturbation quantities have spatial and temporal variation $\sim \exp(i\boldsymbol{k} \cdot \boldsymbol{x} - i\omega t)$, then we are left with the sound wave dispersion relation,

$$\omega^2 = c_s^2 |\boldsymbol{k}|^2, \qquad c_s^2 = \gamma p_0 / \rho_0. \tag{2.33}$$

Here c_s^2 is the adiabatic sound speed. Because internal gravity waves require gradients background state, the uniform background only supports sound waves.

Isothermal Background

A simple nonuniform background is the isothermal atmosphere. We also assume an ideal gas equation of state, so the background density and pressure are proportional to each other. Then the profiles

$$\log \rho_0 = \log p_0 = -\frac{z}{H}, \tag{2.34}$$

satisfy hydrostatic balance when $H = g\rho_0/p_0$. Here H is both the density and pressure scaleheight. These profiles have a non-zero *buoyancy frequency*, also known as the Brunt–Väisälä frequency. The buoyancy frequency is related to internal gravity wave propagation and is given by

$$N^2 = g\frac{ds_0}{dz} = \frac{\gamma - 1}{\gamma}\frac{g}{H} \tag{2.35}$$

Typically quantities like the sound speed and buoyancy frequency are spatially dependent. The isothermal background is a special case in which they are constant, which simplifies calculations.

For the isothermal atmosphere, the linearized perturbation equations are

$$\rho_0 \partial_t \boldsymbol{u}' + \nabla p' = -g\rho' \boldsymbol{e}_z, \tag{2.36a}$$

$$\partial_t p' + \gamma p_0 \nabla \cdot \boldsymbol{u}' + \boldsymbol{u}' \cdot \nabla p_0 = 0, \tag{2.36b}$$

$$\partial_t \rho' + \rho_0 \nabla \cdot \boldsymbol{u}' + \boldsymbol{u}' \cdot \nabla \rho_0 = 0. \tag{2.36c}$$

Because the background density changes with height, the perturbations also have non-sinusoidal vertical structure. If we make the ansatz

$$\boldsymbol{u}',\ p'/p_0,\ \rho'/\rho_0 \sim \exp(i\boldsymbol{k} \cdot \boldsymbol{x} - i\omega t + z/2H), \tag{2.37}$$

then one can show the solutions have both $\boldsymbol{k}$ and ω real. This choice of vertical dependence, $\sim \exp(z/2H)$, means the wave energy flux (proportional to $\rho_0 u_x u_z$) neither increases nor decreases exponentially with height. Then one can find that the dispersion relation is (Vasil et al. 2013)

$$-\frac{\omega^4}{c_s^2} + \omega^2 \left[k_h^2 + k_z^2 + \frac{1}{4H^2} \right] = k_h^2 N^2, \tag{2.38}$$

where we define the horizontal wavenumber by $k_h^2 = k_x^2 + k_y^2$.

This dispersion relation is quadratic in ω^2, which implies two different types of waves (each of which can propagate either up or down). In the limit of high frequency we have sound waves (i.e., pressure waves)

$$\omega_P^2 = \left(|\boldsymbol{k}|^2 + (2H)^{-2} \right) c_s^2, \tag{2.39}$$

and in the limit of low frequency we have internal gravity waves

$$\omega_G^2 = \frac{k_h^2}{|\boldsymbol{k}|^2 + (2H)^{-2}} N^2. \tag{2.40}$$

These are both similar to the wave dispersion relations for a uniform medium (for sound waves), and for an incompressible fluid with uniform buoyancy frequency (for internal gravity waves). In the latter case, the dispersion relation is $\omega_G^2/N^2 = k_h^2/|\boldsymbol{k}|^2$. In both cases, the dispersion relation in an isothermal atmosphere can be recovered by replacing $k_z^2 \to k_z^2 + (2H)^{-2}$. This extra scaleheight factor comes from the vertical dependence of both the background and perturbations.

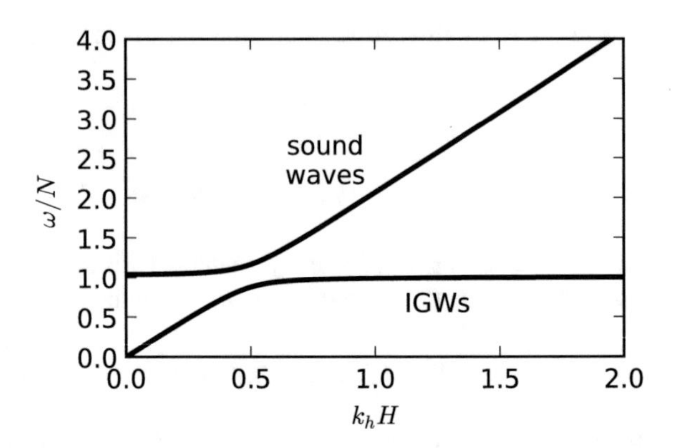

Fig. 2.4 Plot of the dispersion relation, Eq. 2.38, for $k_z H = 1/10$, and $\gamma = 5/3$. The upper curve corresponds to sound waves (Eq. 2.39), and the lower curve corresponds to internal gravity waves (Eq. 2.40)

The full solution of the dispersion relation is

$$\omega^2 = \frac{\omega_P^2}{2}\left[1 \pm \sqrt{1 - \frac{4\omega_G^2}{\omega_P^2}}\right]. \tag{2.41}$$

We plot the solution to this equation in Fig. 2.4. The properties of the two wave branches are related to the Boussinesq and anelastic approximations which can be used to model stellar convection. Neither approximation admits sound wave solutions. It is only possible to formally filter sound waves if there is a gap between the frequencies of the sound waves and of the internal gravity waves and/or convective modes (whose growth/decay rates are identical to the internal gravity wave frequencies). There is a gap between the frequencies if $k_h \gg H^{-1}$, giving the Boussinesq approximation; or if $N^2 \ll g/H$, giving the anelastic approximation. Also, note that the limit $k_h \ll H^{-1}$ also has a gap between frequencies—this corresponds to the *primitive equations*.

General Background

Now we will consider the general case of oscillations of a general background state, such as those derived from the stellar structure equations. We specify the background in terms of the buoyancy frequency, $N^2 = g ds_0/dz$, the density scaleheight, $H = -(d\log\rho_0/dz)^{-1}$, and the adiabatic sound speed $c_s^2 = \Gamma_1 P_0/\rho_0$. The first adiabatic exponent $\Gamma_1 = (d\log p/d\log\rho)_{\rm ad}$ is equal to the ratio of specific heats for an ideal gas, but more generally can be a function of height. Then it becomes convenient to define

$$\Upsilon = \rho_0^{1/2} c_s^2 \nabla \cdot \boldsymbol{u}, \tag{2.42}$$

such that the oscillation equations can be written in the simple form (Deubner and Gough 1984)

$$\partial_z^2 \Upsilon + k_z^2 \Upsilon = 0, \tag{2.43}$$

where

$$k_z^2 = \frac{\omega^2 - \omega_c^2}{c_s^2} + k_h^2 \left(\frac{N^2}{\omega^2} - 1 \right), \tag{2.44}$$

where the *cut-off frequency* ω_c is defined by

$$\omega_c^2 = \frac{c_s^2}{4H^2} \left(1 - 2\frac{dH}{dz} \right). \tag{2.45}$$

Although the background is much more general than the isothermal background, the dispersion relation is very similar. It is straightforward to verify that Eq. 2.44 reduces to Eq. 2.38 when $dH/dz = 0$. The vertical dependence of the eigenfunctions remains the same if c_s is constant.

We can factorize the dispersion relation as

$$c_s^2 k_z^2 = \omega^2 \left(1 - \frac{\omega_+^2}{\omega^2} \right) \left(1 - \frac{\omega_-^2}{\omega^2} \right), \tag{2.46}$$

where

$$\omega_\pm^2 = \frac{1}{2} \left(S^2 + \omega_c^2 \right) \pm \sqrt{\frac{1}{4} (S^2 + \omega_c^2)^2 - N^2 S^2}, \tag{2.47}$$

and

$$S^2 = k_h^2 c_s^2 \tag{2.48}$$

is known as the *Lamb frequency*. In spherical geometry, scalar functions like Υ can be written as a sum of spherical harmonics $Y_{\ell,m}(\theta, \phi)$. The horizontal (angular) laplacian is then $-\ell(\ell + 1)/r^2$. Thus, it is typical to see the Lamb frequency written as

$$S_\ell^2 = \frac{\ell(\ell + 1)c_s^2}{r^2}. \tag{2.49}$$

The advantage of writing the dispersion relation as Eq. 2.46 is it shows that waves are only propagating when ω is either smaller than both $\omega_\pm$, or larger than both $\omega_\pm$. In the limit $S \gg \omega_c$, we have $\omega_+ = S$ and $\omega_- = N$. Thus, waves only propagate if $\omega > S$ (sound waves), or if $\omega < N$ (internal gravity waves). In certain stars, there

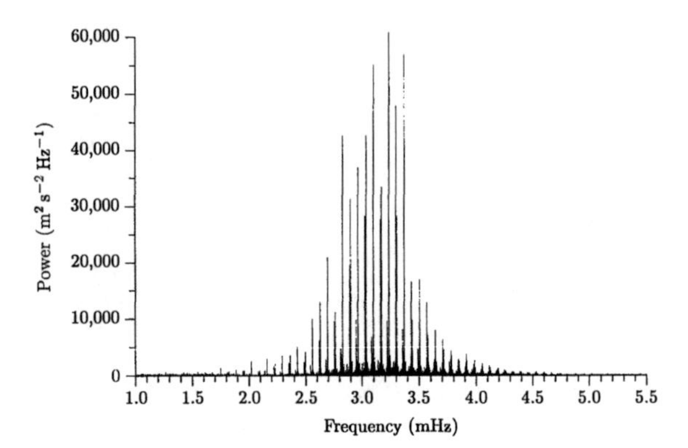

Fig. 2.5 Power in sound wave oscillation modes at the surface of the Sun. The properties of these modes are used to infer the interior structure of the Sun. (Adapted from Chaplin et al. 1996, with permission)

are also "mixed modes," which satisfy $\omega < N$ in certain parts of the star, but $\omega > S$ in other parts. Thus, depending on its position, the same oscillation mode can either act like a sound wave, or an internal gravity wave.

Helioseismology

The surface of the Sun exhibits oscillations at coherent frequencies (Fig. 2.5). These are sound waves excited by convection near the surface; internal gravity waves do not propagate in the surface convection zone (Fig. 2.2), so are much more difficult to detect. The field of *helioseismology* uses these global oscillation modes to infer characteristics of the Sun.

In the preceding discussion, we used cartesian geometry to simplify our analysis. When applying these results to stars, one must translate to spherical geometry. This is relatively straightforward for waves with large radial wavenumber; to lowest order, vertical derivatives only act on the perturbations, so we can neglect any geometry terms. In this case, we can replace $k_z \to k_r$. Furthermore, as described above, it is also customary to replace $k_h^2 \to \ell(\ell+1)/r^2$, where ℓ is the spherical harmonic degree of the angular variation of the wave. This can be made more precise with a WKB analysis of the problem.

Figure 2.5 shows there is a characteristic frequency spacing between the sound waves. The frequency spacing tells us about the average sound speed in the Sun. We can label each oscillation mode by the number of radial nodes, n. Then, assuming $k_r \gg 1/r$, we have

$$\int_{r_-}^{r_+} k_r \, dr = (n + \epsilon)\pi, \tag{2.50}$$

where the limits of the integral $r_\pm$ are the upper and lower turning point of the sound wave, which is where $\omega = \omega_+ \approx S_\ell$. The order unity factor ϵ takes into account the behavior of the modes near the turning points.

If we consider sound waves with $k_r \gg \ell/r$, then we can approximate the dispersion relation as $\omega = k_r c_s$, so we have

$$\omega_n \int_{r_-}^{r_+} \frac{dr}{c_s} = (n + \epsilon)\pi, \tag{2.51}$$

where ω_n is the frequency of the mode with n zeros. If we assume $r_\pm$ do not change significantly with n, we can approximate the integral to be the same for nearby frequencies, and find

$$\Delta\omega = \omega_{n+1} - \omega_n = \pi \left[\int_{r_-}^{r_+} \frac{dr}{c_s} \right]^{-1}. \tag{2.52}$$

Thus, the frequency spacing of sound waves is related to the average inverse sound speed over the wave propagation region.

Although internal gravity waves are not detected in the Sun, they are detected in other stars which have radiative envelopes (e.g., massive stars, Fig. 2.2). The study of global oscillation modes in other stars is known as *asteroseismology*. Because the oscillations are low amplitude, telescopes must measure the brightness of stars very precisely to reliably detect oscillations. There are several recent space missions, e.g., CoRoT, Kepler, and TESS, which all measure the brightness of stars to great precision. These were primarily designed to detect planets orbiting other stars, which can pass between the star and the telescope, causing the brightness of the star to momentarily (and periodicity) dim.

For internal gravity waves, we have $k_r = (N/\omega)\sqrt{\ell(\ell+1)}/r$ i.e., put the (N/ω) in front of the square-root factor. So

$$\frac{\sqrt{\ell(\ell+1)}}{\omega} \int_{r_-}^{r_+} \frac{N\,dr}{r} = (n + \epsilon)\,\pi, \tag{2.53}$$

where now $r_\pm$ are the turning points where $\omega = \omega_- \approx N$. As before, we assume $r_\pm$ do not vary much when ω changes, such that the integral can be approximated as constant for similar ω. Then we find

$$\Delta P = \frac{1}{\omega_{n+1}} - \frac{1}{\omega_n} = \pi\sqrt{\ell(\ell+1)} \int_{r_-}^{r_+} \frac{N\,dr}{r}. \tag{2.54}$$

Thus, internal gravity waves have approximately constant *period* spacing. The period spacing is related to the average buoyancy frequency in the star.

The frequency spacing of sound waves and period spacing of internal gravity waves give simple global measures of the properties of a star. However, there is a more general inverse problem: What is the structure of a star which has oscillation modes which most closely match observations? For distant stars where relatively few oscillation modes are identified, this inversion is ill-posed. However, the many oscillation modes observed in the Sun allow us to gain detailed information about its interior (e.g., Christensen-Dalsgaard 2002).

In particular, helioseismology has allowed us to determine the internal rotation profile of the Sun. As described above, the angular variation of the modes can be decomposed into spherical harmonic functions, $Y_{\ell,m}(\theta, \phi)$. In the absence of rotation, the horizontal derivatives can always be composed to form horizontal laplacians, which depend only on ℓ. Thus, there is a degeneracy between different m modes if there is no rotation. The presence of rotation breaks the degeneracy. If ω_ℓ is the frequency of a mode with no rotation, the frequency of the mode in the limit of weak rotation can be calculated using perturbation theory,

$$\omega_{\ell,m} = \omega_\ell + m \int K_\ell(r, \theta)\Omega(r, \theta) dr d\theta, \tag{2.55}$$

where $K_\ell(r, \theta)$ is a *kernel*, related to the square of the eigenfunction. Eigenfunctions with different radial structure and/or different ℓ have different kernels. If the frequency splitting, $\omega_{\ell,+m} - \omega_{\ell,-m}$, is calculated with many different ℓ and different radial structure, one can infer the rotation profile, $\Omega(r, \theta)$.

We can measure the convective flows in the Sun using *time–distance helioseismology*. The strategy is to consider two points a and b at the surface, and calculate the cross-correlation of, e.g., line-of-sight velocities between these two points,

$$C(a, b, \tau) = \langle u(a, t)\, u(b, t + \tau) \rangle \,. \tag{2.56}$$

The cross-correlation gives you information about sound waves which propagate between a and b in a time τ. If there is no underlying flow, the correlation is symmetric: $C(a, b, \tau) = C(b, a, \tau)$. However, convective flows can doppler-shift the waves, as the waves travel faster from a to b if the flow is properly aligned. By measuring the asymmetry in the cross-correlation for several pairs of points, one can measure the average convective velocity at depth.

This time–distance helioseismology technique was used in Hanasoge et al. (2012) to calculate the power spectrum of convective flows at two depths in the Sun. The measured velocities are orders of magnitude lower than the velocities predicted by simulations of solar convection. The simulation results are similar to the estimates of mixing length theory, which is the basis for convective parameterizations in stellar structure models. It is worrisome that the theory of stellar convection does not reproduce helioseismic observations (although other analyses of the data (Greer et al. 2015) suggest the discrepancy—though present—may be weaker than in Hanasoge et al. 2012). Some have suggested convective flows are weaker than predicted in the Sun because the large density stratification allows narrow downflows to transport the solar luminosity over the entire depth of the solar convection zone (Brandenburg 2016). This differs fundamentally from the local assumption of mixing length theory. The resolution of this "convective conundrum" is a major open problem in solar astrophysics.

Convective–Radiative Interfaces: Convective Overshoot

In the next two sections, we will discuss dynamics which occur at the interface between the convective and radiative zones of stars. These cannot be solved for directly within stellar evolution models, but may lead to important physics which ideally could be parameterized in these 1D models.

This section describes convective overshoot. To fix ideas, consider the solar case with a convection zone above a radiative zone. The convection zone ends when the mean entropy gradient is zero, i.e., there is no buoyancy force on fluid elements. A falling fluid element feels no acceleration at the convective–radiative interface, but it still has a negative velocity, so it will continue to fall into the radiative zone. Once in the radiative zone, there will be a stabilizing buoyancy force which will decelerate the fluid element.

Convective overshoot is important in describing the compositional evolution in stars. Recall that compositional mixing is very efficient within convection zones. Chemical species follow an advection–reaction–diffusion equation

$$\frac{\partial c}{\partial t} + \boldsymbol{u} \cdot \nabla c = \sum_k \omega_k \Delta c^k + D\nabla^2 c. \tag{2.57}$$

The relative importance of the advection term to the diffusion term is parameterized by a *compositional Péclet* number

$$\mathrm{Pe}_C = \frac{UL}{D}, \tag{2.58}$$

where U and L are estimates of the flow's typical velocity and lengthscale, and D is the chemical diffusivity. Stellar convection can have $\mathrm{Pe}_C \gtrsim 10^7$ (e.g., Hughes et al. 2007). Because the convection is efficient, it is thought to mix chemicals to a nearly constant value within the convection zone; this process is parameterized with a turbulent diffusivity, $D_t \sim \mathrm{Pe}_C\, D$.

Now consider a radiative zone adjacent to a convection zone. If there are no bulk fluid motions in the radiative zone, then chemical species diffuse according to their chemical diffusivity D. However, if there is some remnant motions in the radiative zone from convective overshoot, this can lead to extra mixing. The amount of mixing is related to how vigorous the mixing was in the convection zone itself, which is given by D_t. Since $D_t/D \sim \mathrm{Pe}_C \gg 1$, only a small amount of convective overshoot might lead to greatly enhanced mixing in radiative zones.

Convective overshoot is important for many aspects of stellar evolution. When stars have nuclear burning in a convection zone (for instance, in massive main-sequence stars), overshoot can transport fresh fuel into the convection zone. This can extend the main-sequence lifetime of massive stars (e.g., Rosenfield et al. 2017). In less massive, solar-type stars with convective envelopes, convective overshoot can control how much lithium is transported out of the convection zone into the hot core, where it is burnt (Andrássy and Spruit 2013). More recently, asteroseismology has probed the buoyancy frequency structure near the convective–radiative interfaces of massive stars, which is in part set by convective overshoot (Pedersen et al. 2018).

Convective Penetration Length

To understand convective overshoot, we will first estimate the lengthscale over which we expect convection to penetrate into the radiative zone. Define $\ell_{\rm ov}$ to be the *penetration length*. Our first estimate will balance the inertia term in the momentum equation with the buoyancy term,

$$\boldsymbol{u} \cdot \nabla \boldsymbol{u} \sim g \frac{\rho'}{\rho_0}. \tag{2.59}$$

Here ρ' is the difference in density between the convecting fluid element and the background radiative zone after it has moved $\ell_{\rm ov}$ into the radiative zone. Thus, we can estimate

$$\frac{\rho'}{\rho_0} = \ell_{\rm ov} \frac{\partial_z \rho_0}{\rho_0}. \tag{2.60}$$

If we estimate the inertial term as $u_c^2/\ell_{\rm ov}$, and identify $\partial_z \log \rho_0 \sim H^{-1}$ and $c_s^2 \sim gH$, we find

$$\frac{\ell_{\rm ov}}{H} \sim \frac{u_c}{c_s} = \mathrm{Ma}_c. \tag{2.61}$$

Near the base of the solar convection zone, $c_s \sim 200\,\mathrm{km/s}$ (Hughes et al. 2007), and the convective velocities may be $u_c \sim 1 - 100\,\mathrm{m/s}$ (Hanasoge et al. 2012). Thus, we estimate the penetration length to be $10^{-3} - 10^{-5}\,H$. At the base of the solar convection zone, $H \sim 0.1\,R_\star$, so we estimate $\ell_{\rm ov} \lesssim 10^{-4}\,R_\star$. This is a very small fraction of the size of the convection zone, $0.3\,R_\star$.

One problem with this estimate is that the source of buoyancy in stars is the entropy, not density. That is, the pressure gradient contributes to the buoyancy term in Eq. 2.59, such that a more accurate estimate of the buoyancy is

$$\boldsymbol{u} \cdot \nabla \boldsymbol{u} \sim g \frac{s'}{c_p}. \tag{2.62}$$

As above, the naive estimate for s' would be

$$s' = \ell_{\rm ov}\partial_z s_0. \tag{2.63}$$

However, the boundary of the convection zone is exactly defined to be the location where $\partial_z s_0 = 0$. To use Eq. 2.62 we need to specify a model for how the entropy gradient varies from zero near the boundary of the radiative zone.

One simple assumption is that $N^2 = (g/c_p)\partial_z s$ increases linearly away from the convective–radiative interface. Then dN^2/dz is a constant. Assuming $dN^2/dz \sim g/H^2$, we have that $s'/c_p \sim (\ell_{\rm ov}/H)^2$, so

$$\frac{\ell_{\rm ov}}{H} \sim {\rm Ma}_c^{2/3} \tag{2.64}$$

Depending on the value of the convective velocity, one has $\ell_{\rm ov}/H \lesssim 10^{-2}$. However, it's more likely that N^2 has even more curvature (e.g., Christensen-Dalsgaard et al. 2011), in which case we could have an even larger estimate for the overshoot length. Finally, recall that because the compositional Péclet number of stellar convection is very high, even a small amount of fluid motion can lead to enhanced mixing. Thus, there could be important effects of convective overshoot beyond distances of $\ell_{\rm ov}$.

This discussion reveals that the exact structure of N^2 near convective–radiative interfaces likely plays an important role in convective overshoot. What sets this structure in stars? In general, convection occurs in stars because radiative cannot transfer the full luminosity. In solar-type stars with convective envelopes, convection starts because of opacity variations. Thus, using a temperature-dependent opacity is important for deriving a self-consistent convective–radiative interface (Käpylä et al. 2017). For massive main-sequence stars, convection occurs in the core of the star. The size of the convection zone is set by geometry effects. Thus, spherical geometry is key for modeling convective–radiative interfaces for stars with core convection.

Parameterizations of Convective Overshoot

One way to parameterize convective overshoot is to extend the turbulent chemical diffusivity from the convection zone into the radiative zone. One must choose exactly how to perform this extension. A common approach is called *step overshoot*, where the convective turbulent diffusivity at the convective–radiative interface is simply extending into the radiative zone by a penetration length $\ell_{\rm ov}$, which is generally assumed to be some fraction of the pressure scaleheight H_p. This effectively increases the size of the convection zone by $\ell_{\rm ov}$. Observations of eclipsing binary stars can be explained by stellar evolution models assuming step overshoot with $\ell_{\rm ov} \sim 0.1$–$0.2H_p$ (Stancliffe et al. 2015). This is much larger than our estimate in the previous section.

Another approach is *exponential overshoot*, where the turbulent diffusivity is assumed to decay exponentially away from the convection zone. This leads to a continuous profile of turbulent diffusivity. Freytag et al. (1996) found that the velocities decrease exponentially outside a convection zone; however, those simulations have $\mathrm{Pe} < 1$, so are not directly applicable to turbulent stellar convection. As with the step overshoot parameterization, an exponential decay lengthscale is a free parameter of the model, which must be specified. Exponential overshoot has been invoked to explain gravity wave period spacing from asteroseismology (Pedersen et al. 2018).

Not all parameterizations invoke a turbulent diffusivity. Some work instead studies the mass flux of material into the convection zone (e.g., Cristini et al. 2019, and other work by their group). The goal is to determine an *entrainment law* of the form

$$\dot{M} = A\,\mathrm{Ri}^{-b}, \tag{2.65}$$

where $\dot{M}$ is the flux of mass entrained by convective overshoot, and Ri is the *Richardson number*,

$$\mathrm{Ri} = \frac{N^2}{(dU/dz)^2}, \tag{2.66}$$

and A and b are free parameters. Here, N^2 and dU/dz the buoyancy frequency and velocity shear measured near the convective–radiative interface. The motivation behind this entrainment law is that shear flow instabilities will cause mixing into the convection zone, which only occur when the Richardson number is small. This type of parameterization is used in atmospheric sciences to describe the growth of the planetary boundary layer during the day due to solar heating (Mellado 2012). However, it is difficult to use this type of parameterization in stellar evolution models, as the mass and size of the convection zone are outputs of the model, so it is unclear how to impose Eq. 2.65.

Measuring the Turbulent Diffusivity

There has been substantial effort by different groups to determine the effects of convective overshoot from 3D numerical simulations, and use these to develop parameterizations for stellar evolution (e.g., Cristini et al. 2017, 2019; Woodward et al. 2015; Jones et al. 2017; Lecoanet et al. 2016; Rogers and Glatzmaier 2006; Pratt et al. 2017; Korre et al. 2019). One way to study the mixing of material from a radiative zone into a convection zone is to solve an advection–diffusion equation for a passive tracer field representing chemical composition

$$\partial_t c + \boldsymbol{u} \cdot \nabla c = D\nabla^2 c. \tag{2.67}$$

Assume the advective term can be parameterized as a turbulent diffusivity

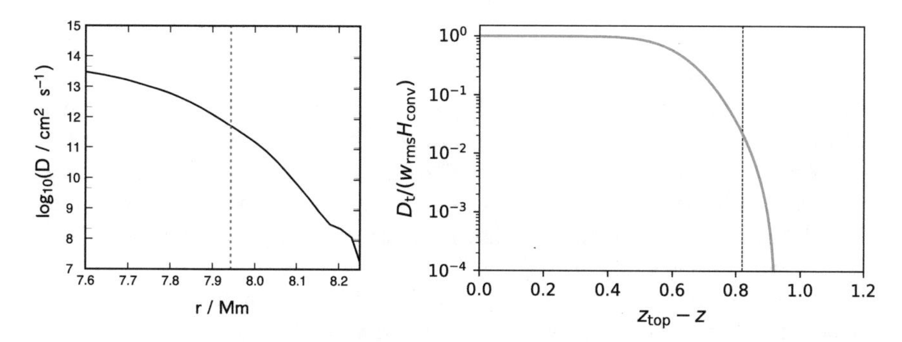

Fig. 2.6 Turbulent diffusivities calculated in convective overshoot simulations in Jones et al. (2017) (left, simulation D1), and Lecoanet et al. (2016) (right, simulation R8). The vertical dashed lines denote the boundary of the convection zone, where $N^2 = 0$. (Left panel adapted from Jones et al. 2017)

$$\boldsymbol{u} \cdot \nabla c \rightarrow -\nabla \cdot (D_t \nabla c), \tag{2.68}$$

and that $D_t \gg D$. The diffusion equation can be reduced to one (spatial) dimension by taking an angular average. Then the change in composition, $\Delta c = c(r, t_e) - c(r, t_s)$, where t_s and t_e are the start and end times, is related to the turbulent diffusivity,

$$D_t = \frac{\int \Delta c \, dr}{\partial_r \overline{c}}, \tag{2.69}$$

where $\overline{c} = \int c \, dt$.

Jones et al. (2017) used this approach to measure convective overshoot. They ran a simulation of convection bounded by two radiative zones. Using Eq. 2.69, they found a turbulent diffusivity that decreased roughly exponentially away from the convection zone (see left panel of Fig. 2.6). However, they did not test if the turbulent diffusivity parameterization was itself valid: it is possible to use Eq. 2.69 to calculate a "turbulent diffusivity" for any solution to Eq. 2.67, even if effect of advection is poorly modeled by diffusion.

To test the turbulent diffusivity approximation directly, Lecoanet et al. (2016) also solved the advection–diffusion equation for a passive tracer (Eq. 2.67) in a simulation with a convection zone above a radiative zone. In contrast to Jones et al. (2017), their calculation is in cartesian geometry. They found that after many convective turnover times, the horizontal average of the tracer field evolves according to the separable solution

$$c(z, t) = c_0 + \exp(-\lambda t) C(z), \tag{2.70}$$

where c_0 is the volume average of the tracer.

Using the turbulent diffusion assumption (Eq. 2.68), one finds

$$-\lambda C = \partial_z \left[(D + D_t)\partial_z C\right], \tag{2.71}$$

which can be solved for D_t, similar to Eq. 2.69. In their simulations, Lecoanet et al. (2016) found the turbulent diffusivity decreases like a Gaussian: $D_t \sim \exp(-(z - z_i)^2)$, where z_i is the height of the convective–radiative interface (Fig. 2.6). This is a much steeper decline than found in Jones et al. (2017). Lecoanet et al. (2016) also verified the diffusion assumption by showing that the horizontal average of the tracer changes with time according to Eq. 2.68.

Why does the turbulent diffusion approximation work in these simulations? One can decompose $c(\boldsymbol{x}, t)$ into a horizontal mean, and a fluctuating component,

$$c(\boldsymbol{x}, t) = \overline{c}(z, t) + \tilde{c}(\boldsymbol{x}, t). \tag{2.72}$$

Then the equations for the mean and fluctuations are

$$\partial_t \overline{c} - D\partial_z^2 \overline{c} = -\partial_z \langle w\tilde{c} \rangle, \tag{2.73}$$

$$\partial_t \tilde{c} - D\nabla^2 \tilde{c} = -w\partial_z \overline{c} - \partial_z \left(w\tilde{c} - \langle w\tilde{c} \rangle\right), \tag{2.74}$$

where $\langle \cdot \rangle$ denotes horizontal average and we assume the fluid is incompressible so w has no horizontal mean. The right-hand side of Eq. 2.73 represents the change in the mean $\overline{c}$ from fluctuations of w and $\tilde{c}$. The first term on the right-hand side of Eq. 2.74 represents the change in the fluctuations $\tilde{c}$ from the fluctuating w interacting with the mean $\overline{c}$. The second term is the change in fluctuations caused by interactions between the fluctuations themselves.

If we neglect this triple-fluctuation term, we have

$$\partial_t \tilde{c} - D\nabla^2 \tilde{c} = -w\partial_z \overline{c}. \tag{2.75}$$

This is known as the quasi-linear approximation (e.g., Marston et al. 2016). Since the left-hand side is linear in $\tilde{c}$, we can invert it by convolving with a Green's function G,

$$\tilde{c} = -\int G(D) * (w\partial_z \overline{c})\, dV dt. \tag{2.76}$$

If we assume the convolution with G acts primarily on w rather than $\overline{c}$, then substituting into Eq. 2.73,

$$\partial_t \overline{c} - D\partial_z^2 \overline{c} = \partial_z \left[\left\langle w \int G(D) * w\, dV dt \right\rangle \partial_z \overline{c}\right]. \tag{2.77}$$

The term in the horizontal average can then be identified as the turbulent diffusivity. Thus, the turbulent diffusivity model holds if the triple-fluctuation term in Eq. 2.74 can be neglected, and if the fluctuations $\tilde{c}$ are linear in $\partial_z \overline{c}$.

Physical Mechanism of Convective Overshoot

We finish this section with a brief discussion of the physical mechanism of convective overshoot. It is difficult for convection to penetrate into radiative zones because the radiative zone is very stably stratified. One way to induce mixing in the radiative zone is via shear flow instabilities (i.e., Kelvin–Helmholtz instabilities). Shear flow instabilities were invoked by Woodward et al. (2015) to explain "cat's eye" rolls found at the convective–radiative interface in simulations. These features were assumed to cause the overshoot mixing.

The stable stratification in the radiative zone can stabilize shear flow instabilities. The Richardson number (Eq. 2.66) is the ratio of stable stratification to destabilizing velocity shear. If the Richardson number is much larger than unity, the flow is typically linearly stable. Woodward et al. (2015) calculated the Richardson number to be approximately 10^3 in near the convective–radiative interface in their simulations, apparently precluding shear flow instabilities.

Another system in which mixing occurs despite strong stable stratification is the air–ocean interface. Wind can excite surface waves to nonlinear amplitudes, at which point they can break and induce mixing. Herault et al. (2018) suggest a similar process might occur in laboratory experiments of a jet of fresh water impinging on a layer of salt water. They find waves excited at their interface, very similar in appearance to Woodward et al. (2015). One possible method for the excitation of ocean waves is the Miles instability (e.g., Hristov et al. 2003), a resonance between the wind velocity and the surface wave phase speed. Whether or not the same instability operates at the boundary of stellar convection zones is a matter of current research.

Convective–Radiative Interfaces: Internal Gravity Wave Generation

Internal gravity waves (IGWs) can propagate in the radiative zones of stars. They are excited by convection, similar to the sound waves observed at the surface of the Sun. But because they do not propagate within a convection zone, they are predominately excited near convective–radiative interfaces. Thus, IGW generation is another phenomenon which occurs near these interfaces.

IGWs are thought to play an important dynamical, as well as diagnostic, role in stellar evolution. IGWs propagate in different parts of stars than sound waves. As will be discussed below, they also carry much more energy than sound waves. Thus, IGWs may contribute to angular momentum transport, mixing, and mass loss in different stars.

IGW Transport and Damping

To simplify our analysis, we will consider a stably stratified, Cartesian background atmosphere with constant buoyancy frequency N^2, under the Boussinesq approximation. Then, the equations of motion are

$$\rho_0 \partial_t \boldsymbol{u} + \nabla p' + g\rho' \boldsymbol{e}_z = -\rho_0 \boldsymbol{u} \cdot \nabla \boldsymbol{u}, \tag{2.78a}$$

$$\partial_t \rho' - \boldsymbol{e}_z \cdot \boldsymbol{u} \frac{\rho_0 N^2}{g} = -\boldsymbol{u} \cdot \nabla \rho', \tag{2.78b}$$

$$\nabla \cdot \boldsymbol{u} = 0. \tag{2.78c}$$

If the momentum equation is multiplied by $\boldsymbol{u}$ and the density equation is multiplied by ρ', the two can be combined into an energy equation

$$\partial_t E + \nabla \cdot \left(\boldsymbol{u} p' + \boldsymbol{u} E\right) = 0, \tag{2.79}$$

where

$$E = \frac{1}{2}\rho_0 |\boldsymbol{u}|^2 + \frac{1}{2}\frac{g^2}{\rho_0 N^2}\rho'^2 \tag{2.80}$$

is the IGW wave energy. The wave energy can be transported by the *wave energy flux*

$$\boldsymbol{F}_w = \boldsymbol{u} p'. \tag{2.81}$$

We will typically only consider the vertical component of the wave flux, which we write as $F_w = u_z p'$. For low-frequency waves with $\omega \ll N$, we can write the wave flux as

$$F_w = \frac{\omega}{k_h}\rho_0 |\boldsymbol{u}_h|^2, \tag{2.82}$$

where $k_h = \sqrt{k_x^2 + k_y^2}$ denotes the horizontal wavenumber, and similarly for $|\boldsymbol{u}_h|^2$.

IGWs can generate mean flows. The horizontal mean of the momentum equation is

$$\partial_t \langle \boldsymbol{u}_h \rangle = -\langle \partial_z (u_z \boldsymbol{u}_h) \rangle, \tag{2.83}$$

where we use $\langle \cdot \rangle$ to denote horizontal average. The correlation $u_z \boldsymbol{u}_h$ is known as the *Reynolds stress*. We can calculate the Reynolds stress by studying an IGW with

$$u_z = A \sin(\boldsymbol{k}_h \cdot \boldsymbol{x} + k_z z - \omega t). \tag{2.84}$$

By incompressibility,

$$\boldsymbol{u}_h = -A\frac{\boldsymbol{k}_h k_z}{k_h^2}\sin(\boldsymbol{k}_h \cdot \boldsymbol{x} + k_z z - \omega t). \tag{2.85}$$

We can rewrite the divergence of the Reynolds stress as

$$u_z \partial_z \boldsymbol{u}_h = -\frac{A^2}{2}\frac{\boldsymbol{k}_h k_z^2}{k_h^2}\sin\left[2\left(\boldsymbol{k}_h \cdot \boldsymbol{x} + k_z z - \omega t\right)\right]. \tag{2.86}$$

Thus, we find that this uniform, planar IGW has

$$\langle u_z \partial_z \boldsymbol{u}_h \rangle = 0, \tag{2.87}$$

and therefore transports no momentum.

However, we have neglected wave damping. In stars, radiative diffusion transports energys, but also damps waves. In this Boussinesq framework, the density equation is modified by the radiative diffusivity

$$\partial_t \rho' - \boldsymbol{e}_z \cdot \boldsymbol{u}\,\frac{\rho_0 N^2}{g} = -\boldsymbol{u} \cdot \nabla\rho' + \chi\nabla^2\rho', \tag{2.88}$$

where $\chi = \rho_0 \kappa$ is the radiative diffusivity and κ is the opacity. This changes the IGW dispersion relation to

$$-i\omega\left(-i\omega - \chi\nabla^2\right)\nabla^2 u_z + N^2\nabla_h^2 u_z = 0. \tag{2.89}$$

To solve this, we introduce the solution ansatz

$$u_z = A\exp\left(i\boldsymbol{k}_h \cdot \boldsymbol{x} + ik_z z - \ell_d^{-1} z - i\omega t\right), \tag{2.90}$$

where the wave is assumed to be propagating upwards, and ℓ_d^{-1} is the inverse damping length. This decomposition assumes k_z and ℓ_d^{-1} are real. Substituting these into the dispersion relation

$$\left[-\omega^2 + i\omega\chi\left(-k^2 - 2ik_z\ell_d^{-1} + \ell_d^{-2}\right)\right]\left(-k^2 - 2ik_z\ell_d^{-1} + \ell_d^{-2}\right) = N^2 k_h^2. \tag{2.91}$$

The dispersion relation specifies the k_z and ℓ_d^{-1} for a wave of a given k_h and ω. It may then appear that Eq. 2.91 is a single equation for two variables, but we can split it into real and imaginary components,

$$\left(\omega^2 - 4\omega\chi k_z\ell_d^{-1}\right)\left(k^2 - \ell_d^{-2}\right) = N^2 k_h^2, \tag{2.92}$$

$$\left(\omega^2 - 2\omega\chi k_z\ell_d^{-1}\right)\left(2k_z\ell_d^{-1}\right) + \omega\chi\left(k^2 - \ell_d^{-2}\right)^2 = 0. \tag{2.93}$$

The $(k^2 - \ell_d^{-2})$ can be eliminated between the equations to arrive at an equation for $k_z\ell_d^{-1}$,

$$\left(\omega^2 - 2\omega\chi k_z\ell_d^{-1}\right)\left(2k_z\ell_d^{-1}\right)\left(\omega^2 - 4\omega\chi k_z\ell_d^{-1}\right)^2 = -\omega\chi N^4 k_h^4 \tag{2.94}$$

In the limit of weak dissipation, where $\omega \gg \chi k_z \ell_d^{-1}$, we have

$$\ell_d^{-1} = -\frac{\chi N^4 k_h^4}{2\omega^5 k_z}. \tag{2.95}$$

Recall that the group and phase velocity of IGWs are perpendicular such that a wave propagating upward has $k_z < 0$ and thus $\ell_d^{-1} > 0$. Waves with $\omega \ll N$ have

$$\ell_d^{-1} = \frac{\chi N^3 k_h^3}{2\omega^4}. \tag{2.96}$$

Damping means the waves now transport momentum. Now the vertical velocity is

$$u_z = A\sin(\phi(\boldsymbol{x}, t))\exp(-\ell_d^{-1}z), \tag{2.97}$$

where for simplicity we define the phase to be

$$\phi(\boldsymbol{x}, t) = \boldsymbol{k}_h \cdot \boldsymbol{x} + k_z z - \omega t \tag{2.98}$$

Using the continuity equation, we have

$$\boldsymbol{u}_h = -A\frac{\boldsymbol{k}_h k_z}{k_h^2}\sin(\phi)\exp(-\ell_d^{-1}z) - A\frac{\boldsymbol{k}_h \ell_d^{-1}}{k_h^2}\cos(\phi)\exp(-\ell_d^{-1}z). \tag{2.99}$$

We can write the divergence of the Reynolds stress as

$$\partial_t\langle\boldsymbol{u}_h\rangle = -\partial_z\langle u_z u_x\rangle = 2\boldsymbol{k}_h\ell_d^{-1}A^2\frac{(-k_z)}{k_h^2}\exp(-2\ell_d^{-1}z). \tag{2.100}$$

Recall that for an upward propagating IGW, $k_z < 0$, so the momentum deposition is in the $\boldsymbol{k}_h$ direction.

In stars, thermal diffusion is primarily due to radiation, and can be quite an important effect. For example, consider the diffusion of waves excited by convection in the solar convection zone. Assume most energetic waves will have frequencies similar to the convective frequency (the convection has a period of about a month), and horizontal wavenumber $k_h = \sqrt{\ell(\ell+1)}/(0.7R_\odot)$, with $\ell = 3$. Then using $N \sim 2\,\text{mHz}$ (e.g., Deubner and Gough 1984), and $\chi \sim 10^7\,\text{cm}^2/\text{s}$ (e.g., Hughes et al. 2007), we find that

$$\frac{\ell_d}{R_\odot} \sim 3 \times 10^{-5}. \tag{2.101}$$

These waves are so strongly damped because their vertical wavenumber is $k_h N/\omega \sim 4 \times 10^{-5} R_\odot^{-1}$ because N/ω is very large. But because ℓ_d^{-1} depends strongly on ω/N, waves with frequencies $\sim$10 times larger than the convective frequency have $\ell_d \sim R_\odot$. This demonstrates the importance of understanding the frequency dependence of waves excited by convection.

Spectrum of IGWs Excited by Convection

There are several proposed mechanisms for IGW generation by convection (Ansong and Sutherland 2010). Two that are particularly relevant for stellar convection are the mechanical oscillator effect and deep forcing. The *mechanical oscillator* mechanism assumes IGWs are generated by motions of the convective–radiative interface, which are distorted by the convection. The *deep forcing* mechanism assumes waves are excited deep within the convection zone by thermal or Reynolds stresses, and then travel evanescently to the convective–radiative interface where they begin to propagate.

Lecoanet et al. (2015) tested both of these mechanisms in a series of numerical simulations. First, they solved the Boussinesq equations with a nonlinear equation of state which self-consistently leads to a lower convective region, and an upper radiative region in their simulation. They compared the waves generated in this "full" simulation to waves in two simplified simulations testing the two physical mechanisms. In the mechanical oscillator simulation, the linearized wave equations were solved only in the radiative zone, while the bottom boundary was forced with the movement of the convective–radiative interface from the full simulation. In the deep forcing simulation, the linearized wave equations were solved in the full domain, but a source term related to the Reynolds stresses from the full simulation was included within the convection zone. The deep forcing simulation was able to reproduce the waves in the full simulation, whereas the interface forcing simulation was not.

The deep forcing mechanism can be used to make quantitative predictions about the spectrum of IGWs generation by stellar convection. We can rewrite the fluid equations as

$$\partial_t^2 \nabla^2 \xi_z + N^2(z) \nabla_h^2 \xi_z \equiv L\xi_z = S, \tag{2.102}$$

where ξ_z is the vertical displacement associated with the wave, and S is related to the Reynolds stresses $\boldsymbol{F}$ by,

$$S = -\nabla^2 F_z + \partial_z \nabla \cdot \boldsymbol{F}, \qquad \boldsymbol{F} = \nabla \cdot (\boldsymbol{u}\boldsymbol{u}). \tag{2.103}$$

The only approximation we have made here is to neglect the contributions of thermal stresses to S. Then the linear wave operator L can be inverted in Eq. 2.102 using a Green's function G,

$$\xi_z = \int G * S\, dV dt. \tag{2.104}$$

The Green's function depends on the exact structure of the buoyancy frequency near the convective–radiative interface (Lecoanet and Quataert 2013).

The principal difficulty is in modeling the convective Reynolds stresses, S. We will follow the analysis of Lecoanet and Quataert (2013), and assume the flow can be decomposed into "eddy" structures, each of which has a typical size h and coherence time τ_h. We assume the coherence time is the eddy's turnover time, $\tau_h \sim h/u_h$, where u_h is the velocity of the eddy. To recover the Kolmogorov $k^{-5/3}$ energy spectrum, we require $u_h \sim h^{1/3}$, so $\tau_h \sim h^{2/3}$. We will first calculate the wave excitation by eddies of a given size, and then sum over all eddy sizes to calculate the total wave spectrum.

Which waves does an eddy of size h and coherence time τ_h couple to? The length and timescales of the wave must match those of the eddy in Eq. 2.104 for the convolution to be nonzero. This is because the Green's function has the same temporal and horizontal dependence in the convection zone as the wave has in the radiative zone. If the horizontal wavenumber of the wave, k_h, is much larger than $1/h$, then there are many wave oscillations over the eddy, and the convolution will be exponentially small. Thus, there is negligible wave excitation for $k_h h > 1$. If $k_h h < 1$, multiple eddies fit across a single wavelength of the wave. Most of the time these eddies will roughly cancel out, but sometimes they will be in phase with the wave, and lead to coherent excitation. Although this may sound rare, it will occur roughly $\sqrt{k_h h}$ of the time. Thus, wave excitation is much more efficient for $k_h h < 1$ than for $k_h h > 1$. The same argument holds for the wave frequency ω: there is negligible wave excitation for $\omega\tau_h > 1$, but strong wave excitation for $\omega\tau_h < 1$.

To calculate the wave excitation by eddies of size h, we assume the eddies are isotropic, space-filling, and have random phases. This is enough to estimate the S term in Eq. 2.104. We can calculate the differential wave energy flux,

$$\frac{dF_w}{d\log k_h\, d\log\omega\, d\log h} \sim \rho_0 \left(\frac{h}{\tau_h}\right)^3 \frac{\omega}{N_0}(k_h h)^4, \tag{2.105}$$

where N_0 is a characteristic value of the buoyancy frequency in the radiative zone. The wave energy flux tells you the energy input into the waves with frequencies around ω and horizontal wavenumbers around k_h. This does not, by itself, give the wave amplitude. The wave amplitude is set by a balance between energy gain by convection, and energy loss by linear or nonlinear damping processes.

Now we will add up the effects of all eddies. This requires describing how energy is injected into the Kolmogorov cascade. We assume that the largest convective features are "eddies" with size H, the scaleheight, and frequency ω_c, the convective frequency. Then we assume there are eddies of every size $h < H$, each of which has a coherence time

$$\tau_h \omega_c = \left(\frac{h}{H}\right)^{2/3}. \quad (2.106)$$

Of course, convection is neither isotropic, nor homogeneous, so the Kolmogorov theory does not necessarily apply. However, laboratory experiments show a robust $k^{-5/3}$ spectrum at small scales (e.g., Lohse and Xia 2010, and references within). The largest structures in convection can organize into plumes, which may not satisfy our assumptions of the eddy cascade. But we will show below that it is precisely the waves generated by small-scale eddies which are most relevant in stars.

Consider a wave with frequency ω and horizontal wavenumber k_h. Equation 2.105 shows that the flux decreases rapidly as τ_h and h become small. Thus, the excitation is dominated by the largest eddy that can couple to the wave, via the conditions $k_h h < 1$ and $\omega\tau_h < 1$. We will focus on low k_h waves; these are waves where, if $\tau_{h^*} = \omega^{-1}$, then $k_h h^* < 1$. That is, the coupling to eddies is determined by the frequency, not the wavenumber. Then, using that the excitation is dominated by eddies with sizes around h^* satisfying $\tau_{h^*} = \omega^{-1}$, we find

$$\frac{dF_w}{d\log k_h\, d\log\omega} \sim \rho_0 u_c^3 \frac{\omega_c}{N_0} (k_h H)^4 \left(\frac{\omega}{\omega_c}\right)^{-13/2}. \quad (2.107)$$

Note this expression is only for $k_h h^* < 1$. The main predictions of this theory are the steep $\omega^{-13/2}$ and k_h^4 power laws. There is also an overall wave excitation efficiency, which is $\omega_c/N_0 \sim \mathrm{Ma}_c$, the convective Mach number. IGW excitation is more dynamically important than sound wave excitation, because the efficiency factor for sound waves scales like $\mathrm{Ma}_c^{15/2}$ (Goldreich and Kumar 1990), and $\mathrm{Ma}_c \ll 1$ for efficient stellar convection.

There have been several efforts to recover the spectrum in Eq. 2.107 in numerical simulations (e.g., Alvan et al. 2014; Rogers and Glatzmaier 2006; Rogers et al. 2013; Edelmann et al. 2019). This is difficult, as the theory is strictly only valid when the waves are excited by three-dimensional, turbulent convection, with frequencies satisfying $\omega \ll N$, and $\omega_c \ll \omega$, and for wavenumbers $k_h \gg H^{-1}$. Nevertheless, recent Boussinesq simulations in cartesian geometry were able to recover the predicted $\omega^{-13/2}$ and k_h^4 power laws (Fig. 2.7). Couston et al. (2018) was able to measure these power laws even in simulations which were only moderately turbulent (i.e., with hardly any "inertial range"). It is possible that an alternative theory makes the same power law predictions, without any appeal to the Kolmogorov cascade. It remains to be seen if the prediction (or generalizations, e.g., Kumar et al. 1999) is also valid for stratified convection in spherical geometry. If taken at face value, the theory suggests high-frequency waves are excited by small eddies, which should not be strongly affected by stratification and/or sphericity.

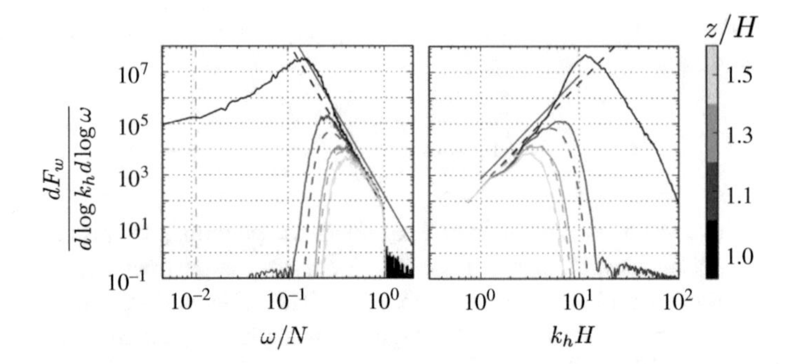

Fig. 2.7 Differential wave flux measured from simulations of wave generation by convection in Couston et al. (2018). The left panel shows the wave flux at fixed k_h, and the right panel shows the wave flux at fixed ω. The solid lines show the wave flux at different heights, where the radiative zone is above $z/H = 1$; the dashed lines are the theory (Eq. 2.107), including the effects of wave damping. The blue lines show the $\omega^{-13/2}$ and k_h^4 power laws. (Adapted from Couston et al. 2018, with permission)

Dynamical Influence of IGW

Using the IGW dispersion relation, one can show that the momentum flux is related to the energy flux via

$$\boldsymbol{u}_h u_z = \frac{\boldsymbol{k}_h}{\omega} F_w. \tag{2.108}$$

In stars, the *angular momentum flux*, is similarly related to the wave energy flux (Kumar and Quataert 1997),

$$\dot{J} = \frac{m}{\omega} F_w, \tag{2.109}$$

where m is the azimuthal wavenumber of the wave (whose angular dependence can be expanded in spherical harmonic functions $Y_{\ell,m}(\theta, \phi)$). To estimate the size of this angular momentum flux, we can calculate how long it would take to make an order unity change in the rotation of the solar radiative zone,

$$t_w \sim \frac{I_{rad}\Omega}{\dot{J}}, \tag{2.110}$$

where I_{rad} is the moment of inertia and Ω is the rotation rate. If we use $\dot{J}$ from waves with $\omega \sim \omega_c$, the convective frequency, and $m = 3$, we find $t_w \sim 10^5$ yrs, which is extremely short. However these waves, which are most efficiently excited by convection, also damp very quickly, and thus do not influence most of the radiative zone.

Low frequency waves carry the largest angular momentum flux, but decay very quickly, whereas very high frequency waves have low angular momentum flux, even if they can propagate through the entire star. The wave angular momentum flux as a function of radius is

$$\dot{j} \sim \rho_0 u_c^3 \frac{\omega_c}{N_0} \frac{m}{\omega} (k_\perp H)^4 \left(\frac{\omega}{\omega_c}\right)^{-13/2} \exp\left(-\int_{r_{\rm conv}}^{r} 2\ell_d^{-1}(r';\omega)dr'\right). \qquad (2.111)$$

One can then check that the waves which transport the most angular momentum to a radius r satisfies

$$\int_{r_{\rm conv}}^{r} \ell_d^{-1}(r';\omega)\, dr' = \frac{2}{7.5}, \qquad (2.112)$$

The frequency of this wave is

$$\omega^* = \left[\frac{4}{7.5}\int_{r_{\rm conv}}^{r} \chi k_h^3 N^3\, dr'\right]^{1/4}. \qquad (2.113)$$

For solar convection, $\omega^* \sim 5\omega_c$ (Fuller et al. 2014). This wave has a wave energy flux smaller than waves with frequency ω_c by a factor of $5^{-13/2} \sim 3 \times 10^{-5}$. Using Eq. 2.110, we find the typical time to substantially change the rotation of the solar interior is about

$$t_w \sim 10^{10}\,\text{yrs}, \qquad (2.114)$$

about the lifetime of the Sun. This suggests IGW transport of angular momentum may be marginally important for setting the rotation rate in the Sun's interior. Note that these estimates depend sensitively on the wave spectrum.

Wave transport of energy has been hypothesized to lead to mass loss at the end of massive stars' lives (Quataert and Shiode 2012; Shiode and Quataert 2014). About a year before a massive star explodes as a supernova, the star starts to burn silicon in its core. These nuclear reactions release tremendous energy, and efficiently excite IGWs. These waves can propagate to the surface of the star, where they reach high enough amplitudes that they break, depositing their energy and angular momentum. This energy/momentum injection in the outer layers of the star may be strong enough to launch winds. The presence of such winds could help explain certain observations of the subsequent supernova (e.g., Smith et al. 2007).

Conclusion

In this chapter, we first reviewed the theory of stellar structure. Energy generated by nuclear reactions in the core of a star is transported outward by a combination of radiation and convection. If the entire stellar luminosity can be transported by radiation, that region of the star is stably stratified. Otherwise, convection supplements the transport, and that region of the star is convective. Almost all stars have a convective region and stably stratified, radiative region.

Because the stellar structure equations are one-dimensional, they cannot self-consistently include the effects of convection. Convection in stars is complicated by the presence of large density contrasts across the convection zone. One way to model this strongly stratified convection is the anelastic equations. Simulations of stratified convection exhibit large asymmetries between upward and downward fluids which are absent from Boussinesq convection. Some properties of these convective flows can be compared to observations of sound waves or internal gravity waves at the surface of stars, through the techniques of helio- or asteroseismology.

Furthermore, because most stars have both convective and radiative zones, important dynamics occur at the interface between these regions. Here we described convective overshoot and the generation of internal gravity waves by convection. Other important problems include the generation of baroclinic flows (Jermyn et al. 2018), and the generation of waves at the interface by tides (Goodman and Dickson 1998). These could be related to mixing processes that occur in the radiative zones of stars (e.g., Pedersen et al. 2018).

However, the greatest challenge in stellar fluid dynamics is how to apply the insights of multidimensional numerical simulations to one-dimensional stellar evolution models. The large separation between dynamic/convective timescales and evolutionary timescales makes it impossible to simulate stellar evolution in three dimensions. Thus, the multi-dimensional fluid dynamical processes described in this chapter must be parameterized for use in stellar evolution models. These can then be tested against astrophysical observations to validate the underlying insights into the fluid dynamics of stars.

References

Alvan, L., Brun, A. S., & Mathis, S. (2014). Theoretical seismology in 3D: Nonlinear simulations of internal gravity waves in solar-like stars. *Astronomy & Astrophysics*, *565*, A42. https://doi.org/10.1051/0004-6361/201323253.

Andrássy, R., & Spruit, H. C. (2013). Overshooting by convective settling. *Astronomy & Astrophysics*, *559*, A122. https://doi.org/10.1051/0004-6361/201321793.

Ansong, J. K., & Sutherland, B. R. (2010). Internal gravity waves generated by convective plumes. *Journal of Fluid Mechanics*, *648*, 405. https://doi.org/10.1017/S0022112009993193.

Augustson, K. C., Brun, A. S., & Toomre, J. (2016). The magnetic furnace: Intense core dynamos in B stars. *The Astrophysical Journal*, *829*, 92. https://doi.org/10.3847/0004-637X/829/2/92.

Boyd, J. P. (2011). Chebyshev spectral methods and the Lane-Emden problem. *Numerical Mathematics: Theory, Methods and Applications*, *4*(2), 142–157. https://doi.org/10.1017/S100489790000057X.

Brandenburg, A. (2016). Stellar mixing length theory with entropy rain. *The Astrophysical Journal*, *832*, 6. https://doi.org/10.3847/0004-637X/832/1/6.

Brown, B. P., Miesch, M. S., Browning, M. K., Brun, A. S., & Toomre, J. (2011). Magnetic cycles in a convective dynamo simulation of a young solar-type star. *The Astrophysical Journal, 731*, 69. https://doi.org/10.1088/0004-637X/731/1/69.

Browning, M. K., Brun, A. S., & Toomre, J. (2004). Simulations of core convection in rotating a-type stars: Differential rotation and overshooting. *The Astrophysical Journal, 601*(1), 512–529. https://doi.org/10.1086/380198.

Burns, K. J., Vasil, G. M., Oishi, J. S., Lecoanet, D., & Brown, B. P. (2019). Dedalus: A flexible framework for numerical simulations with spectral methods. *arXiv e-prints*. arXiv:1905.10388.

Cantiello, M., & Braithwaite, J. (2011). Magnetic spots on hot massive stars. *Astronomy & Astrophysics, 534*, A140. https://doi.org/10.1051/0004-6361/201117512.

Chaplin, W. J., Elsworth, Y., Howe, R., Isaak, G. R., McLeod, C. P., Miller, B. A., et al. (1996). BiSON performance. *Solar Physics, 168*, 1–18. https://doi.org/10.1007/BF00145821.

Christensen-Dalsgaard, J. (2002). Helioseismology. *Reviews of Modern Physics, 74*, 1073–1129. https://doi.org/10.1103/RevModPhys.74.1073.

Christensen-Dalsgaard, J., Monteiro, M. J. P. F. G., Rempel, M., & Thompson, M. J. (2011). A more realistic representation of overshoot at the base of the solar convective envelope as seen by helioseismology. *Monthly Notices of the Royal Astronomical Society, 414*, 1158–1174. https://doi.org/10.1111/j.1365-2966.2011.18460.x.

Clune, T. C., Elliott, J. R., Miesch, M. S., Toomre, J., & Glatzmaier, G. A. (1999). Computational aspects of a code to study rotating turbulent convection in spherical shells. *Parallel Computing, 25*(4), 361–380. ISSN 0167-8191, https://doi.org/10.1016/S0167-8191(99)00009-5.

Couston, L.-A., Lecoanet, D., Favier, B., & Le Bars, M. (2018). The energy flux spectrum of internal waves generated by turbulent convection. *Journal of Fluid Mechanics, 854*, R3. https://doi.org/10.1017/jfm.2018.669.

Cristini, A., Meakin, C., Hirschi, R., Arnett, D., Georgy, C., Viallet, M., et al. (2017). 3D hydrodynamic simulations of carbon burning in massive stars. *Monthly Notices of the Royal Astronomical Society, 471*, 279–300. https://doi.org/10.1093/mnras/stx1535.

Cristini, A., Hirschi, R., Meakin, C., Arnett, D., Georgy, C., & Walkington, I. (2019). Dependence of convective boundary mixing on boundary properties and turbulence strength. *Monthly Notices of the Royal Astronomical Society, 484*, 4645–4664. https://doi.org/10.1093/mnras/stz312.

Denissenkov, P. A., Herwig, F., Truran, J. W., & Paxton, B. (2013). The C-flame quenching by convective boundary mixing in super-AGB stars and the formation of hybrid C/O/Ne white dwarfs and SN progenitors. *The Astrophysical Journal, 772*, 37. https://doi.org/10.1088/0004-637X/772/1/37.

Deubner, F.-L., & Gough, D. (1984). Helioseismology: Oscillations as a diagnostic of the solar interior. *Annual Review of Astronomy and Astrophysics, 22*, 593–619. https://doi.org/10.1146/annurev.aa.22.090184.003113.

Edelmann, P. V. F., Ratnasingam, R. P., Pedersen, M. G., Bowman, D. M., Prat, V., & Rogers, T. M. (2019). Three-dimensional simulations of massive stars: I. Wave generation and propagation. *arXiv e-prints*.

Featherstone, N. A., & Hindman, B. W. (2016a). The spectral amplitude of stellar convection and its scaling in the high-Rayleigh-number regime. *The Astrophysical Journal, 818*, 32. https://doi.org/10.3847/0004-637X/818/1/32.

Featherstone, N. A., & Hindman, B. W. (2016b). The emergence of solar supergranulation as a natural consequence of rotationally constrained interior convection. *The Astrophysical Journal, 830*, L15. https://doi.org/10.3847/2041-8205/830/1/L15.

Fields, C. E., Timmes, F. X., Farmer, R., Petermann, I., Wolf, W. M., & Couch, S. M. (2018). The impact of nuclear reaction rate uncertainties on the evolution of core-collapse supernova progenitors. *The Astrophysical Journal Supplement Series, 234*, 19. https://doi.org/10.3847/1538-4365/aaa29b.

Freytag, B., Ludwig, H.-G., & Steffen, M. (1996). Hydrodynamical models of stellar convection. The role of overshoot in DA white dwarfs, A-type stars, and the sun. *Astronomy & Astrophysics, 313*, 497–516.

Fuller, J., Lecoanet, D., Cantiello, M., & Brown, B. (2014). Angular momentum transport via internal gravity waves in evolving stars. *The Astrophysical Journal*, *796*, 17. https://doi.org/10.1088/0004-637X/796/1/17.

Goldreich, P., & Kumar, P. (1990). Wave generation by turbulent convection. *The Astrophysical Journal*, *363*, 694–704. https://doi.org/10.1086/169376.

Goodman, J., & Dickson, E. S. (1998). Dynamical tide in solar-type binaries. *The Astrophysical Journal*, *507*, 938–944. https://doi.org/10.1086/306348.

Greer, B. J., Hindman, B. W., Featherstone, N. A., & Toomre, J. (2015). Helioseismic imaging of fast convective flows throughout the near-surface shear layer. *The Astrophysical Journal*, *803*, L17. https://doi.org/10.1088/2041-8205/803/2/L17.

Hanasoge, S. M., Duvall, T. L., & Sreenivasan, K. R. (2012). Anomalously weak solar convection. *Proceedings of the National Academy of Science*, *109*, 11928–11932. https://doi.org/10.1073/pnas.1206570109.

Hansen, C. J., Kawaler, S. D., & Trimble, V. (2004). *Stellar interiors: Physical principles, structure, and evolution*. Berlin: Springer.

Herault, J., Facchini, G., & Le Bars, M. (2018). Erosion of a sharp density interface by a turbulent jet at moderate Froude and Reynolds numbers. *Journal of Fluid Mechanics*, *838*, 631–657. https://doi.org/10.1017/jfm.2017.891.

Hotta, H., Rempel, M., & Yokoyama, T. (2014). High-resolution calculations of the solar global convection with the reduced speed of sound technique. I. The structure of the convection and the magnetic field without the rotation. *The Astrophysical Journal*, *786*, 24. https://doi.org/10.1088/0004-637X/786/1/24.

Hristov, T. S., Miller, S. D., & Friehe, C. A. (2003). Dynamical coupling of wind and ocean waves through wave-induced air flow. *Nature*, *422*, 55–58. https://doi.org/10.1038/nature01382.

Hughes, D. W., Rosner, R., & Weiss, N. O. (2007). *The solar tachocline*. Cambridge: Cambridge University Press. ISBN 9781139462587, https://books.google.com/books?id=1WD5xUp8fWIC.

Jermyn, A. S., Tout, C. A., & Chitre, S. M. (2018). Enhanced rotational mixing in the radiative zones of massive stars. *Monthly Notices of the Royal Astronomical Society*, *480*, 5427–5446. https://doi.org/10.1093/mnras/sty1831.

Jiang, Y.-F., Cantiello, M., Bildsten, L., Quataert, E., Blaes, O., & Stone, J. (2018). Outbursts of luminous blue variable stars from variations in the helium opacity. *Nature*, *561*, 498–501. https://doi.org/10.1038/s41586-018-0525-0.

Jones, C. A., Boronski, P., Brun, A. S., Glatzmaier, G. A., Gastine, T., Miesch, M. S., et al. (2011). Anelastic convection-driven dynamo benchmarks. *Icarus*, *216*(1), 120–135. ISSN 0019-1035, https://doi.org/10.1016/j.icarus.2011.08.014, http://www.sciencedirect.com/science/article/pii/S0019103511003319.

Jones, S., Andrassy, R., Sandalski, S., Davis, A., Woodward, P., & Herwig, F. (2017). Idealized hydrodynamic simulations of turbulent oxygen-burning shell convection in 4π geometry. *Monthly Notices of the Royal Astronomical Society*, *465*, 2991–3010. https://doi.org/10.1093/mnras/stw2783.

Käpylä, P. J., Rheinhardt, M., Brandenburg, A., Arlt, R., Käpylä, M. J., Lagg, A., et al. (2017). Extended subadiabatic layer in simulations of overshooting convection. *The Astrophysical Journal*, *845*, L23. https://doi.org/10.3847/2041-8213/aa83ab.

Kippenhahn, R., & Weigert, A. (1994). *Stellar structure and evolution*. Berlin, Heidelberg: Springer.

Korre, L., Garaud, P., & Brummell, N. H. (2019). Convective overshooting and penetration in a Boussinesq spherical shell. *Monthly Notices of the Royal Astronomical Society*, *484*, 1220–1237. https://doi.org/10.1093/mnras/stz047.

Kumar, P., & Quataert, E. J. (1997). Angular momentum transport by gravity waves and its effect on the rotation of the solar interior. *The Astrophysical Journal*, *475*, L143–L146. https://doi.org/10.1086/310477.

Kumar, P., Talon, S., & Zahn, J.-P. (1999). Angular momentum redistribution by waves in the sun. *The Astrophysical Journal*, *520*, 859–870. https://doi.org/10.1086/307464.

Langer, N. (2012). Presupernova evolution of massive single and binary stars. *Annual Review of Astronomy and Astrophysics*, *50*, 107–164. https://doi.org/10.1146/annurev-astro-081811-125534.

Lecoanet, D., & Quataert, E. (2013). Internal gravity wave excitation by turbulent convection. *Monthly Notices of the Royal Astronomical Society*, *430*, 2363–2376. https://doi.org/10.1093/mnras/stt055.

Lecoanet, D., Brown, B. P., Zweibel, E. G., Burns, K. J., Oishi, J. S., & Vasil, G. M. (2014). Conduction in low mach number flows. I. Linear and weakly nonlinear regimes. *The Astrophysical Journal*, *797*, 94. https://doi.org/10.1088/0004-637X/797/2/94.

Lecoanet, D., Le Bars, M., Burns, K. J., Vasil, G. M., Brown, B. P., Quataert, E., et al. (2015). Numerical simulations of internal wave generation by convection in water. *Physical Review E*, *91*(6), 063016. https://doi.org/10.1103/PhysRevE.91.063016.

Lecoanet, D., Schwab, J., Quataert, E., Bildsten, L., Timmes, F. X., Burns, K. J., et al. (2016). Turbulent chemical diffusion in convectively bounded carbon flames. *The Astrophysical Journal*, *832*, 71. https://doi.org/10.3847/0004-637X/832/1/71.

Lohse, D., & Xia, K.-Q. (2010). Small-scale properties of turbulent Rayleigh-Bénard convection. *Annual Review of Fluid Mechanics*, *42*, 335–364. https://doi.org/10.1146/annurev.fluid.010908.165152.

Marston, J. B., Chini, G. P., & Tobias, S. M. (2016). Generalized quasilinear approximation: Application to zonal jets. *Physical Review Letters*, *116*(21), 214501. https://doi.org/10.1103/PhysRevLett.116.214501.

Mellado, J. P. (2012). Direct numerical simulation of free convection over a heated plate. *Journal of Fluid Mechanics*, *712*, 418–450. https://doi.org/10.1017/jfm.2012.428.

Mosumgaard, J. R., Ball, W. H., Aguirre, V. S., Weiss, A., & Christensen-Dalsgaard, J. (2018). Stellar models with calibrated convection and temperature stratification from 3D hydrodynamics simulations. *Monthly Notices of the Royal Astronomical Society*, *478*, 5650–5659. https://doi.org/10.1093/mnras/sty1442.

Ogura, Y., & Phillips, N. A. (1962). Scale analysis of deep and shallow convection in the atmosphere. *Journal of the Atmospheric Sciences*, *19*(2), 173–179. https://doi.org/10.1175/1520-0469(1962)019<C0173:SAODAS>2.0.CO;2.

Paxton, B., Bildsten, L., Dotter, A., Herwig, F., Lesaffre, P., & Timmes, F. (2011). Modules for experiments in stellar astrophysics (MESA). *The Astrophysical Journal Supplement Series*, *192*, 3. https://doi.org/10.1088/0067-0049/192/1/3.

Paxton, B., Cantiello, M., Arras, P., Bildsten, L., Brown, E. F., Dotter, A., et al. (2013). Modules for experiments in stellar astrophysics (MESA): Planets, oscillations, rotation, and massive stars. *The Astrophysical Journal Supplement Series*, *208*, 4. https://doi.org/10.1088/0067-0049/208/1/4.

Paxton, B., Marchant, P., Schwab, J., Bauer, E. B., Bildsten, L., Cantiello, M., et al. (2015). Modules for experiments in stellar astrophysics (MESA): Binaries, pulsations, and explosions. *The Astrophysical Journal Supplement Series*, *220*, 15. https://doi.org/10.1088/0067-0049/220/1/15.

Paxton, B., Schwab, J., Bauer, E. B., Bildsten, L., Blinnikov, S., Duffell, P., et al. (2018). Modules for experiments in stellar astrophysics (MESA): Convective boundaries, element diffusion, and massive star explosions. *The Astrophysical Journal Supplement Series*, *234*, 34. https://doi.org/10.3847/1538-4365/aaa5a8.

Pedersen, M. G., Aerts, C., Pápics, P. I., & Rogers, T. M. (2018). The shape of convective core overshooting from gravity-mode period spacings. *Astronomy & Astrophysics*, *614*, A128. https://doi.org/10.1051/0004-6361/201732317.

Pratt, J., Baraffe, I., Goffrey, T., Constantino, T., Viallet, M., Popov, M. V., et al. (2017). Extreme value statistics for two-dimensional convective penetration in a pre-main sequence star. *Astronomy & Astrophysics*, *604*, A125. https://doi.org/10.1051/0004-6361/201630362.

Quataert, E., & Shiode, J. (2012). Wave-driven mass loss in the last year of stellar evolution: Setting the stage for the most luminous core-collapse supernovae. *Monthly Notices of the Royal Astronomical Society*, *423*, L92–L96. https://doi.org/10.1111/j.1745-3933.2012.01264.x.

Rogers, T. M., & Glatzmaier, G. A. (2006). Angular momentum transport by gravity waves in the solar interior. *The Astrophysical Journal*, *653*, 756–764. https://doi.org/10.1086/507259.

Rogers, T. M., Lin, D. N. C., McElwaine, J. N., & Lau, H. H. B. (2013). Internal gravity waves in massive stars: Angular momentum transport. *The Astrophysical Journal*, *772*, 21. https://doi.org/10.1088/0004-637X/772/1/21.

Rosenfield, P., Girardi, L., Williams, B. F., Clifton Johnson, L., Dolphin, A., Bressan, A., et al. (2017). A new approach to convective core overshooting: Probabilistic constraints from color-magnitude diagrams of LMC clusters. *The Astrophysical Journal*, *841*, 69 (2017). https://doi.org/10.3847/1538-4357/aa70a2.

Shiode, J. H., & Quataert, E. (2014). Setting the stage for circumstellar interaction in core-collapse supernovae. II. Wave-driven mass loss in supernova progenitors. *The Astrophysical Journal*, *780*, 96. https://doi.org/10.1088/0004-637X/780/1/96.

Smith, N. (2014). Mass loss: Its effect on the evolution and fate of high-mass stars. *Annual Review of Astronomy and Astrophysics*, *52*, 487–528. https://doi.org/10.1146/annurev-astro-081913-040025.

Smith, N., Li, W., Foley, R. J., Wheeler, J. C., Pooley, D., Chornock, R., et al. (2007). SN 2006gy: Discovery of the most luminous supernova ever recorded, powered by the death of an extremely massive star like η carinae. *The Astrophysical Journal*, *666*, 1116–1128. https://doi.org/10.1086/519949.

Stancliffe, R. J., Fossati, L., Passy, J.-C., & Schneider, F. R. N. (2015). Confronting uncertainties in stellar physics: Calibrating convective overshooting with eclipsing binaries. *Astronomy & Astrophysics*, *575*, A117. https://doi.org/10.1051/0004-6361/201425126.

Sukhbold, T., & Woosley, S. E. (2014). The compactness of presupernova stellar cores. *The Astrophysical Journal*, *783*, 10. https://doi.org/10.1088/0004-637X/783/1/10.

Townsend, R. H. D., Goldstein, J., & Zweibel, E. G. (2018). Angular momentum transport by heat-driven g-modes in slowly pulsating B stars. *Monthly Notices of the Royal Astronomical Society*, *475*, 879–893. https://doi.org/10.1093/mnras/stx3142.

Vasil, G. M., Lecoanet, D., Brown, B. P., Wood, T. S., & Zweibel, E. G. (2013). Energy conservation and gravity waves in sound-proof treatments of stellar interiors. II. Lagrangian constrained analysis. *The Astrophysical Journal*, *773*, 169. https://doi.org/10.1088/0004-637X/773/2/169.

Viallet, M., Baraffe, I., & Walder, R. (2013). Comparison of different nonlinear solvers for 2D time-implicit stellar hydrodynamics. *Astronomy & Astrophysics*, *555*, A81. https://doi.org/10.1051/0004-6361/201220725.

Woodward, P. R., Herwig, F., & Lin, P.-H. (2015). Hydrodynamic simulations of H entrainment at the top of He-shell flash convection. *The Astrophysical Journal*, *798*, 49. https://doi.org/10.1088/0004-637X/798/1/49.

Chapter 3
Internal Waves in the Atmosphere and Ocean: Instability Mechanisms

Bruce R. Sutherland

Abstract This chapter summarizes and extends part of the lectures on internal gravity waves in the atmosphere and ocean by focusing upon the various instabilities associated with vertically propagating internal waves, including breaking due to convective overturning and shear, parametric subharmonic instability, and modulational instability associated with spatially localized wave packets.

Overview

Internal gravity waves move within a fluid whose density changes with height, and are driven by buoyancy forces. The waves can be subdivided broadly into two classes: interfacial and vertically propagating waves. Interfacial internal waves and modes move horizontally being confined by localized stratification and/or by horizontal boundaries at the top and bottom of the domain. Vertically propagating internal waves move upward and downward in continuously varying stratification. Here we will focus on the latter case, referring to them simply as "internal waves". (For an introduction to the theory for propagation, generation, and breaking of interfacial and internal waves, see the textbook by Sutherland 2010). The aim of this chapter is to examine some of the stability properties of vertically propagating internal waves. There are a myriad of mechanisms through which energy and momentum on the scale of the waves can be transferred to different scales. Understanding these processes is essential to improve predictions of the general circulation of the atmosphere and of mixing in the ocean. Particularly in the latter case, internal waves are an important conduit through which energy injected into the ocean at large scales (through surface winds or tidal flow over bottom topography) is ultimately dissipated, thus providing closure to the ocean energy budget. The use of linear theory is well established to estimate the breaking of waves due to overturning or (in the atmosphere) due to

B. R. Sutherland (✉)
Departments of Physics and of Earth and Atmospheric Sciences, University of Alberta, Edmonton, AB, Canada
e-mail: bsuther@ualberta.ca

M. Le Bars and D. Lecoanet (eds.), *Fluid Mechanics of Planets and Stars*,
CISM International Centre for Mechanical Sciences 595,
https://doi.org/10.1007/978-3-030-22074-7_3

interactions with the synoptic-scale winds near a critical level, which is where the ground-based horizontal phase speed of the waves matches the wind speed (Lindzen 1981). Because of the interest in mixing induced by oceanic internal waves, several new breaking mechanisms have only recently come under investigation. These include the breakdown of a wave beam reflecting from bottom topography with a critical slope (e.g., see Chalamalla and Sarkar 2016), breaking waves exerting drag upon oceanic eddies (Trossman et al. 2013; Marshall et al. 2017), or waves simply being trapped and eventually absorbed within an anticyclonic eddy (Danioux et al. 2015). These and other energy and momentum cascade processes are discussed in a recent review paper (Sutherland et al. 2019).

Although an entire book could be devoted to internal wave breaking, this chapter focuses specifically upon the stability of internal waves in the absence of topography, background flows, or other waves. Even with these restrictions, there are still many mechanisms through which internal waves may become unstable, whether through overturning, shear instability, parametric subharmonic instability, or modulational instability. In the section "Dispersion and Polarization Relations", we review the linear theory for small-amplitude, plane-periodic Boussinesq, and anelastic internal waves. Breaking conditions (whether due to convective overturning or shear instability) are then reviewed in the section "Breaking Conditions". In the section "Triad Resonant Instability" reviews the theory for triad resonance instability (TRI) with discussion of its adaptation to confined modes and internal wave beams. Moderately large amplitude quasi-monochromatic wave packets may be susceptible to modulational instability, as discussed in the section "Modulational Stability and Instability". Finally, future directions for research are discussed in the section "Future Directions".

Dispersion and Polarization Relations

The motion of internal waves is given by the laws of conservation of mass, momentum, and internal energy with additional approximations to filter the dynamics of sound waves and to account for the magnitude of background density variations in the vertical. In the ocean or in the atmosphere over small ($\lesssim 5\,\text{km}$) vertical distances, the Boussinesq approximation is used: this amounts to treating density as a constant, ρ_0, where it multiplies the material derivative in the momentum equations; density variations are included only as they influence buoyancy. For internal waves with vertical structure or propagating over vertical distances larger than about $10\,\text{km}$, it is necessary to use the anelastic approximation. In this case, the background density profile $\bar{\rho}(z)$ multiplies the material derivative in the momentum equations. The way in which sound is filtered is different under these two approximations: a Boussinesq fluid is assumed to be incompressible; for an anelastic gas, the background density variations are accounted for in the mass conservation equation.

Explicitly, ignoring viscosity and dissipative effects, the equations are given by

$$\begin{array}{ll}
\text{Boussinesq equations} & \text{anelastic equations} \\
\nabla \cdot \vec{u} = 0 & \nabla \cdot (\bar{\rho}\vec{u}) = 0 \\
\frac{Du}{Dt} - fv = -\frac{1}{\rho_0}\frac{\partial p}{\partial x} & \frac{Du}{Dt} - fv = -\frac{\partial}{\partial x}\left(\frac{p}{\bar{\rho}}\right) \\
\frac{Dv}{Dt} + fu = -\frac{1}{\rho_0}\frac{\partial p}{\partial y} & \frac{Dv}{Dt} + fu = -\frac{\partial}{\partial y}\left(\frac{p}{\bar{\rho}}\right) \\
\frac{Dw}{Dt} = -\frac{1}{\rho_0}\frac{\partial p}{\partial z} - b & \frac{Dw}{Dt} = -\frac{\partial}{\partial z}\left(\frac{p}{\bar{\rho}}\right) - b \\
\frac{Db}{Dt} = -N^2 w & \frac{Db}{Dt} = -N^2 w.
\end{array} \tag{3.1}$$

Here we will examine internal waves in otherwise stationary fluid so that $\vec{u} = (u, v, w)$ represents the velocity fluctuations due to waves. We also assume that the fluid is uniformly stratified, meaning that the buoyancy frequency, N, is a constant. The buoyancy frequency represents the frequency (in radians per second) of vertical oscillations in the fluid which results when fluid is carried upward and so finds itself more dense than its surroundings or it is carried downward and finds itself lighter than its surroundings. In the ocean, over vertical domains less than about a kilometer (so that the thermodynamic effects of pressure upon density and temperature can be ignored), the buoyancy frequency is given by $N = [-(g/\rho_0)\, d\bar{\rho}/dz]^{1/2}$. The ocean is uniformly stratified where the density increases linearly with depth, corresponding to an approximately linear decrease in temperature and/or linear increase in salinity provided that these changes are not so large that the nonlinear equation of state of sea water needs to be taken into account. In the atmosphere, the buoyancy frequency is given generally by $N = [(g/\bar{\theta})\, d\bar{\theta}/dz]^{1/2}$, in which g is gravity and $\bar{\theta}(z)$ is the vertical profile of the background potential temperature. Explicitly, $\bar{\theta} \equiv \bar{T}(\bar{p}/p_0)^{-\kappa}$, in which $\bar{T}(z)$ and $\bar{p}(z)$, respectively, are the background temperature and pressure, p_0 is a reference pressure (typically taken to be that at the ground), and $\kappa \simeq 2/7$ for Earth's atmosphere, which is composed primarily of the diatomic molecules of nitrogen and oxygen. The atmosphere is uniformly stratified where the background temperature is constant with height in which case the potential temperature increases exponentially with height and the corresponding density decreases exponentially with height according to $\bar{\rho}(z) = \rho_0 \exp(-z/H_\rho)$, with $H_\rho = R_a T/g$ being the density scale height for air with gas constant $R_a = 287\,\text{J/kg K}$. Internal waves exist on sufficiently small spatial scales that the influence of the Earth's rotation can be treated as approximately uniform with latitude, if its influence is considered at all. Thus, the Coriolis parameter, f, is taken to be constant. Finally, b represents the buoyancy associated with perturbations in the fluid: for the ocean $b \equiv -g\rho/\rho_0$; for the atmosphere $b \equiv g\theta/\bar{\theta}$.

What is clear from (3.1) is that the anelastic equations reduce to the Boussinesq equations by taking $\bar{\rho} \to \rho_0$. This is one reason why one can perform laboratory experiments in tanks filled with stratified salt solutions and yet still make predictions about stratified air flow on relatively small vertical scales in the atmosphere.

The equations for small-amplitude waves are produced by neglecting the advective terms in the material derivatives appearing in (3.1) so that $D/Dt \equiv \partial/\partial t + \vec{u} \cdot \nabla \to$

$\partial/\partial t$. In the Boussinesq approximation, this gives five coupled linear equations in five unknowns all having constant coefficients. The structure and dynamics of these waves are correspondingly straightforward to derive. For small-amplitude anelastic waves, the exponentially decreasing background density $\bar{\rho}$ leads to a moderately more involved analysis. Nonetheless, after combining the coupled equations to a single equation in one unknown, the mass streamfunction, the result is a partial differential equation with constant coefficients, and so is straightforward to solve. These procedures are discussed in more detail below.

Boussinesq Internal Waves

After linearizing the Boussinesq equations in (3.1), we seek plane-periodic solutions with the understanding from the superposition principle that any solution can be written as a sum of waves. Thus, we seek solutions of the form

$$(u, v, w, b, p) = (A_u, A_v, A_w, A_b, A_p) \exp[\imath(kx + \ell y + mz - \omega t)], \tag{3.2}$$

in which it is understood that the actual field is the real part of the right-hand side expression. Each of the constant amplitudes (A_u, etc) is possibly complex valued such that its magnitude represents the actual amplitude and its argument represents the phase.

Substituting (3.2) into (3.1) thus results in an algebraic matrix eigenvalue problem whose eigenvalue gives the dispersion relation (an equation relating the frequency, ω, to the wavenumber $\vec{k} = (k, \ell, m)$) and whose eigenvector gives the polarization relations (the relationship between the amplitudes of each field). Explicitly, the dispersion relationship for Boussinesq internal waves is

$$\omega^2 = \frac{N^2(k^2 + \ell^2) + f^2 m^2}{k^2 + \ell^2 + m^2} = N^2 \cos^2\Theta + f^2 \sin^2\Theta, \tag{3.3}$$

where in the last expression we have defined $\Theta \equiv \tan^{-1}(m/k_h)$ with $k_h \equiv (k^2 + \ell^2)^{1/2}$ being the horizontal wavenumber. Thus, Θ represents the angle from the horizontal of the wavenumber vector or, equivalently, the angle from the vertical of lines of constant phase. Supposing that the maximum vertical displacement, A_0, is prescribed and the phase is arbitrarily set so that the vertical displacement amplitude $A_\eta = A_0$ is real valued, then the polarization relations can be written to express how the amplitudes of the remaining fields depend upon A_0. For mathematical convenience, we further simplify the results by supposing the x-axis points in the horizontal direction of propagation of the waves. (Such a transformation is possible because the full Eq. (3.1) is invariant upon rotation about the z-axis provided that the Coriolis parameter, f, is constant.) In effect, this allows us to set $\ell = 0$ so that $k_h = k$. The results are given in Table 3.1 and illustrated in Fig. 3.1.

Table 3.1 Polarization relations for Boussinesq internal waves giving (left) the possibly complex amplitudes and (right) the actual fields supposing the vertical displacement amplitude A_0 is real valued

$A_\eta = A_0$	$\Rightarrow$	$\eta = A_0 \cos(kx + mz - \omega t)$
$A_u = \imath\omega \tan\Theta\, A_0$	$\Rightarrow$	$u = -\omega\,(m/k)\,A_0 \sin(kx + mz - \omega t)$
$A_v = f \tan\Theta\, A_0$	$\Rightarrow$	$v = f\,(m/k)\,A_0 \cos(kx + mz - \omega t)$
$A_w = -\imath\omega\, A_0$	$\Rightarrow$	$w = \omega\, A_0 \sin(kx + mz - \omega t)$
$A_b = -N^2 A_0$	$\Rightarrow$	$b = -N^2 A_0 \cos(kx + mz - \omega t)$

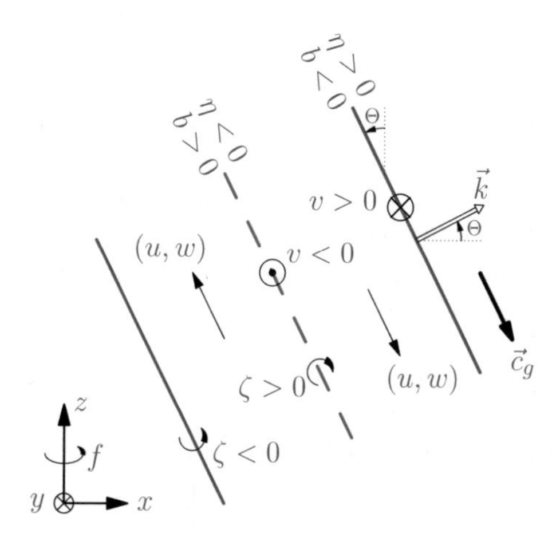

Fig. 3.1 Schematic illustrating the polarization relations associated with internal waves in the x–z plane having phase (group) velocity directed upward (downward) to the right. The Coriolis parameter is taken to be positive. Circles with crosses inside indicate vectors pointing into the page, whereas circles with filled circles inside indicate vectors pointing out of the page. The spanwise vorticity is indicated by ζ

Of course another approach to come up with these results is to combine the five coupled linearized Boussinesq equations in (3.1) to form a single differential equation. Assuming the waves have no structure in the y-direction, this is done most straightforwardly by defining the streamfunction ψ implicitly through $u = -\partial_z\psi$ and $w = \partial_x\psi$, thus automatically satisfying the condition for incompressibility. The terms involving pressure are eliminated by taking the spanwise component of the curl of the momentum equations and so arriving at an equation for the spanwise vorticity, $\zeta = \partial_z u - \partial_x w = -\nabla^2\psi$. Finally, buoyancy can be eliminated by using the internal energy equation. If this procedure is followed retaining the nonlinear terms, the resulting fully nonlinear equation for the evolution of the streamfunction is

$$[\partial_{tt}\nabla^2 + N^2\partial_{xx} + f^2\partial_{zz}]\psi = \nabla \cdot [\partial_t(\zeta\vec{u}) - \partial_x(b\vec{u}) + f\partial_z(v\vec{u})]. \tag{3.4}$$

For small-amplitude waves, the right-hand side can be neglected and the resulting linear partial differential equation can immediately be seen to give the dispersion relation (3.3).

Anelastic Internal Waves

Because of the z-dependent coefficients involving $\bar{\rho}$ in the anelastic equations given in (3.1), one cannot start by assuming solutions with vertical periodicity in z. Instead, after linearizing it is necessary to combine the equations to form one equation in one unknown. As with Boussinesq waves, we simplify the algebra by orienting the x-axis in the horizontal direction of propagation of the waves, thus eliminating terms involving y-derivatives in (3.1). From the expression for mass conservation of anelastic waves, $\nabla \cdot (\bar{\rho}\vec{u}) = 0$, we define the mass streamfunction, Ψ, implicitly so that $u = -(1/\bar{\rho})\partial_z \Psi$ and $w = (1/\bar{\rho})\partial_x \Psi$. We then proceed as described above for Boussinesq flow by taking the spanwise curl of the momentum equations to give an evolution equation for the spanwise vorticity,

$$\zeta = \partial_z u - \partial_x w = -(1/\bar{\rho})[\nabla^2 \Psi + (1/H_\rho)\partial_z \Psi], \tag{3.5}$$

in which $H_\rho \equiv -(\bar{\rho}'/\bar{\rho})^{-1}$ is the density scale height and, for simplicity, the influence of the Earth's rotation has been neglected. Eliminating buoyancy from the resulting equations and neglecting background rotation gives

$$\frac{\partial^2}{\partial t^2}\nabla^2 \Psi + \frac{1}{H_\rho}\frac{\partial^3 \Psi}{\partial t^2 \partial z} + N^2 \frac{\partial^2 \Psi}{\partial x^2} = 0. \tag{3.6}$$

This reduces to the linearized Boussinesq equation (3.4) with f = 0 by taking the limit $H_\rho \to \infty$.

Because both N and H_ρ are constant in an isothermal, uniformly stratified atmosphere it is possible to find solutions to (3.6) in the form of an exponential: $\Psi = A_\Psi \exp[\imath(kx + Mz - \omega t)]$, in which the z-dependent background suggests that M is not necessarily real valued. Indeed, by substituting into (3.6), we find $M = m + \imath/(2H_\rho)$, in which m is the usual real-valued vertical wavenumber. Thus, the mass streamfunction has the form of a exponentially decaying oscillator with height. The corresponding velocity fields increase exponentially with height according to

$$\begin{aligned} u &= -\frac{1}{\rho_0}(\imath m - 1/2H_\rho)\, A_\Psi\, e^{z/2H_\rho}\, e^{\imath(kx+mz-\omega t)} \\ w &= \frac{1}{\rho_0}(\imath k)\, A_\Psi\, e^{z/2H_\rho}\, e^{\imath(kx+mz-\omega t)}, \end{aligned} \tag{3.7}$$

and the dispersion relation for anelastic waves not influenced by the Earth's rotation is given by $\omega^2 = N^2 k^2\, [k^2 + m^2 + (4H_\rho)^{-2}]^{-1/2}$. While it may seem unphysical that u and w appear to be unbounded, the exponential increase in these fields follows directly from the physical requirement that momentum is conserved, which is to say that in the absence of breaking the horizontally averaged vertical flux of horizontal momentum, $\bar{\rho}\,\langle uw \rangle$, must be constant. The velocities increase because the background density decreases with height. In reality, waves do not grow to infinite amplitude: upward-propagating waves eventually grow to such large amplitude that weakly nonlinear effects (ignored in the derivation of (3.6)) begin to play a role and eventually fully nonlinear effects of convective overturning and turbulent breakdown occur.

Breaking Conditions

Although linear theory strictly makes reliable predictions only for waves of sufficiently small amplitude, it is insightful to extrapolate the results for waves that are of such large amplitude that they are overturning or otherwise lead to fine-scale instabilities associated with breaking waves (Lindzen 1981). While we can imagine that anelastic waves eventually grow to such large amplitude that they break, the breaking itself is expected to occur over small vertical scales compared with the density scale height. This gives some justification for using the polarization relations for Boussinesq waves to examine breaking.

Overturning

Internal waves are overturning if the amplitude is so large that the waves carry dense fluid over less dense fluid rendering the fluid locally to be unstably stratified. Mathematically, this means that the total squared buoyancy frequency, $N_T^2 \equiv N^2 + \Delta N^2$, composed of the background squared buoyancy frequency plus the change due to waves $\Delta N^2 \equiv \partial b/\partial z$, is negative somewhere in the flow field. Thus, the condition for the critical vertical displacement amplitude A_{OT} of waves on the cusp of overturning is $\min\{[\Delta N^2/N^2]\}|_{A=A_{\mathrm{OT}}} = -1$. Using the polarization relations in Table 3.1 gives

$$A_{\mathrm{OT}} = \left|\frac{1}{m}\right|. \tag{3.8}$$

Convection

Just because the waves are overturning, it does not mean they will necessarily convectively breakdown. This is because convective instability itself takes time to grow. If the growth rate is not sufficiently fast, the waves may pull back the overturned fluid so that the region becomes stably stratified again.

Thus, an assessment of whether convection actually occurs is found by comparing the growth rate, σ_c, for convection with the frequency, ω, of the waves. The growth rate itself is given by the maximum value of $(-N_T^2)^{1/2}$ in which the discriminant is evaluated where the local total stratification is unstable ($N_T^2 < 0$). The critical vertical displacement amplitude, A_{CV}, of overturning waves that are susceptible to convection is given implicitly by the condition $\sigma_c = \omega$. Using the polarization and dispersion relations, we find

$$A_{\mathrm{CV}} = \left|\frac{1}{m}\right| \left(1 + \frac{\omega^2}{N^2}\right), \tag{3.9}$$

in which the frequency is given by (3.3). In the low-frequency limit, appropriate for near-inertial internal waves, $A_{\mathrm{CV}} \simeq A_{\mathrm{OT}}$.

Shear Instability

For near-inertial internal waves, which have low frequency close to f, their motion is close to horizontal ($\Theta \lesssim 90°$). Thus, even if the waves are not overturning, it is possible for shear instability to develop and ultimately result in the breakdown of the waves. A necessary condition for the linear instability of a parallel stratified shear flow is that the minimum gradient Richardson number, Ri_g, is below $1/4$. For the case of internal waves moving in the x–z plane ($\ell = 0$), Ri_g is the ratio of the local stratification, N_T^2, to the square of the local vertical shear $\partial u/\partial z$. For small amplitude, non-overturning waves, it turns out that this has a minimum close to where the stratification is unaffected by the waves ($\Delta N^2 = 0$) but where the shear is strongest (see Sect. 4.6.3 of Sutherland 2010). Thus, the critical vertical displacement amplitude for waves satisfying $\min\{\mathrm{Ri}_g\} = 1/4$ is

$$A_{\mathrm{SHR}} = 2\frac{N|k|}{\omega m^2}\sqrt{1 - \left(\frac{Nk}{\omega m}\right)^2}. \tag{3.10}$$

This criterion is met for amplitudes below the overturning threshold ($A_0 = A_{\mathrm{OT}}$) for internal waves having $|k/m| < \sqrt{f/N} \simeq 0.1$, in which the last value is estimated from characteristic values in the atmosphere and ocean of $f \sim 10^{-4}\mathrm{s}^{-1}$ and $N \sim 10^{-2}\mathrm{s}^{-1}$.

Summary of Breaking Instabilities

Figure 3.2 plots the critical amplitudes A_{OT}, A_{CV}, and A_{SHR} against relative wave numbers. Also shown in Fig. 3.2a is the prediction for instability due to

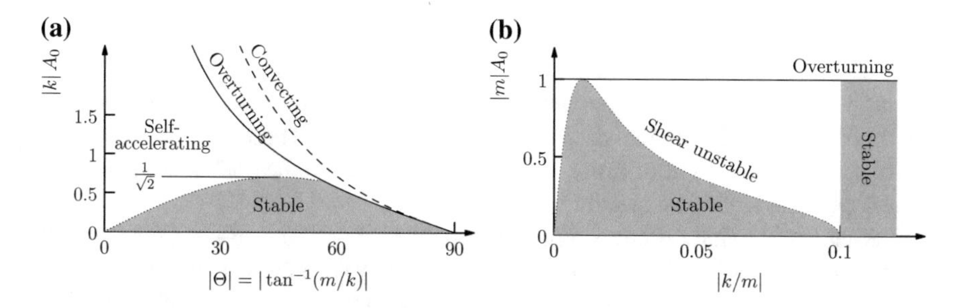

Fig. 3.2 Breaking regimes heuristically predicted from linear theory showing (**a**) the critical amplitudes associated with overturning and convective instability as well as self-acceleration and (**b**) the critical amplitude associated with stratified shear instability of near-inertial waves. (Adapted with permission from Fig. 4.18 of Sutherland 2010)

self-acceleration, a weakly nonlinear mechanism to be discussed in the section "Modulational Stability and Instability". Focusing on the high-frequency non-hydrostatic waves in Fig. 3.2a, it appears that the waves become unstable to self-acceleration before reaching overturning amplitudes. This is bourne out in simulations of horizontally periodic vertically localized wave packets though fully localized non-hydrostatic wave packets may achieve higher amplitudes before being driven to overturn by the effects of self-acceleration (e.g., see Sutherland 2001). Focusing on the near-inertial waves in Fig. 3.2b, it is clear that the waves become unstable due to shear instability before reaching overturning amplitudes if their wavenumber is such that $|k/m| < (f/N)^{1/2}$. Surprisingly, if $|k/m| \lesssim (f/N)^{1/2}$ then even waves with very small amplitude are susceptible to shear instability.

Triad Resonant Instability

Parametric subharmonic instability (PSI) or, more generally, triad resonant instability (TRI) is considered as a potentially important and pervasive mechanism for the breakdown of internal waves, particularly in the ocean where anelastic growth cannot contribute to drive the waves to breaking through one of the mechanisms described above. The general theory was developed for plane periodic waves (Hasselmann 1967) and applied specifically to internal waves by several researchers (see reviews by Phillips (1981); Staquet and Sommeria (2002); Dauxois et al. (2018)). Mathematically, TRI is a consequence of weakly nonlinear interactions between a triplet of waves including a "parent wave" with wavenumber $\vec{k}_0$ and a pair of "sibling waves" with wave numbers $\vec{k}_\pm$. Through the nonlinear advective terms in the equations of motion, energy can pass between the parent and siblings provided they satisfy the resonance conditions

$$\begin{aligned} \vec{k}_0 &= \pm\vec{k}_+ \pm \vec{k}_- \\ \omega_0 &= \pm\omega_+ \pm \omega_-, \end{aligned} \tag{3.11}$$

in which the frequencies ω_0 and $\omega_\pm$ are given by the corresponding dispersion relations. Even for waves restricted to the x–z plane, there are a rich number of solutions to these three equations in four unknowns. Of these it remains to determine for which the amplitudes of the sibling waves indeed grow. These are the waves that are expected to grow out of noise, extracting energy from the parent wave with the fastest growing siblings expected to be dominant.

Taking a somewhat different approach from that of previous researchers (Benielli and Sommeria 1998; Bourget et al. 2013), we begin by working with the fully nonlinear Boussinesq equations in the form given by (3.4). The parent wave is characterized in terms of the streamfunction in which extracting the real part of the complex exponential is made explicit:

$$\psi_0 = \frac{1}{2} a_0 \exp[\imath(k_0 x + m_0 z - \omega_0 t)] + \text{c.c.}, \tag{3.12}$$

in which c.c. denotes the complex conjugate. Likewise, the sibling waves have streamfunction

$$\psi_\pm = \frac{1}{2} a_\pm \exp[\imath(k_\pm x + m_\pm z - \omega_\pm t)] + \text{c.c.}. \tag{3.13}$$

The polarization relations can be used to find corresponding expressions for the vorticity, velocity, and buoyancy associated with the parent and sibling waves. In all these expressions, the amplitudes a_0 and $a_\pm$ are considered to be functions of time allowing for the possible growth of the sibling waves at the expense of the parent.

Putting these results into (3.4) together with the resonance conditions (3.11) gives a triplet of equations expressing how two of the waves interact to influence the third wave. In particular, the interaction between the parent and the "+" sibling to influence the "−" sibling is expressed through

$$\begin{aligned} &[\partial_{tt}\nabla^2 + N^2\partial_{xx} + f^2\partial_{zz}]\psi_- = \\ &\quad \nabla \cdot [\partial_t(\zeta_0\vec{u}_+ + \zeta_+\vec{u}_0) - \partial_x(b_0\vec{u}_+ + b_+\vec{u}_0)] + f\partial_z(v_0\vec{u}_+ + v_+\vec{u}_0)]. \end{aligned} \tag{3.14}$$

With use of (3.12) and (3.13) and the corresponding polarization relations, this can then be written as an equation for the time evolution of the amplitude a_-. In doing so, it is convenient to assume that the growth rate is small compared to the wave frequency so that $\|d_{tt}a_-\| \ll \omega_- \|d_t a_-\|$ and the term $d_t(a_0 a_+^\star)$ arising on the right-hand side of (3.14) is considered to be negligible.

The resulting equation can be written $d_t a_- = I_- a_0 a_+^\star$, in which I_- is an expression involving the wave numbers of the parent wave and the "+" sibling. Similarly, the evolution equation for the amplitude of the "+" sibling can be written $d_t a_+ = I_+ a_0 a_-^\star$. Combining these results gives

$$\frac{d^2}{dt^2} a_+ = I_+ I_-^\star \, |a_0|^2 \, a_+. \tag{3.15}$$

Thus, through the process of passing energy from one sibling to the other via the parent, it is possible for both siblings to grow in amplitude. Assuming the change in a_0 is small while the sibling amplitudes are small, so that $|a_0|$ can be treated as a constant in (3.15), the growth is exponential with growth rate, σ_p, given by the positive real part of $\sqrt{I_+ I_-^\star}\,|a_0|$. (As $|a_\pm|$ grows to be comparable to $|a_0|$, the waves can saturate or, if $|a_0|$ is sufficiently large, additional nonlinear effects can lead to energy being transferred to a broader frequency spectrum possibly leading to overturning.)

If rotation is neglected, then $I_\pm$ is pure real. In this case, it is possible for there to be no growth of sibling waves if their wave numbers are such that $I_+ I_- < 0$. However, numerous studies have shown growth for a range of sibling waves in two dimensions (Mied 1976) and for a wide range in three dimensions (Klostermeyer 1991; Lombard and Riley 1996).

Although the resonance conditions (3.11) strictly require the waves to be perfectly periodic in infinite space, sibling waves may also grow from a bounded parent mode in uniform stratification if the wave numbers of the sibling waves are sufficiently small that a wavelength can fit inside the domain. This has been observed in laboratory experiments (Bouruet-Aubertot et al. 1995; Benielli and Sommeria 1998), as shown in Fig. 3.3a.

TRI has also been observed for horizontally propagating vertically confined internal waves having a mode-1 vertical structure (Joubaud et al. 2012), as shown in Fig. 3.3b. Recent advances have been made in the study of TRI for internal wave beams (Bourget et al. 2013; Karimi and Akylas 2014; Dauxois et al. 2018), showing that sibling waves can grow within and emerge from a parent beam if there are a sufficient number of wavelengths of parent waves within the beam. For the interpretation of experiments showing TRI occurring within an internal wave beam (Fig. 3.3c), Bourget et al. (2013) included viscosity in their theoretical analysis.

Modulational Stability and Instability

If the amplitude of the waves varies in space, as surely occurs for naturally occurring internal waves, then there is another mechanism for amplitude growth or decay in time resulting from the waves inducing a flow that then Doppler shifts the waves themselves. If resulting in amplitude growth, the waves are said to be modulationally unstable. Otherwise, they are modulationally stable, meaning that their amplitude decreases faster than linear theory for dispersion would predict. The structure of the induced flow and its consequent impact back upon the waves depends non-trivially upon the dimensionality of the amplitude envelope, which describes the space- and time-dependent amplitude of the waves.

Conceptually, it is most straightforward to understand the physics of induced flows by considering a one-dimensional wave packet, which is spanwise uniform and horizontally periodic so that the amplitude envelope varies spatially only in the vertical. Thus, the vertical displacement field for a 1D Boussinesq wave packet is

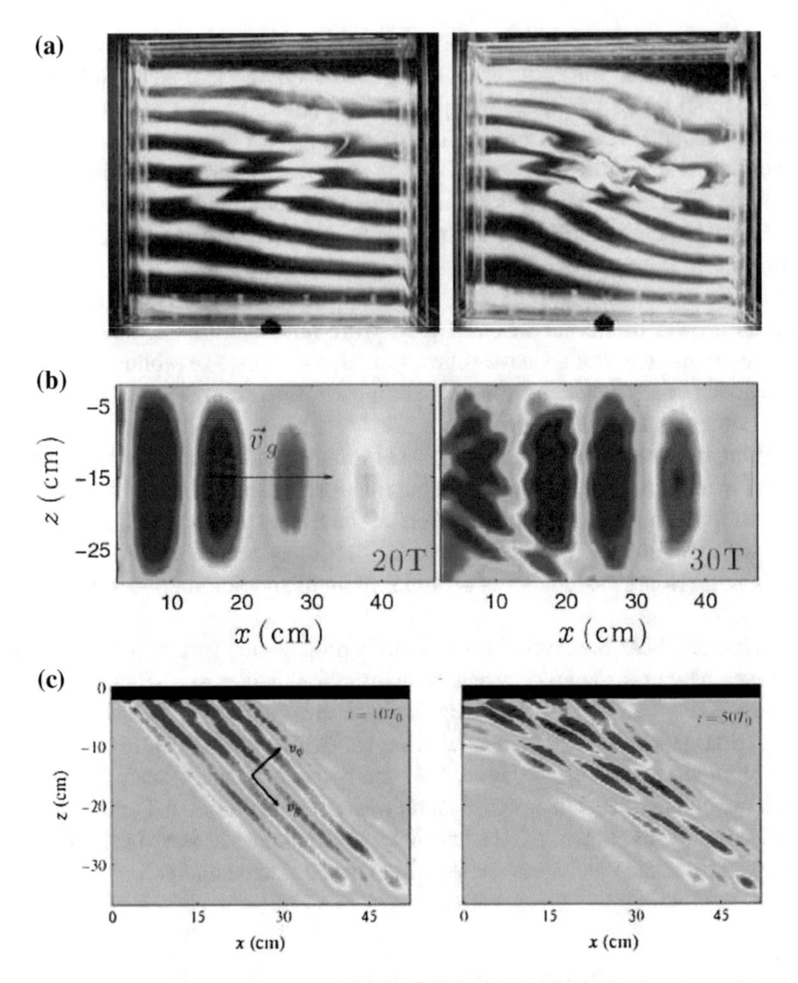

Fig. 3.3 Observations in laboratory experiments of TRI occurring for internal waves as (**a**) a mode in an oscillating square tank (adapted with permission from Fig. 6 of Bouruet-Aubertot et al. 1995), (**b**) a horizontally propagating vertically confined mode (adapted with permission from Fig. 1 of Joubaud et al. 2012), and (**c**) a down and rightward propagating beam (adapted with permission from Fig. 2 of Bourget et al. 2013)

written

$$\eta(x, z, t) = \frac{1}{2}A(Z, T)e^{\imath(kx+mz-\omega t)} + \text{c.c.}. \tag{3.16}$$

Assuming the modulated waves are quasi-monochromatic, the variables Z and T, respectively, represent slow variations in the vertical and in time as compared with the vertical wavelength and period of the waves.

Using the polarization relations in Table 3.1, the leading-order expressions for the corresponding horizontal and vertical velocity fields are

$$\begin{aligned} u_1(x,z,t) &= \tfrac{1}{2}\imath(\omega m/k)\, A(Z,T)\, e^{\imath(kx+mz-\omega t)} + \text{c.c.} \\ w_1(x,z,t) &= -\tfrac{1}{2}\imath\omega\, A(Z,T)\, e^{\imath(kx+mz-\omega t)} + \text{c.c.}, \end{aligned} \tag{3.17}$$

in which the subscripts emphasize that these fields, arising from linear theory, are proportional to the amplitude. The mean momentum flux, $F_z = \rho_0 \langle u_1 w_1 \rangle$, is straightforwardly computed from the cross-terms of the product, $u_1 w_1$, for which the complex conjugate cancels the complex exponential leaving only the slow variations. Explicitly,

$$F_z = -\rho_0 \frac{\omega^2 m}{2k} |A|^2. \tag{3.18}$$

Rightward propagating wave packets with upward group velocity, c_{gz}, have $k > 0$ and $m < 0$, and so $F_z > 0$ for such waves as expected.

Being horizontally periodic and incompressible, one can write the x-momentum equation in flux-form and average over one horizontal wavelength. For conceptual convenience we neglect rotation and so find

$$\frac{\partial}{\partial t}\left(\rho_0 \langle u \rangle\right) = -\frac{\partial}{\partial z} F_z. \tag{3.19}$$

The key here is to recognize that u on the left-hand side is not the u_1 appearing in (3.17). Instead it is a quantity proportional to the amplitude squared and which is independent of x. For convenience, we will write $u_2 \equiv \langle u \rangle$ on the left-hand side of (3.19). Thus, $\partial_t u_2$ represents the horizontal acceleration of the background resulting from the divergence of the momentum flux.

For plane waves the amplitude is constant and, consequently, so is the momentum flux, F_z. Hence, there is no force or corresponding flow induced by plane waves until they break or otherwise dissipate. However, even without dissipation, modulated internal wave packets do induce a (time-changing) flow directly as a consequence of the vertical variation in F_z. An example with an upward-propagating wave packet having a Gaussian amplitude envelope is illustrated in Fig. 3.4. The momentum flux is largest where the amplitude is largest at the vertical center of the wave packet. Corresponding to the decrease in the momentum flux going upward from this point there is a time-transient acceleration of the background flow. Likewise, the momentum flux increases going upward from below the wave packet to its center and so there the background flow decelerates after the center of the wave packet passes. The net increase in speed due to the acceleration occurring on the leading flank of the wave packet is canceled by the net decrease occurring on the trailing flank. And so the speed of the background flow remains the same from before to after the passage of the wave packet. However, a floating particle (a passive tracer) will find itself displaced rightward from its initial position.

A formula for the induced flow itself can be found if one assumes the flow is steady in a frame of reference moving with the group velocity of the wave packet. That is, we suppose $u_2(z,t) = u_2(\tilde{z})$ in which $\tilde{z} \equiv z - c_{gz}t$. Likewise, we know that the amplitude envelope translates at the group velocity and assume that time variations in the translating frame due to wave packet dispersion occur on much

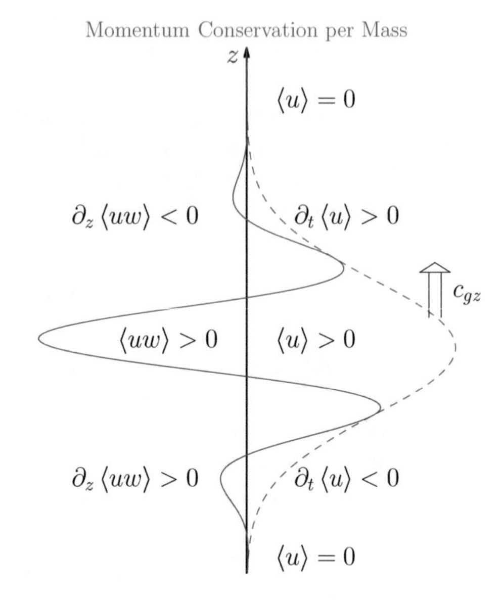

Fig. 3.4 Schematic illustrating the physical mechanism for the transient acceleration of the mean flow due to the divergence of the momentum flux associated with a wave packet. The curved solid line represents vertical variations of, e.g., displacement along some vertical cut; the dashed line represents the amplitude envelope, which translates vertically upward at the group velocity c_{gz}

slower time scales. Thus, we can approximate $A(Z,T) \simeq A(\tilde{z})$ in the expression for F_z in (3.18). So, changing co-ordinates in (3.19) to the translating frame gives $-c_{gz}\partial_{\tilde{z}}(\rho_0 u_2) = -\partial_{\tilde{z}} F_z$. Integrating both sides and using (3.18) gives

$$u_2 = -\frac{1}{2}\frac{1}{c_{gz}}\frac{\omega^2 m}{k}|A|^2 = \frac{1}{2}N|\vec{k}||A|^2, \tag{3.20}$$

in which $|\vec{k}| = (k^2+m^2)^{1/2}$. In the case illustrated by Fig. 3.4, the induced flow is a vertically translating squared Gaussian. A parcel at a fixed point in space accelerates rightward as the leading edge of the wave packet approaches and then decelerates as the squared Gaussian flow passes, but overall is displaced rightward as given by the vertical integral of u_2.

Now being aware that a modulated 1D wave packet induces a flow that translates vertically with the wave packet, it is anticipated that this flow, if sufficiently large, could act significantly to Doppler shift the waves, just as a time-independent background shear flow can act to cause wave reflection or breaking near a critical level. This is indeed the case. As a crude heuristic, an estimate for the critical amplitude at which this Doppler shifting might force the waves to overturn is given by the condition for "self-acceleration", which assumes the waves break if the maximum of the induced flow exceeds the horizontal group velocity of the waves. The critical amplitude for self-acceleration is $A_{SA} = \sqrt{2}\,m/|\vec{k}|$, which is plotted in Fig. 3.2a.

In a more rigorous examination, we note that the effect of a background horizontal flow, U, acting upon waves is given by the advection operator $U\partial_x$. So, in the absence of other influences, the evolution of the amplitude envelope due to the leading-order effect of Doppler shifting of the waves is represented by $\partial_t A = -\imath k U A$. Including the effects of the vertical translation of the wave packet at the group velocity as well as the dispersion of the waves, which can be important at this order, the evolution of the amplitude envelope of a wave packet as influenced by the flow it induces is given by

$$\partial_t A = -c_{gz}\partial_z A + \imath \frac{1}{2}\omega_{mm}\partial_{zz} A - \imath \frac{1}{2} N k |\vec{k}| |A|^2 A. \tag{3.21}$$

This is a nonlinear Schrödinger (NLS) equation (normally written in a frame translating with the wave packet at its vertical group velocity, in which case the first term on the right-hand side disappears and z is replaced with $\tilde{z} = z - c_{gz}t$). An example of the solution of the NLS equation for a modulationally unstable wave packet is shown in Fig. 3.5.

The NLS equation predicts whether the amplitude initially grows or decays depending upon the product of the sign of the induced flow and the sign of $\omega_{mm} = \partial_m c_{gz}$: modulational stability occurs if $U\omega_{mm} > 0$ (the group velocity increases with increasing wavenumber); otherwise, the wave packet is modulationally unstable. (For the general theory, see Whitham 1974 or Sect. 4.2.4 of Sutherland 2010.) For non-rotating internal waves, the maximum upward group velocity occurs for waves with $m = m_c = -k/\sqrt{2}$. For larger m (with smaller magnitude), c_{gz} is smaller and

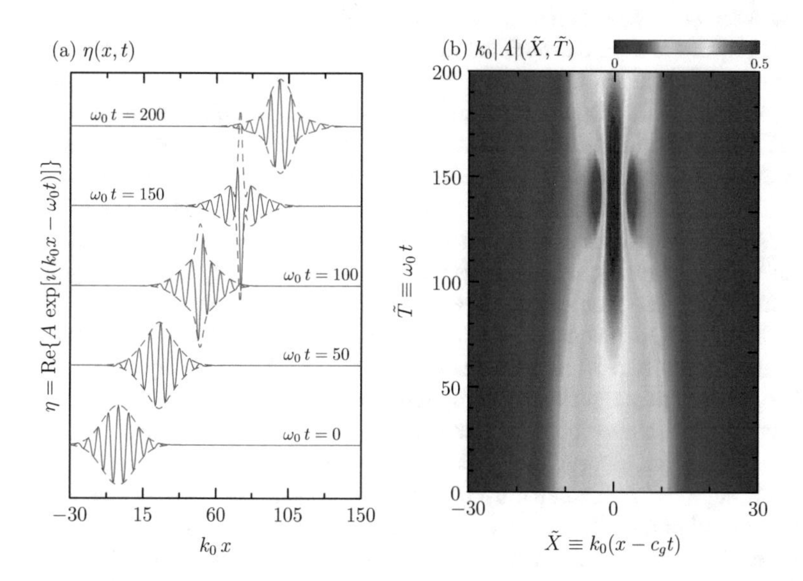

Fig. 3.5 Evolution of a modulationally unstable Gaussian wave packet showing (**a**) the displacement due to waves (solid blue lines) and the amplitude envelope (dashed red lines) at five different times during the evolution as indicated and (**b**) the evolution of the amplitude envelope in a frame moving with the group velocity of the wave packet, with magnitude indicated by the color scale. Note that after the initial amplitude growth the wave packet then broadens and its amplitude decays. Over longer times the growth and decay process repeats. (Adapted with permission from Fig. 4.4 of Sutherland (2010))

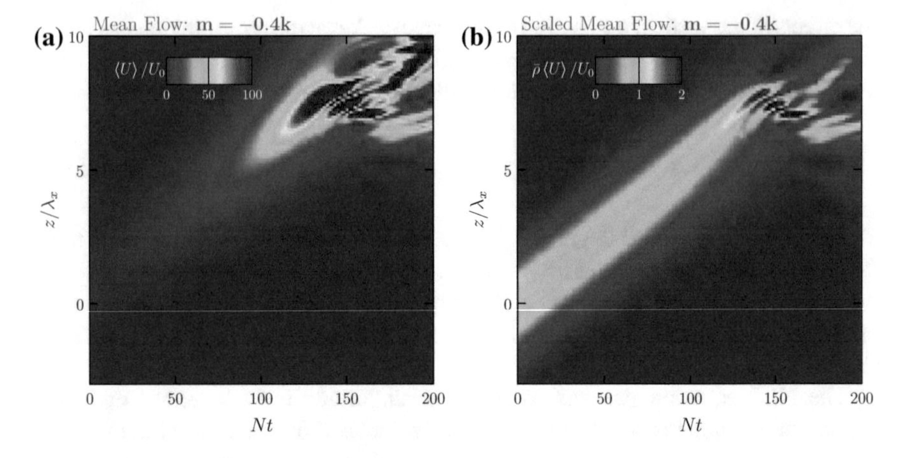

Fig. 3.6 Breakdown of an anelastic nonlinear 1D packet showing (**a**) induced flow that grows exponentially with height, (**b**) induced momentum that would be unchanging in the absence of weakly nonlinear effects

so $\omega_{mm} < 0$: the waves are modulationally unstable. Conversely, more hydrostatic waves with $|m| > k/\sqrt{2}$ are modulationally stable.

The physics of modulational stability and instability are clear by considering the effect of the induced flow that Doppler shifts the waves. For non-hydrostatic waves with $|m/k| < 1/\sqrt{2}$, Doppler shifting acts to reduce the effective frequency of the waves which, in turn, reduces their vertical group velocity. Thus, the trailing edge of the wave packet catches up with the leading edge causing the amplitude to increase while the vertical extent of the packet decreases. On the other hand, if $|m/k| > 1/\sqrt{2}$ then the effective decrease in the wave frequency leads to an increase in the vertical group velocity causing the wave packet to spread vertically and its amplitude to decrease faster than would occur due to linear dispersion alone.

Although the NLS equation predicts that a modulationally unstable wave packet will periodically narrow and then spread (see Fig. 3.5), in reality other nonlinear effects may come into play as the amplitude peaks. An example is shown in Fig. 3.6 illustrating the evolution of the induced flow associated with an anelastic wave packet. As the wave packet narrows and peaks, the wave packet becomes convectively unstable. Although this is predicted by linear theory to happen for a 1D anelastic wave packet, the addition of the weakly nonlinear effect of modulational instability shows that breaking occurs at much lower altitudes (Dosser and Sutherland 2011).

Conversely, modulationally stable 1D anelastic wave packets spread as they grow in amplitude and weakly nonlinear effects become important. Thus, overturning is retarded and simulations show that they break at much higher altitudes than predicted by linear theory. The overturning heights found in simulations are compared with those predicted by linear theory in Fig. 3.7.

What has been presented so far relates to 1D wave packets whose amplitude envelope varies spatially only in the z-direction. The structure of the flow changes

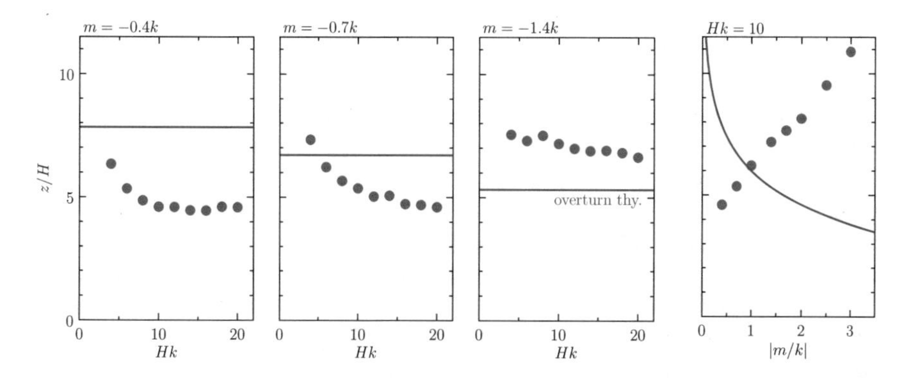

Fig. 3.7 Overturning heights of 1D anelastic wave packets measured in fully nonlinear numerical simulations and compared with theory, showing the altitude of overturning anelastic waves predicted by linear theory (solid blue lines) and fully nonlinear numerical simulation results for the altitude of the first occurrence of overturning (red dots). (Copyright © American Meteorological Society. Adapted with permission from Fig. 6 of Dosser and Sutherland 2011)

qualitatively for wave packets that are modulated in both the x- and z-directions. In this case, the induced flow resulting from the divergence of the momentum flux is itself horizontally divergent and so an order amplitude-squared pressure gradient is established to ensure the fluid remains incompressible. The result is that a spanwise infinite wave packet that is otherwise horizontally and vertically localized induces a flow in the form of a long wave provided the frequency of the long wave is larger than the Coriolis frequency (Bretherton 1969; Tabaei and Akylas 2007; van den Bremer and Sutherland 2014). Unlike the induced flow for 1D wave packets, this flow is positive over the leading flank of the wave packet and is negative over the trailing flank. Thus, whatever the wavenumber of the waves in the wave packet, the waves are always modulationally unstable with the leading edge narrowing and peaking if $|m/k| < 1/\sqrt{2}$ and the trailing edge narrowing and peaking otherwise. Evidently from numerical simulations of 2D anelastic wave packets, this results in overturning occurring at altitudes not too different from those predicted by linear theory (Gervais et al. 2018).

The flow induced by 3D wave packets is entirely different again. In this case, provided the spanwise extent of the wave packet is not too wide, the response to the horizontal divergence of the momentum flux is to create a circulation that goes horizontally around the wave packet in what is termed the Bretherton flow (Bretherton 1969; Tabaei and Akylas 2007; Bühler 2014; van den Bremer and Sutherland 2018). The induced flow over the extent of the wave packet is unidirectional as in the case of 1D wave packets, although the magnitude of the flow is smaller. Thus, while the effects of modulational instability and stability are again anticipated, their influence is expected to be not so pronounced for 3D as for 1D wave packets. Examination of these effects is currently under investigation.

Future Directions

Here we have presented a brief overview of some of the instability mechanisms associated with internal waves in uniformly stratified fluid. While plane-periodic internal waves are an exact solution of the fully nonlinear Euler equations, their stability properties are by no means trivial. Even at small amplitudes, periodic waves, modes, and beams can eventually break down due to resonant-triad wave interactions. If, as in the atmosphere, the waves grow sufficiently rapidly with height that TRI can be ignored, then the waves can break either due to convection or shear instability. And the height at which this breakdown occurs can be preconditioned by the weakly nonlinear effects of modulational stability and instability.

Despite interest in internal waves for their momentum transport in the atmosphere and their energy transport in the ocean, several fundamental aspects of their instability properties remain to be explored. Of these, this author believes the most important avenues for exploration include studying the influence of nonuniform stratification and spatial confinement upon TRI and examining the influence of background rotation upon wave-induced flows and their corresponding weakly nonlinear influence upon the modulational stability/instability and ultimate breaking of wave packets.

References

Benielli, D., & Sommeria, J. (1998). Excitation and breaking of internal gravity waves by parametric instability. *Journal of Fluid Mechanics*, *374*, 117–144.

Bourget, B., Dauxois, T., Joubaud, S., & Odier, P. (2013). Experimental study of parametric subharmonic instability for internal plane waves. *Journal of Fluid Mechanics*, *723*, 1–20.

Bouruet-Aubertot, P., Sommeria, J., & Staquet, C. (1995). Instabilities and breaking of standing internal gravity waves. *Journal of Fluid Mechanics*, *285*, 265–301.

Bühler, O. (2014). *Waves and mean flows* (2nd ed.). Cambridge: Cambridge University Press.

Bretherton, F. P. (1969). Momentum transport by gravity waves. *Quarterly Journal Royal Meteorological Society*, *95*(404), 213–243.

Chalamalla, V. K., & Sarkar, S. (2016). PSI in the case of internal wave beam reflection at a uniform slope. *Journal of Fluid Mechanics*, *789*, 347–367.

Danioux, E., Vanneste, J., & Bühler, O. (2015). On the concentration of near-inertial waves in anticyclones. *Journal of Fluid Mechanics*, *773*, R2. https://doi.org/10.1017/jfm.2015.252.

Dauxois, T., Joubaud, S., Odier, P., & Venaille, A. (2018). Instabilities of internal gravity wave beams. *Annual Reviews of Fluid Mechanics*, *50*, 131–156.

Dosser, H. V., & Sutherland, B. R. (2011). Anelastic internal wavepacket evolution and stability. *Journal of the Atmospheric Science*, *68*, 2844–2859.

Gervais, A., Swaters, G. E., van den Bremer, T. S., & Sutherland, B. R. (2018). Evolution and stability of two-dimensional anelastic internal gravity wavepackets. *Journal of the Atmospheric Sciences*, *75*, 3703–3724. https://doi.org/10.1175/JAS-D-17-0388.1.

Hasselmann, K. (1967). A criterion for non-linear wave stability. *Journal of Fluid Mechanics*, *30*, 737–739.

Joubaud, S., Munroe, J., Odier, P., & Dauxois, T. (2012). Experimental parametric subharmonic instability in stratified fluids. *Physics of Fluids*, *24*, 041703.

Karimi, H. H., & Akylas, T. R. (2014). Parametric subharmonic instability of internal waves: Locally confined beams versus monochromatic wavetrains. *Journal of Fluid Mechanics*, *757*, 381–402.

Klostermeyer, J. (1991). Two-dimensional and three-dimensional parametric instabilities in finite amplitude internal gravity waves. *Geophysical Astrophysical Fluid Dynamics*, *61*, 1–25.
Lindzen, R. S. (1981). Turbulence and stress owing to gravity wave and tidal breakdown. *Journal of Geophysical Research*, *86*, 9707–9714.
Lombard, P. N., & Riley, J. J. (1996). On the breakdown into turbulence of propagating internal waves. *Dynamics of Atmospheres and Oceans*, *23*, 345–355.
Marshall, D. P., Ambaum, M. H. P., Maddison, J. R., Munday, D. R., & Novak, L. (2017). Eddy saturation and frictional control of the Antarctic circumpolar current. *Geophysical Research Letters*, *44*(1), 286–292. https://doi.org/10.1002/2016gl071702.
Mied, R. R. (1976). The occurrence of parametric instabilities in finite-amplitude internal gravity waves. *Journal of Fluid Mechanics*, *78*, 763–784.
Phillips, O. M. (1981). Wave interactions - the evolution of an idea. *Journal of Fluid Mechanics*, *106*, 215–227.
Staquet, C., & Sommeria, J. (2002). Internal gravity waves: From instabilities to turbulence. *Annual Review of Fluid Mechanics*, *34*, 559–593.
Sutherland, B. R. (2001). Finite-amplitude internal wavepacket dispersion and breaking. *Journal of Fluid Mechanics*, *429*, 343–380.
Sutherland, B. R. (2010). *Internal gravity waves*. Cambridge: Cambridge University Press.
Sutherland, B. R., Achatz, U., Caulfield, C. P., & Klymak, J. M. (2019). Recent progress in modeling imbalance in the atmosphere and ocean. *Physical Review Fluids*, *4*(010501), 1–22. https://doi.org/10.1103/PhysRevFluids.4.010501.
Tabaei, A., & Akylas, T. R. (2007). Resonant long-short wave interactions in an unbounded rotating stratified fluid. *Studies in Applied Mathematics*, *119*, 271–296.
Trossman, D. S., Arbic, B. K., Garner, S. T., Goff, J. A., Jayne, S. R., Metzger, E. J., et al. (2013). Impact of parameterized lee wave drag on the energy budget of an eddying global ocean model. *Ocean Modelling*, *72*, 119–142. https://doi.org/10.1016/j.ocemod.2013.08.006.
van den Bremer, T. S., & Sutherland, B. R. (2014). The mean flow and long waves induced by two-dimensional internal gravity wavepackets. *Physics of Fluids*, *26*(106601), 1–23. https://doi.org/10.1063/1.4899262.
van den Bremer, T. S., & Sutherland, B. R. (2018). The wave-induced flow of internal gravity wavepackets with arbitrary aspect ratio. *Journal of Fluid Mechanics*, *834*, 385–408. https://doi.org/10.1017/jfm.2017.745.
Whitham, G. B. (1974). *Linear and nonlinear waves*. New York: Wiley.

Chapter 4
Rotational Dynamics of Planetary Cores: Instabilities Driven By Precession, Libration and Tides

Thomas Le Reun and Michael Le Bars

Abstract In this chapter, we explore how gravitational interactions drive turbulent flows inside planetary cores and provide an interesting alternative to convection to explain dynamo action and magnetic fields around terrestrial bodies. In the first section, we introduce tidal interactions and their effects on the shape and rotation of astrophysical bodies. A method is given to derive the primary response of liquid interiors to these tidally-driven perturbations. In the second section, we detail the stability of this primary response and demonstrate that it is able to drive resonance of inertial waves. As the instability mechanism is introduced, we draw an analogy with the parametric amplification of a pendulum whose length is periodically varied. Lastly, we present recent results regarding this instability, in particular its nonlinear saturation and its ability to drive dynamo action. We present how it has proved helpful to explain the magnetic field of the early Moon.

Introduction: From Planetary Magnetic Fields to Core Turbulence

In addition to the Earth, several terrestrial bodies of the Solar System are known to be presently protected from solar radiation by an intense magnetic field, or present evidence of a past one. For instance, flybys operated by probes equipped with magnetometers have revealed the presence of a magnetic field surrounding Mercury as well as Ganymede and Io, two of Jupiter's largest moons (Ness et al. 1975; Kivelson et al. 1996a; Sarson et al. 1997; Showman et al. 1999; Kivelson et al. 2002). Magnetized rock samples from Mars and the Earth's Moon have also revealed the existence of a past intense magnetic field (Stevenson 2001; Garrick-Bethell et al. 2009; Le

T. Le Reun · M. Le Bars (✉)
CNRS, Aix Marseille University, Centrale Marseille, IRPHE, Marseille, France
e-mail: lebars@irphe.univ-mrs.fr

M. Le Bars and D. Lecoanet (eds.), *Fluid Mechanics of Planets and Stars*,
CISM International Centre for Mechanical Sciences 595,
https://doi.org/10.1007/978-3-030-22074-7_4

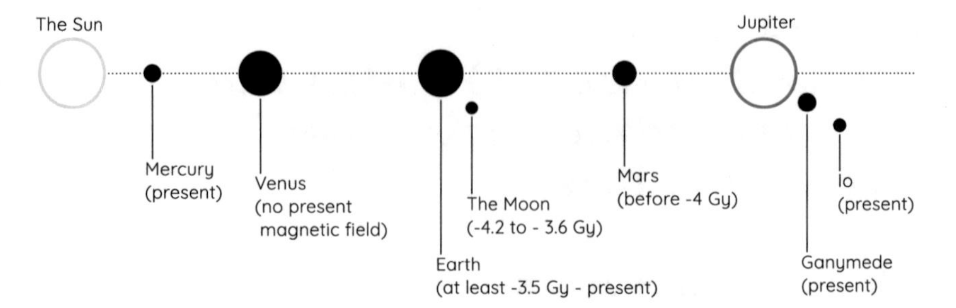

Fig. 4.1 A summary of the terrestrial bodies (in black) of the Solar System known for having a past or present dynamo. The relative size of these planets and moons is respected (apart for Jupiter and the Sun). Venus is given for comparison as it is not surrounded by a magnetic field although it is of similar size to the Earth. It is not known whether Venus had a magnetic field in the past. We report the estimated period of existence of these fields when it is known, based on Garrick-Bethell et al. (2009), Tarduno et al. (2010), Stevenson (2001)

Bars et al. 2011). A summary of what is presently known about magnetic fields of terrestrial planets in the Solar System is given in Fig. 4.1. Beyond the Solar System, magnetic fields are also expected in extra-solar planets, where they constitute one of the key ingredients for habitability.

As first conjectured by Larmor (1919), the magnetic field of a planet originates in the turbulent motion of liquid iron in its core (Olson 2015). Following the seminal works of Roberts (1968) and Busse (1970), it has been shown that buoyancy-driven flows such as thermal and solutal convection in cores provoke turbulent stirring and dynamo action (Glatzmaiers and Roberts 1995). This convective motion is driven by the secular cooling of a planet, by radiogenic heating, and by latent heat and potential energy release during its core solidification. While the energy budget to sustain the present day magnetic field of the Earth is closed—even if still partly controversial, see e.g. Labrosse (2015)—, the Earth early dynamo and the dynamo in smaller bodies remain largely unexplained.

As an illustration, let us assume that the main source for dynamo lies in the initial thermal energy of the body acquired during its formation. This initial thermal energy can be estimated assuming that it is roughly tantamount to the loss of gravitational potential energy from a dispersed cloud to an aggregated body, and is proportional to R^5 where R is the radius of the planet at the end of its formation. As the secular heat loss scales like the surface of a planet, we can infer that a typical planetary cooling time grows like R^3. As a consequence, the core of relatively small planetary bodies such as Ganymede, Mercury or the early Moon cools down very quickly and should not be able to sustain turbulent convective motion and long-term dynamo action. Even for larger planets such as the Earth, the initial temperature to maintain a dynamo all along their lifetime should be extremely hot, in possible contradiction with the presence of a solid mantle at the beginning of their existence (Andrault et al. 2016).

However, initial heat is not the only source of energy available to drive fluid motion. In particular, a huge amount of mechanical energy is stored in the rotational dynamics of planetary systems (Le Bars et al. 2015). If it is possible to convert this mechanical energy into turbulent kinetic energy inside a planetary core, it provides an interesting alternative to convective instabilities to drive planetary dynamos. In the ideal case of a perfectly spherical planet with uniform rotation, this conversion cannot happen: the liquid core follows the terrestrial planet in its solid-body rotation. However, tidal interactions between astrophysical bodies result in periodic alteration of their shape, of the direction of their rotation axis and of their rotation rate, which can then force fluid motion inside their cores.

This idea that tidal interactions force core turbulence was first introduced by Malkus in three seminal articles (Malkus 1963, 1968, 1989), but was largely dismissed by geophysicists for decades. As noted by Kerswell, this was mainly due to a misunderstanding regarding the nature of the flow excited by tides (Kerswell 1996, 2002). Tidal interactions are of small amplitudes and their direct forcing only generates small departures from the solid-body rotation of the fluid core. Alone, these small perturbations are not powerful enough to sustain a magnetic field. However, these periodic perturbations are able to excite resonant instabilities which can then break down into bulk-filling turbulence.

While the flow directly created by tidal perturbations is purely laminar and of low amplitude in the first place, the excited resonant instabilities are responsible for converting the huge rotational kinetic energy into turbulence, and possibly dynamo action.

Flows driven by tidal instabilities in a geophysical context have benefited from extensive investigation over the past two decades. Theoretical and experimental studies have revealed that these instabilities, for the most part, rely on the interplay between inertial waves—which exist in any rotating fluid because of the restoring action of the Coriolis force—and the harmonic forcing. The underlying mechanism is a sub-harmonic resonance called the elliptical instability (Kerswell 1993; Le Dizès 2000; Lacaze et al. 2005; Le Bars et al. 2007; Cébron et al. 2014; Grannan et al. 2014; Lemasquerier et al. 2017). In particular, this research has clarified the conditions for such an instability to develop in terms of tidal forcing versus viscous damping inside planetary cores (see for instance Cébron et al. 2012b). Yet there is still much to understand about tidally-driven instabilities; in particular, the comprehension of their saturation into bulk-filling turbulence remains a challenging problem although of crucial importance to predict the resulting dynamo action. Such a turbulence bears many particularities compared to classical turbulence. In addition to being strongly influenced by rotation, this turbulence is forced at low amplitude—tidal perturbations are weak forcings—in a very small dissipation regime—because of the massive size of the considered bodies. This problem is thus difficult to study as such regimes are far beyond the reach of any numerical simulation or laboratory experiment. Only careful extrapolations to planetary cores can be drawn from present knowledge. Nevertheless, significant steps have been made over the past few years

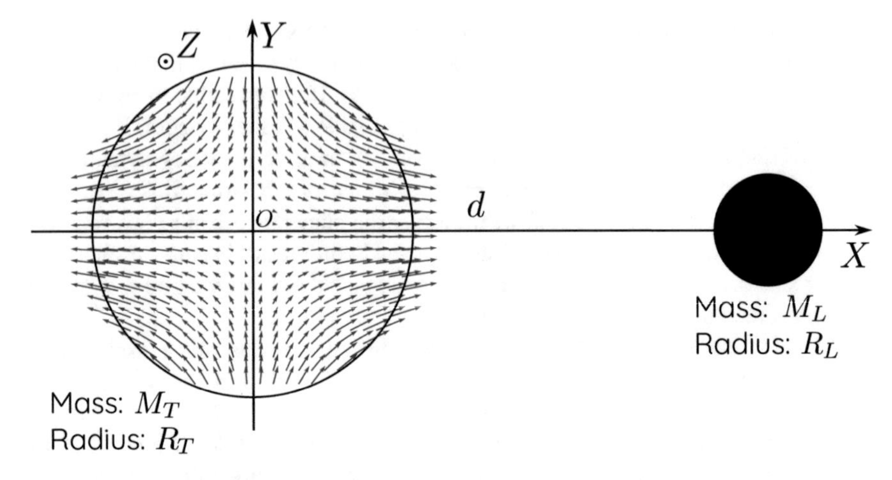

Fig. 4.2 Schematic diagram of two astrophysical bodies such as the Earth and the Moon, assuming that they are perfectly spherical and homogeneous. The tidal force field induced by the Moon is represented by the red arrows in the Earth. This field is invariant under rotation around the X axis. Note that the Earth induces a similar field inside the Moon

(see for instance Favier et al. 2015; Grannan et al. 2017; Le Reun et al. 2017), one of the most striking results being the evidence of a fully turbulent kinematic dynamo driven by tidal forcing in a planetary-relevant ellipsoidal geometry (Reddy et al. 2018).

In this chapter, we aim at providing a simple understanding of tidally-driven instabilities. In the first section, we dwell on the primary response of planetary cores to tidal forcings. The second section is devoted to the parametric instabilities which develop on this primary response; in particular, we draw an analogy with the resonance of length-varying pendulums and exhibit the mathematical underpinnings of tidally-driven instabilities. In the last section, we review a few recent results and challenges brought by the study of mechanical forcing in planetary cores.

Tidal Forcings in Planetary Cores: The Primary Response to Tides

The aim of this section is to review the basic effects of tides on planetary cores. We introduce the principal perturbations to the rotational motion of planets induced by tides and present the method to infer the primary response of a fluid cavity to those perturbations.

The Shape of a Planet Undergoing Tidal Distortion

As seen in chap. 1 of this book, the gravitational interaction between two astrophysical bodies results in a force field distorting them called "tides". For each body, it reflects the difference between the total gravitational attraction which drives its motion and the local attraction. Considering a planet T and a moon L separated by a distance d taken constant in first approximation—see Fig. 4.2—, the tidal potential writes at lowest order:

$$U_{\text{tides}} = \frac{GM_T}{R_T^3}\frac{M_L}{M_T}\left(\frac{R_T}{d}\right)^3\left(r^2 - (\boldsymbol{e}_X \cdot \boldsymbol{r})^2\right) \tag{4.1}$$

with $r^2 = (X^2 + Y^2 + Z^2)$, $\boldsymbol{r} = X\boldsymbol{e}_X + Y\boldsymbol{e}_Y + Z\boldsymbol{e}_Z$ and G the gravitational constant—the remaining variables are defined on Fig. 4.2. The tidal force field is represented in Fig. 4.2 and bears two important symmetries: it is invariant by rotation around the planet-moon axis (OX) and by reflection relative to the (YOZ) plane. The deformation induced by the tidal potential (4.1) can be analytically determined, see for instance discussions in Barker (2016) and Barker et al. (2016). At the lowest order, it can be shown that the planet adopts an ellipsoidal shape; the combination of both equatorial—or rotational—and tidal bulges forces the three axes of this ellipsoid to have different lengths.

In the following, we assume that the outer boundaries of planetary cores are ellipsoidal. For simplicity's sake, we do not take into account the possible presence of a solid inner core. In a general context, as the rigidities of a solid iron inner core and of the rocky mantle are different, the liquid iron domain would be a shell confined between two ellipsoids which are not necessarily homothetic. Such a geometry does not change the overall dynamics but makes its analysis more difficult—see for instance Lemasquerier et al. (2017). In addition, the case we consider is an actual situation encountered in young planets as solid inner cores only crystallize later after their formation. For instance, the Earth is known to have been surrounded by a magnetic field since at least 3.5 Gy although the inner core is only around 1 Gy old or less (Labrosse 2015).

Flow Driven By Differential Spin and Orbit

In a configuration similar to the Earth–Moon system, the spin of the Earth is not synchronized with the orbit of the Moon. Because the Earth rotates every day, but the Moon's orbit is 27 days long, and because of the symmetry of the tidal field, the solid part of the Earth is subject to slightly less than two tidal rises per day. Focusing on the core, while the liquid iron rotates at the same rate as the Earth, its outer shape bears a tidal bulge which follows the orbiting motion of the Moon.

Let us assume a simple case where those two rotations take place in the same plane, the corresponding situation being depicted in Fig. 4.3. In such a configuration,

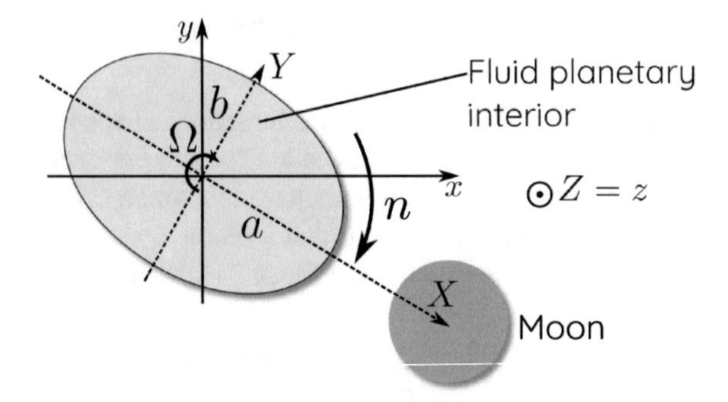

Fig. 4.3 Schematic diagram of a moon orbiting around a planet. We define two systems of axes: $(OXYZ)$ follows the revolution of the moon (hence of the ellipsoidal shape of the planet) and rotates at rate $n\boldsymbol{e}_z$; $(Oxyz)$ tracks the rotation of the solid mantle of the planet and rotates at rate $\Omega\boldsymbol{e}_z$. We assume that the two rotation motions occur in the same plane for simplicity's sake. The fluid envelop is ellipsoidal: its axes have lengths a and b in the equatorial plane, and c along the axis of rotation

the flow inside the core is not a solid-body rotation. We aim at determining the inviscid flow created by the differential rotation of the planet and its moon; it must satisfy the Euler equation and the non-penetration boundary conditions at the edge of the ellipsoidal container with axes lengths a, b and c—see Fig. 4.3. We use the method introduced by Hough (1895). First we note that the derivation is facilitated when carried out in the frame of reference $(OXYZ)$ in which the boundary shape does not change over time. Then, we look for a uniform vorticity solution which directly satisfies the boundary conditions. This is achieved by first using a rescaled system of coordinates $\tilde{\boldsymbol{X}} = (\tilde{X}, \tilde{Y}, \tilde{Z})$ which transforms the ellipsoid into a sphere, the velocity $\boldsymbol{U} = (U, V, W)$ being also transformed accordingly:

$$\begin{cases} \tilde{X} = X/a \\ \tilde{Y} = Y/b \\ \tilde{Z} = Z/c \end{cases} \text{and} \quad \begin{cases} \tilde{U} = U/a \\ \tilde{V} = V/b \\ \tilde{W} = W/c \end{cases} \tag{4.2}$$

and then looking for a vector $\boldsymbol{\omega}(t)$ such that:

$$\tilde{\boldsymbol{U}} = \boldsymbol{\omega}(t) \times \tilde{\boldsymbol{X}}. \tag{4.3}$$

With this ansatz, the transformed velocity field is at each time t a solid-body rotation which necessarily satisfies the non-penetration boundary conditions in the sphere. As a consequence, the flow $\boldsymbol{U}$ transformed back into the original coordinates also satisfies the boundary conditions. Such a flow is divergence-free and has a uniform vorticity $\boldsymbol{\varpi}$ which writes:

$$\boldsymbol{\varpi} = \boldsymbol{\nabla} \times \boldsymbol{U} = \begin{bmatrix} \left(\frac{c}{b} + \frac{b}{c}\right) \omega_x \\ \left(\frac{c}{a} + \frac{a}{c}\right) \omega_y \\ \left(\frac{a}{b} + \frac{b}{a}\right) \omega_z \end{bmatrix}. \quad (4.4)$$

We look for a steady solution in the frame orbiting with the Moon at a rate $n\boldsymbol{e}_z$—see Fig. 4.3. Taking into account the Coriolis acceleration due to the rotation of the frame of reference, the stationary vorticity equation derived from the Euler equation reads:

$$(\boldsymbol{U} \cdot \boldsymbol{\nabla})\boldsymbol{\varpi} = ((\boldsymbol{\varpi} + 2n\boldsymbol{e}_z) \cdot \boldsymbol{\nabla})\, \boldsymbol{U}. \quad (4.5)$$

As $\boldsymbol{\varpi}$ is space-independent, the left hand side of this equation vanishes. The equations on the components of the vorticity are therefore:

$$\begin{cases} \left[\frac{1}{c}\left(\frac{a}{b} + \frac{b}{a}\right) - \frac{1}{b}\left(\frac{c}{a} + \frac{a}{c}\right)\right] \omega_y \omega_z + 2\frac{1}{c} n \omega_y = 0 \\ \left[\frac{1}{a}\left(\frac{c}{b} + \frac{b}{c}\right) - \frac{1}{c}\left(\frac{a}{b} + \frac{b}{a}\right)\right] \omega_x \omega_z - 2\frac{1}{a} n \omega_x = 0 \\ \left[\frac{1}{b}\left(\frac{c}{a} + \frac{a}{c}\right) - \frac{1}{a}\left(\frac{c}{b} + \frac{b}{c}\right)\right] \omega_x \omega_y = 0. \end{cases} \quad (4.6)$$

The last equation prescribes either ω_x or ω_y to be equal to 0. Let us assume $\omega_y = 0$ (a similar reasoning is possible for $\omega_x = 0$): the first equation is then directly satisfied. The second equation either prescribes a ω_z proportional to n for a nonzero ω_x, or a zero ω_x. However, the former possibility does not account for the global rotation of the planet at rate Ω, which must be part of the solution. The physical solution is therefore obtained for $\omega_x = 0$[1]. To constrain the value of ω_z, we consider that the vorticity of the fluid must match the planetary vorticity, i.e.:

$$\left(\frac{a}{b} + \frac{b}{a}\right) \omega_z = 2(\Omega - n) \iff \omega_z = \frac{2ab}{a^2 + b^2}(\Omega - n) \quad (4.7)$$

which finally gives the following steady flow driven by tides:

$$\begin{bmatrix} U \\ V \\ W \end{bmatrix} = \frac{2ab}{a^2 + b^2}\,(\Omega - n) \begin{bmatrix} -\frac{a}{b} Y \\ \frac{b}{a} X \\ 0 \end{bmatrix}. \quad (4.8)$$

[1]This condition is in any case the only one when we consider the viscous problem where the flow solution of (4.6) must reconnect to the rotating solid mantle through a thin boundary layer, see for instance Tilgner (2007b).

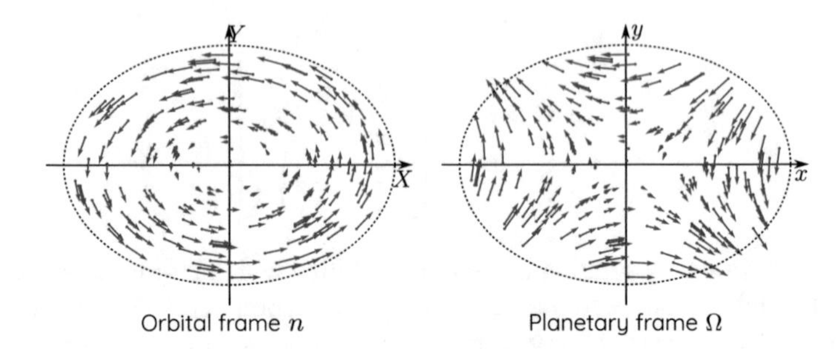

Fig. 4.4 Tidal flow velocity field seen from the orbital frame rotating at rate n (left) and from the planetary frame rotating at rate Ω (right). The arrows scale is not the same on the two figures. $\Omega - n$ is positive: the fluid moves counterclockwise in the orbital frame and the elliptical bulge moves clockwise when seen from the planetary frame. Planetary frame highlights the strain field which perturbs the solid-body rotation of the fluid

Lastly, we can introduce the ellipticity of the deformation $\beta = (a^2 - b^2)/(a^2 + b^2)$; the base flow then writes into a simpler and more compact form:

$$\boldsymbol{U} = (\Omega - n) \begin{bmatrix} 0 & -1-\beta & 0 \\ 1-\beta & 0 & 0 \\ 0 & 0 & 0 \end{bmatrix} \begin{bmatrix} X \\ Y \\ Z \end{bmatrix}. \tag{4.9}$$

Note that this last form is quite meaningful as it corresponds to the superposition of a circular vortex and a strain, which is a configuration known to be unstable (Pierrehumbert 1986; Bayly 1986; Waleffe 1990). Moreover, in the frame rotating with the planet, the flow velocity, denoted $\boldsymbol{U}^{\Omega}$, is:

$$\boldsymbol{U}^{\Omega} = -\beta\gamma \begin{bmatrix} \sin(2\gamma t) & \cos(2\gamma t) & 0 \\ \cos(2\gamma t) & -\sin(2\gamma t) & 0 \\ 0 & 0 & 0 \end{bmatrix} \begin{bmatrix} x \\ y \\ z \end{bmatrix} \quad \text{with: } \gamma = \Omega - n \tag{4.10}$$

which can be retrieved from (4.9) via a rotation of coordinates and a velocity composition. The two fields (4.9) and (4.10) are shown in Fig. 4.4. This last way of writing the tidal flow is even more meaningful as it highlights the time periodicity of the tidal excitation: the tidal frequency is twice the differential rotation between the planet and the moon, which reflects the symmetry of the tidal bulge respective to the plane (OYZ). For $n \ll \Omega$, this is tantamount to undergoing two tidal rises a day as on the Earth where $n/\Omega \sim 1/27$. Besides, the tidal flow amplitude is proportional to the ellipticity of the deformation. Tidal excitation is a small perturbation to the planetary solid-body rotation, of relative amplitude 10^{-7} on Earth for instance. Nevertheless, we show later in this chapter that this repetitive excitation, although of small amplitude, is able to excite turbulent flows.

Perturbation of the Rotation Rate: Libration

In the previous section, we have introduced the tidal potential and its consequences on the shape of planetary cores. Tides not only induce distortion, but they also alter the rotation of planets. We review in the next two sections typical perturbations of the rotation rate and rotation axis, and combine them with the tidal distortion to derive the corresponding core flows.

Libration of moons Physical longitudinal libration, hereafter called libration, is the oscillation of the rotation rate of an astrophysical body without change in its rotation axis. This kind of motion is excited by tidal interaction between the considered body's tidal bulge and its parent planet or star. One common situation where libration is observed is presented in Fig. 4.5: a moon is synchronized in a spin-orbit resonance along an elliptical orbit, meaning that its orbital rate matches its mean rotation rate (like our Moon, which always shows us the same side). This is due to tidal dissipation inside the rocky mantle of such bodies which despins them from any initial rotation rate into this particular equilibrium (Rambaux and Castillo-Rogez 2013). Moreover, because of the rigidity of the moon, its tidal bulge is in general not exactly aligned with the parent body. The Fig. 4.5 presents the extreme situation where the bulge is frozen and follows the rotation of the moon instead of staying aligned with the planet. This is for instance the case of the Earth's Moon, whose large fossil bulge has not relaxed and is not induced by tidal interaction anymore, but still persists. As depicted in Fig. 4.5, this misalignment and the difference of gravitational attraction between the two sides of the bulge create a torque which tends to accelerate the rotation rate at ② and to decelerate the rotation rate at ④. The rotation rate is therefore perturbed around a mean with the same period as the orbit.

The situation shown in Fig. 4.5 is not the only one leading to libration oscillations. In planetary systems with many satellites such as the Jovian and Saturnian systems, a body not only interacts with its parent planet but also with all the other moons. Some of them are in what is called a Laplace resonance, which is a stable situation where the orbit rates of several moons are multiples of one another (Rambaux and Castillo-Rogez 2013).

Libration-driven flows We seek now to determine the flow driven by libration and assume for simplicity's sake a purely rigid tidal bulge. The total rotation rate including libration can be written as:

$$\boldsymbol{\Omega} = \Omega_0 \left(1 + \varepsilon \sin(\omega_\ell t)\right) \boldsymbol{e}_z \tag{4.11}$$

where ε is the relative variation of the rotation rate and ω_ℓ is the libration frequency. Observed from the frame rotating at rate Ω_0, the moon and its bulge oscillate with a frequency ω_ℓ and with an amplitude angle $\varepsilon\Omega_0/\omega_\ell$. Although in the example of spin-orbit synchronization presented in Fig. 4.5 $\omega_\ell = \Omega_0$, the libration frequency can take any value due to tidal interactions with many bodies as in the Jovian and Saturnian systems. To determine the libration-driven flow, we use the same method as for the tidal base flow: we look for a flow $\tilde{\boldsymbol{U}} = \boldsymbol{\omega}(t) \times \tilde{\boldsymbol{X}}$—see definition (4.3)—in

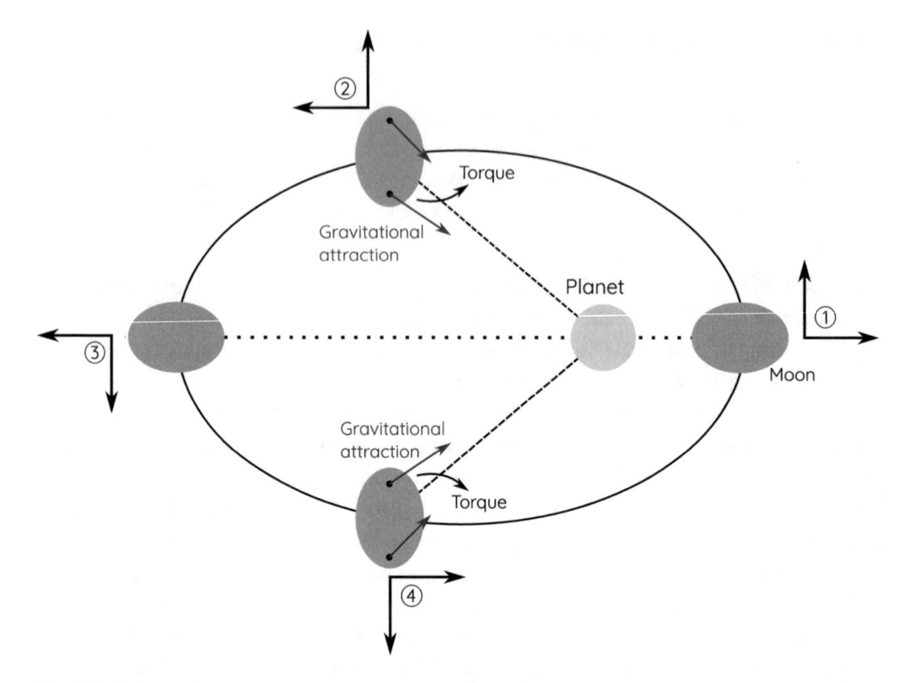

Fig. 4.5 Schematic diagram of the excitation of libration on a moon with an eccentric orbit in spin-orbit resonance. Its bulge is frozen and follows the rotation of the moon instead of staying aligned towards the parent planet. The four snapshots separate the orbit and the rotation in four equal periods

the frame where the boundary stands still, i.e. the librating frame. We use the same notation as before: in the frame $(OXYZ)$ the boundary is still—therefore it is the librating frame—and $(Oxyz)$ is the mean rotation frame. In the $(OXYZ)$ frame the equation for the vorticity ϖ writes:

$$\partial_t \varpi + (\boldsymbol{u} \cdot \nabla)\varpi = ((\varpi + 2\boldsymbol{\Omega}) \cdot \nabla)\,\boldsymbol{u} + 2\dot{\boldsymbol{\Omega}} \tag{4.12}$$

where we include the time dependence of the flow—which oscillates at the libration frequency—and a last term corresponding to the Poincaré's acceleration. In its expanded form, the Eq. (4.12) yields:

$$\left\{\begin{aligned}
\left(\frac{c}{b}+\frac{b}{c}\right)\dot{\omega}_x &= \left[\frac{a}{c}\left(\frac{a}{b}+\frac{b}{a}\right)-\frac{a}{b}\left(\frac{c}{a}+\frac{a}{c}\right)\right]\omega_y\omega_z + 2\frac{a}{c}\Omega\omega_y \\
\left(\frac{c}{a}+\frac{a}{c}\right)\dot{\omega}_y &= \left[\frac{b}{a}\left(\frac{c}{b}+\frac{b}{c}\right)-\frac{b}{c}\left(\frac{a}{b}+\frac{b}{a}\right)\right]\omega_x\omega_z - 2\frac{b}{c}\Omega\omega_x \\
\left(\frac{a}{b}+\frac{b}{a}\right)\dot{\omega}_z &= \left[\frac{c}{b}\left(\frac{c}{a}+\frac{a}{c}\right)-\frac{c}{a}\left(\frac{c}{b}+\frac{b}{c}\right)\right]\omega_x\omega_y + 2\dot{\Omega}.
\end{aligned}\right. \tag{4.13}$$

Assuming ω_x and ω_y are zero, the last equation relates the temporal variation of ω_z to the Poincaré acceleration:

$$\left(\frac{a}{b}+\frac{b}{a}\right)\dot{\omega}_z = 2\dot{\Omega} \iff \omega_z = \frac{2ab}{a^2+b^2}\Omega_0\varepsilon\sin(\omega_\ell t)+\overline{\text{cst}}. \tag{4.14}$$

Considering a synchronized body in the librating frame, the constant component of the vorticity must be equal to zero. The resulting base flow can be written in the librating frame in terms of axes lengths or the ellipticity β defined earlier:

$$\begin{aligned} \boldsymbol{U} &= \frac{2ab}{a^2+b^2}\Omega_0\varepsilon\sin(\omega_\ell t)\begin{bmatrix} 0 & -a/b & 0 \\ b/a & 0 & 0 \\ 0 & 0 & 0 \end{bmatrix}\begin{bmatrix} X \\ Y \\ Z \end{bmatrix} \\ &= \Omega_0\varepsilon\sin(\omega_\ell t)\begin{bmatrix} 0 & -1-\beta & 0 \\ 1-\beta & 0 & 0 \\ 0 & 0 & 0 \end{bmatrix}\begin{bmatrix} X \\ Y \\ Z \end{bmatrix} \end{aligned}$$

The libration flow written in the mean rotation frame—which requires doing velocity composition and coordinates change—writes, at the lowest order in ε:

$$\boldsymbol{U}^{\Omega} = \Omega_0\varepsilon\beta\sin(\omega_\ell t)\begin{bmatrix} 0 & 1 & 0 \\ 1 & 0 & 0 \\ 0 & 0 & 0 \end{bmatrix}\begin{bmatrix} x \\ y \\ z \end{bmatrix}. \tag{4.15}$$

Note that at $t=\pi/(2\omega_\ell)$ the structure of the flow is exactly similar to (4.9): snapshots of the velocity in the libration case are the sames as Fig. 4.4 left for the librating frame and Fig. 4.4 right for the mean rotation frame. Generally, seen from the frame where the bulge is stationary, a fluid particles rotates around elliptical streamlines in the tidal case and oscillates along elliptical streamlines in the libration case. Seen from the mean rotation frame, the strain field in the tidal case rotates with a constant amplitude, whereas the strain field in the libration case has a stationary spatial structure but an oscillating amplitude.

Perturbations of the Rotation Axis: Precession

Precession and equatorial bulge The spin of planets around their axes is in general predominant compared to any other motion in the barycentric frame of reference. Gravitational interactions can force the spinning axis to change over time, this motion been decomposed into precession and nutation—see Fig. 4.6 left. We rather dwell hereafter on precession, which is a direct consequence of the existence of an equatorial bulge induced by the planet spin and the associated centrifugal force. As depicted in Fig. 4.6 right, a neighboring body exerts a net gravitational torque which tends to align the equatorial bulge with the body. A gyroscopic effect then applies and con-

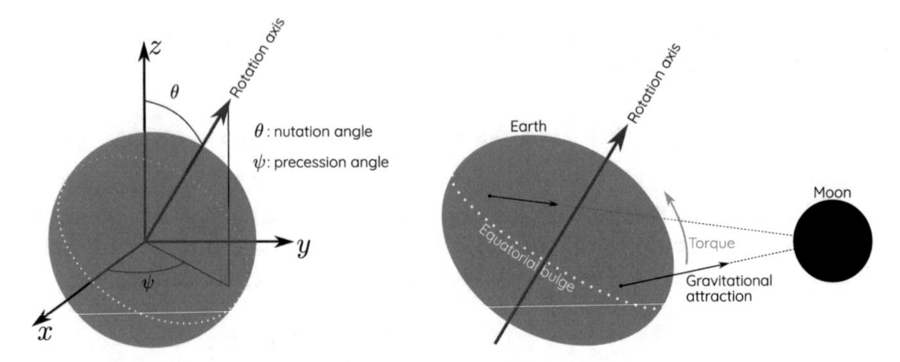

Fig. 4.6 **Left**: illustration of the two motions of the rotation axis which can be decomposed in nutation—variations of θ—and precession—variations of ψ. (xOy) is the ecliptic plane. **Right**: schematic diagram of the cause of the Earth's precession, which can be applied to other bodies. The gravitational force of a companion body—here the Earth's moon—creates a torque which tends to align the equatorial bulge created by rotation with the Moon. The gyroscopic effect due to rapid spinning converts this torque into a precession of the rotation axis

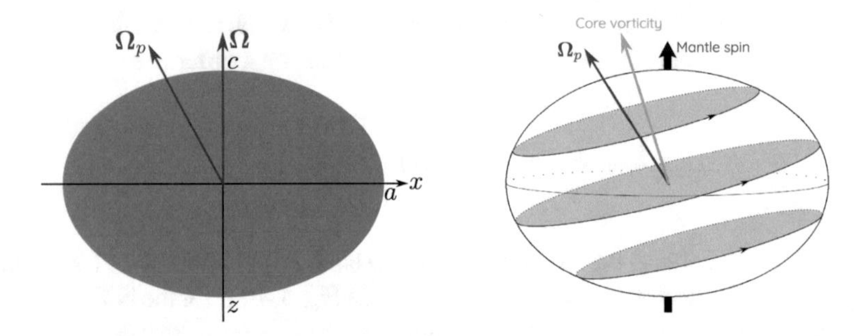

Fig. 4.7 **Left**: schematic cartoon of a precessing spheroid and the different rotation vectors as seen from the precession frame of reference. The fast rotation vector $\boldsymbol{\Omega}$ of the mantle surrounding the liquid core is along the z axis. The frame rotates at rate $\Omega_p \boldsymbol{e}_z$ so that the precession and the rotation vectors remain in the same plane. **Right**: visualization of the precession-driven flow whose vorticity is along an intermediate direction between the precession vector and the mantle fast spinning

verts this torque into precession, i.e. constant increase of the angle ψ with constant θ. This is the case for the Earth which precesses over a period of 26000 years with an angle $\theta \sim 23.5°$ because of gravitational interactions with the Moon and the Sun.

Precession-driven flow As for libration and tides, we would like to compute the base flow driven in a precessing planetary core. To fully expand the vorticity equation, we assume for simplicity's sake that the core is spheroidal, i.e. that $a = b$. As seen earlier, tides generally impose $a \neq b$: this case is also tractable, yet more complicated (Cébron et al. 2010b). With this symmetry, we carry out the computation of the precession-driven flow in the precessing frame, which is the frame in which the rotation axis of the solid mantle is constant equal to $\Omega \mathbf{e}_z$—see Fig. 4.7. This frame rotates at the precession rate Ω_p around the z axis so that both the rotation vector $\boldsymbol{\Omega}$ and the precession vector $\boldsymbol{\Omega}_p$ remain in the same plane (Oxz), as shown in Fig. 4.7. Note also that because we assume $a = b$, the topology of the core outer boundary remains unchanged although it rapidly rotates around the z axis.

We then use exactly the same method as previously and the equations to solve to determine the vector $\boldsymbol{\omega}$ are:

$$\begin{cases} \left(\frac{c}{a} + \frac{a}{c}\right)\dot{\omega}_x = \left[\frac{a}{c} - \frac{c}{a}\right]\omega_y\omega_z + 2\frac{a}{c}\Omega_{pz}\omega_y \\ \left(\frac{c}{a} + \frac{a}{c}\right)\dot{\omega}_y = \left[\frac{c}{a} - \frac{a}{c}\right]\omega_x\omega_z - 2\frac{a}{c}\Omega_{pz}\omega_x + 2\Omega_{px}\omega_z \\ 2\dot{\omega}_z = -2\frac{c}{a}\Omega_{px}\omega_y. \end{cases} \tag{4.16}$$

Note that these equations include a Coriolis force associated to the precession rate. We look for a steady solution, and the last equation implies that $\omega_y = 0$, which is also consistent with the first one: the rotation axis of the liquid core is in the same (Oxz) plane as the precession axis (see Fig. 4.7 right). The second equation yields:

$$\frac{\omega_x}{\omega_z} = \frac{\Omega_{px}}{\frac{a}{c}\Omega_{pz} + \frac{a^2 - c^2}{2ac}\omega_z}. \tag{4.17}$$

As noted in Tilgner (2007b) and Le Bars et al. (2015), the inviscid theory does not allow to relate the z component of the core vorticity to the mantle's spin in order to close the problem: this can be achieved only via the introduction of viscous effects. The interested reader can refer to the key study of Busse (1968) and discussions in the reviews of Tilgner (2007b) or Le Bars et al. (2015). Here, we simply give first order considerations in the limit of small precession rates. In the absence of precession ($\Omega_p = 0$), one expect the fluid to rotate with the mantle, hence $\omega_z = \Omega$. By symmetry between prograde and retrograde precession for the vertical vorticity at small precession rate, one can also expect the first correction of ω_z to appear at order 2 only, i.e. $\omega_z = \Omega + O(\Omega_p^2)$. Then the horizontal vorticity is given at first order by

$$\frac{\omega_x}{\Omega} = \frac{\Omega_{px}}{\frac{a}{c}\Omega_{pz} + \frac{a^2 - c^2}{2ac}\Omega} \tag{4.18}$$

and the base flow written the precessing frame of reference is:

$$\boldsymbol{U}_b^p = \begin{bmatrix} -\Omega Y \\ \Omega X \\ \frac{c}{a}\frac{\Omega_{px}}{\frac{a}{c}\Omega_{pz} + \frac{a^2 - c^2}{2ac}\Omega}\Omega \end{bmatrix}. \tag{4.19}$$

It is represented schematically in Fig. 4.7 right. For instance, in the case of the Earth which has a rotation axis tilted by 23.5° compared to the ecliptic plane, a polar flattening of about $1/400$ and a precession period of about 26000 years, the angle between the mantle's and the core's spinning axes is about 0.001°. We note that Eq. (4.18) obtained in a very simplified approach predicts a "resonance" (i.e. a divergence of ω_x) for specific combinations of shape and precession parameters. A full calculation including viscosity proves the existence of a range of phenomena including resonance and bistability (Cébron 2015).

Instabilities Driven By Mechanical Forcings: From Parametric Resonance to Turbulence

In the preceding section, we have determined the primary response of liquid cores to tidal deformation and rotation perturbations such as libration and precession. We show in this section that the repetitive action of these forcings leads to bulk instabilities. The core destabilization is due to the parametric resonance of inertial waves with the periodic primary flow. This section begins with a presentation of parametric resonance through the example of the Botafumeiro, a famous length-varying pendulum. We then introduce inertial waves and the possibility of resonance with tidal perturbations.

Parametric Sub-harmonic Resonance of a Pendulum

O Botafumeiro One of the most striking example of parametric resonance that can be found around the globe is probably *O Botafumeiro* in Santiago de Compostella's Cathedral. It is a 54 kg thurible—a metal censer—that hangs on the top of the cathedral's dome. The length of the rope can be changed over time around its mean length of 21.5 m by a group of holders. To spread incense in the cathedral, the holders first let the thurible swing with a small angle. Each time the censer goes up, i.e. twice per period, they slightly pull down the rope to shorten the swing length, and let it increase again as the swing goes down. With this twice per period excitation, they manage to swing the censer with very large amplitude in a short time, up to a height of 20.6 m. The velocity at the lowest point of the oscillation reaches 68 km/h.

Mechanical study of the length-varying pendulum As shown in Fig. 4.8, we model *O Botafumeiro* by a pendulum whose length ℓ varies over time:

$$\ell(t) = \ell_0(1 + \eta \sin(\omega t)) \tag{4.20}$$

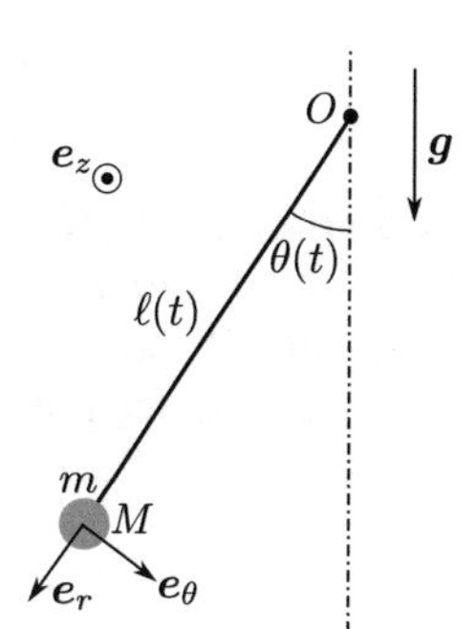

Fig. 4.8 Model for *O Botafumeiro*: a length-varying pendulum. The function $\ell(t)$ is prescribed to periodically vary around a mean length ℓ_0 according to (4.20)

where η is a small parameter accounting for the small modulation of the relative length of the swing. We study the motion of the point object M of mass m at the end of the rope of length $\ell(t)$. We assume that the referential bound to O is Galilean and that the rope is mass-less. The velocity $\boldsymbol{v}$ of M writes:

$$\boldsymbol{v} = \ell\dot{\theta}\boldsymbol{e}_\theta + \dot{\ell}\boldsymbol{e}_r \tag{4.21}$$

and the angular momentum L_z of M with respect to axis $(O, \boldsymbol{e}_z)$ is:

$$L_z = m\ell^2\dot{\theta}. \tag{4.22}$$

As gravity applies a torque $-m\ell g \sin\theta$, the conservation of angular momentum yields:

$$\ddot{\theta} + \frac{2\dot{\ell}}{\ell}\dot{\theta} + \frac{g}{\ell}\sin\theta = 0. \tag{4.23}$$

Introducing the ansatz (4.20) and $\omega_0^2 = g/\ell_0$ finally leads to:

$$\ddot{\theta} + \eta\frac{2\omega\cos(\omega t)}{1+\eta\sin(\omega t)}\dot{\theta} + \frac{\omega_0^2}{1+\eta\sin(\omega t)}\sin\theta = 0, \tag{4.24}$$

where η and ω are two control parameters, the relative variation of the rope length and the frequency of these variations.

We can already qualitatively predict at which frequency the length must be varied for optimal oscillation by looking at Eq. (4.23): indeed, comparing to the classical pendulum equation where ℓ is constant but still accounting for $\dot{\ell}$, the second term in (4.23) appears in place of a viscous damping of type $\nu\dot{\theta}$. Here however the sign of the viscosity depends on the sign of $\dot{\ell}$. In particular, shortening the length of the rope—i.e. $\dot{\ell} < 0$—corresponds to a "negative viscosity", which encourages the motion; $\dot{\ell} > 0$ corresponds to a classical positive damping. For the holders to input maximum energy into the system, the length must thus be shortened when the angular velocity $|\dot{\theta}|$ is maximum, i.e. when the pendulum is at its lowest position, and increased when $\dot{\theta} = 0$, i.e. when it is at its highest position. Therefore, two antagonistic moves must be operated during half the period of the pendulum: the frequency of the excitation must be the double of the free oscillation. This will be formally proved in the following.

Asymptotic analysis To analytically study the length-varying pendulum described by (4.24) and get the main physical properties of this system, we carry out an asymptotic study in the limit of small length variations $\eta \ll 1$ and small angle $\theta \ll 1$. The Eq. (4.24) can then be expanded into:

$$\ddot{\theta} + \omega_0^2 \theta = \eta \left[-2\omega \cos(\omega t)\dot{\theta} + \omega_0^2 \sin(\omega t)\theta \right] \tag{4.25}$$

where the length variation appears at order one as a forcing. The system has two typical time scales. The first one is $1/\omega_0$, the natural or free period of pendulum oscillation. This is the fast timescale of the system. We further assume here that the forcing period $1/\omega$ is of the same order of magnitude as $1/\omega_0$, because as seen just before, this is where interesting physics is expected. Then because of the forcing, the system also evolves over a slow timescale $1/(\eta\omega_0)$. We thus look for a solution to (4.25) where these two timescales are included and decoupled (Strogatz 2016), i.e. a function θ which depends on $\tau = t$ and $T = \eta t$ such that:

$$\theta(\tau, T) = (f_0(\tau) + \eta f_1(\tau))\, F(T). \tag{4.26}$$

In this two-timing framework, the total time derivatives are expanded as partial derivatives according to:

$$\begin{cases} \frac{\mathrm{d}}{\mathrm{dt}} = \dfrac{\partial}{\partial \tau} + \eta \dfrac{\partial}{\partial T} \\ \frac{\mathrm{d}^2}{\mathrm{dt}^2} = \dfrac{\partial^2}{\partial \tau^2} + 2\eta \dfrac{\partial^2}{\partial T \partial \tau} \end{cases} \tag{4.27}$$

where we have kept only the terms up to order 1. Taking into account the ansatz (4.26) and (4.27) yields to an order zero and an order one equations such that:

$$\begin{cases} \frac{\mathrm{d}^2 f_0}{\mathrm{d}\tau^2} + \omega_0^2 f_0 = 0 \\ \frac{\mathrm{d}^2 f_1}{\mathrm{d}\tau^2} + \omega_0^2 f_1 = -2\omega \cos(\omega\tau) f_0' + \omega_0^2 f_0 \sin(\omega\tau) - 2\dfrac{f_0' F'}{F}. \end{cases} \tag{4.28}$$

The solution to the first equation is straightforward, and using complex solutions:

$$f_0(\tau) = A e^{i\omega_0 \tau} + B e^{-i\omega_0 \tau}. \tag{4.29}$$

We then input this solution in the right hand side (RHS) of the second equation in (4.28) and expand it to find the following Fourier decomposition:

$$\begin{aligned}\frac{d^2 f_1}{d\tau^2} + \omega_0^2 f_1 \;=\;& e^{i(\omega_0+\omega)\tau}\left[-Ai\omega_0\omega + A\frac{\omega_0^2}{2i}\right]\\ &+ e^{i(\omega_0-\omega)\tau}\left[-Ai\omega_0\omega - A\frac{\omega_0^2}{2i}\right]\\ &+ e^{i(-\omega_0+\omega)\tau}\left[Bi\omega_0\omega + B\frac{\omega_0^2}{2i}\right]\\ &+ e^{-i(\omega_0+\omega)\tau}\left[Bi\omega_0\omega - B\frac{\omega_0^2}{2i}\right]\\ &+ e^{i\omega_0\tau}\left[-2iA\frac{F'}{F}\omega_0\right] + e^{-i\omega_0\tau}\left[2iB\frac{F'}{F}\omega_0\right].\end{aligned} \tag{4.30}$$

The frequency of the two terms appearing in the last line is the same as the eigen frequency of the harmonic oscillator in the left hand side (LHS) of (4.30) and should give rise to divergence of f_1. Those contributions to the LHS are called "secular terms" as they excite a long-term growth of the solution. However, the Taylor expansion in the ansatz (4.26) requires f_1 to remain bounded over time for the calculation to remain valid: secular terms must therefore be canceled.

In general, a first possibility is to impose $F' = 0$, but this cannot explain the amplitude increase observed in the case of *O Botafumeiro*. A more interesting solution arises when the excitation frequency ω is adequately chosen. When

$$\omega = 2\omega_0$$

the second and third lines then also have a frequency of $\pm\omega_0$ and give a more complex condition for f_1 to remain bounded. With this particular condition, canceling the secular terms leads to

$$\begin{cases} -\dfrac{3}{2}\omega_0 A + 2\dfrac{F'}{F}B = 0 \\ -2\dfrac{F'}{F}A + \dfrac{3}{2}\omega_0 B = 0. \end{cases} \tag{4.31}$$

In order to avoid the simple solution $A = B = 0$ which does not model the amplitude growth, the determinant of the above system must be zero. Canceling the determinant imposes:

$$F(T) \propto e^{\pm\frac{3}{4}\omega_0 T}. \tag{4.32}$$

Under these conditions, it is straightforward that for the exponentially growing branch $A = B$. The total solution at lowest order, in terms of time t and only considering the growing solution for a pendulum released at $t = 0$ from angle θ_0 with no initial velocity, is therefore:

$$\theta(t) = \theta_0 \cos(\omega_0 t) \exp\left(\frac{3}{4}\eta\omega_0 t\right). \tag{4.33}$$

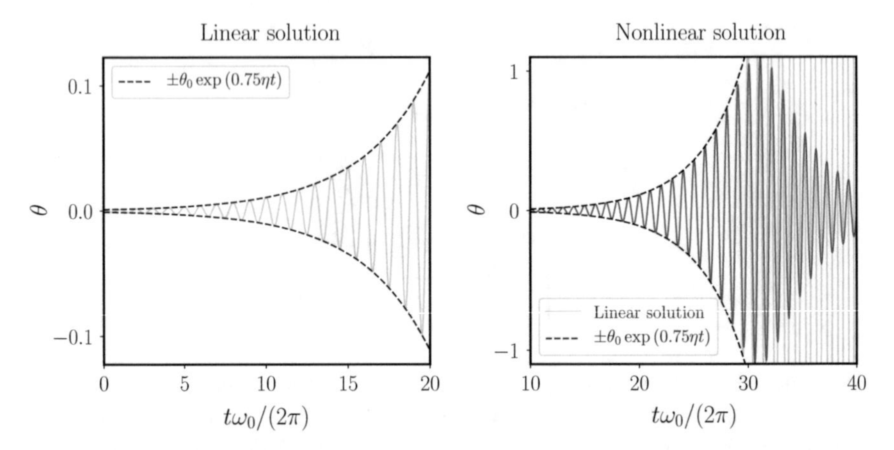

Fig. 4.9 **Left**: linear solution corresponding to the expression (4.33) for $\eta = 5 \times 10^{-2}$, $\omega_0 = 1$, $\omega = 2$ and $\theta_0 = 1 \times 10^{-3}$. **Right**: fully nonlinear solution of equation (4.24) (in red) and comparison with the linear solution (in gray)

Such a resonance process is called "parametric sub-harmonic resonance" as it happens when the excitation frequency is twice the free oscillation frequency. From a tiny perturbation, provided it is repetitive and has the adequate frequency, it leads to a drastic increase of the oscillations' amplitude. This solution is represented in Fig. 4.9 left and superimposed to the fully nonlinear solution obtained via numerical resolution of Eq. (4.24) in Fig. 4.9 right. The two solutions are in very good agreement at early time, but our linear approach does not capture the collapse of the amplitude observed in the numerical solution. As we show in the following paragraph, this is due to nonlinear effects which are not accounted for in our theoretical approach.

Saturation of the resonance As noted in Fig. 4.9 right, the exponential growth computed in the preceding paragraph must come to an end. Here, the nonlinearity—the $\sin\theta$ term in (4.23)—causes a collapse of the oscillations. Indeed, the period of free pendulum oscillations then depends on the amplitude: it increases as the amplitude increases. At second order in θ, the relation between period P and amplitude θ_m is given by Borda's formula (Guéry-Odelin and Lahaye 2010)

$$P(\theta_m) = \frac{2\pi}{\omega_0}\left(1 + \frac{\theta_m^2}{16}\right). \tag{4.34}$$

At early times, when the amplitude of the oscillations remains small, the excitation frequency ω matches the resonance condition $\omega = 2\omega_0$. As the amplitude increases, the frequency of the pendulum decreases and the oscillator is detuned from the excitation. This reverses the energy transfer from the parametric excitation to the pendulum. As the excitation and oscillation are out of phase, energy is pumped back from the pendulum to the forcing—the negative viscosity effect becomes a positive viscosity. The consequence is the global decrease of the amplitude. Note lastly that in the case of *O Botafumeiro*, the holders can tune the excitation to the actual period of the thurible, hence maintain large amplitude oscillations over a long time.

Oscillators in Planetary Cores: Inertial Waves

In the following, we draw an analogy between the length-varying pendulum and tidally-driven mechanical forcings in planetary cores. We identify what can be regarded as a pendulum inside planetary cores and what acts as a slight and repetitive perturbation.

The oscillating eigenmodes of rotating fluids are the so-called inertial waves. They are caused by the restoring action of the Coriolis force. This section is a short reminder of how to derive their governing equation and to infer their dispersion relation.

Let us derive a wave equation from the governing equations of incompressible rotating fluids. We consider the linear limit of the Euler equation for the velocity $\boldsymbol{u}$ in the rotating frame of reference:

$$\partial_t \boldsymbol{u} + 2\boldsymbol{\Omega} \times \boldsymbol{u} = -\boldsymbol{\nabla} p. \tag{4.35}$$

The equation governing the vorticity $\boldsymbol{\varpi} = \boldsymbol{\nabla} \times \boldsymbol{u}$ reads:

$$\partial_t \boldsymbol{\varpi} = 2\,(\boldsymbol{\Omega} \cdot \boldsymbol{\nabla})\,\boldsymbol{u}. \tag{4.36}$$

We assume $\boldsymbol{\Omega} = \Omega \boldsymbol{e}_z$ and take the curl the vorticity equation (4.36) to obtain an equation on the velocity only:

$$\partial_t\,(\boldsymbol{\nabla} \times (\boldsymbol{\nabla} \times \boldsymbol{u})) \;=\; 2\Omega \partial_z \boldsymbol{\varpi}. \tag{4.37}$$

We differentiate this last equation over time and then substitute (4.36) into (4.37) which gives:

$$\partial_{tt}\,(\boldsymbol{\nabla} \times (\boldsymbol{\nabla} \times \boldsymbol{u})) = 4\Omega^2 \partial_{zz} \boldsymbol{u}. \tag{4.38}$$

Considering incompressibility, we finally retrieve the Poincaré equation of rotation flows:

$$\partial_{tt} \boldsymbol{\nabla}^2 \boldsymbol{u} + 4\Omega^2 \partial_{zz} \boldsymbol{u} \;=\; 0. \tag{4.39}$$

Considering the divergence of the Navier–Stokes equation, one can easily find a similar equation on the pressure field p.

In a hypothetical infinite medium which satisfies translational invariance, the Poincaré equation admits plane waves solutions. Assuming that $\boldsymbol{u}$ takes the form of a plane wave of vector $\boldsymbol{k}$ and frequency ω:

$$\boldsymbol{u}(\boldsymbol{r}, t) \;=\; \boldsymbol{u}_0 e^{i(\boldsymbol{k}\cdot\boldsymbol{r} - \omega t)}, \tag{4.40}$$

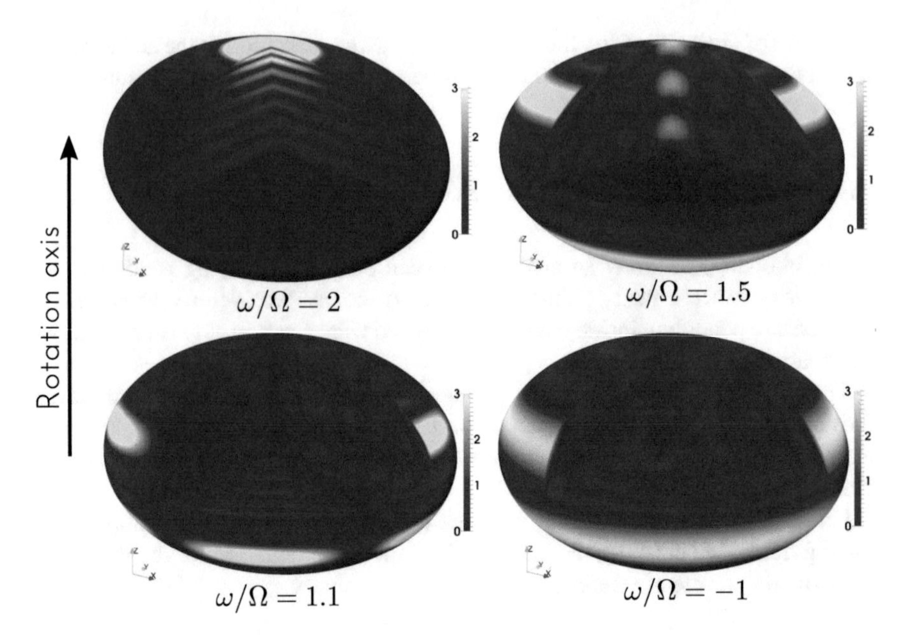

Fig. 4.10 Four examples of the velocity field of inertial modes computed in an ellipsoid with axes $a = 1$, $b = 0.86$ and $c = 0.57$ including viscous boundary layers, performed by Vidal and Cébron (2017). It shows equatorial and meridional cuts and the amplitude map at the surface of the ellipsoid. This figure is adapted from Vidal and Cébron (2017) Fig. 6

$\boldsymbol{r}$ being the position, the dispersion relation of inertial waves writes:

$$\omega^2 = 4\Omega^2 \frac{k_z^2}{\boldsymbol{k}^2} \qquad i.e. \qquad \omega = \pm 2\Omega \cos\xi \tag{4.41}$$

where ξ is the angle between the rotation axis and the wave vector. Note that this dispersion relation is peculiar in the sense that wave frequency is not related to the wavelength but to the direction of the wavevector only. Moreover, the frequency of the waves is bounded between -2Ω and 2Ω.

In the case of bounded geometry such as spheres or ellipsoids, the translational invariance is lost, but oscillatory solutions to the Poincaré equation can still be found. They are called inertial modes and their derivation for the special case of the sphere can be found in Greenspan (1968). Their frequencies remain bounded between -2Ω and 2Ω (Greenspan 1968). An example of inertial modes computed in an ellipsoidal container is given in Fig. 4.10. In the case of periodic mechanical forcing of planets, these inertial modes play the role of the excited oscillators.

Parametric Excitation: The Case of Tidally-Driven Instabilities

In section "Flow Driven by Differential Spin and Orbit", we derived the primary response of a fluid planetary interior to tidal distortion—see for instance the expressions (4.9) and (4.10); nothing has been said yet regarding the stability over time of this flow. This section shows that a tidal flow can excite a parametric resonance of inertial waves. We give the conditions under which a resonance can happen, and we briefly present a few ideas on how to quantify its grow rate.

We investigate the time evolution of perturbations to the tidal base flow, an instability being characterized by an exponential growth of these perturbations. In the frame rotating with the planet—see section "Flow Driven by Differential Spin and Orbit"—, we write the total flow as:

$$\boldsymbol{U} = \boldsymbol{U}^{\Omega} + \boldsymbol{u} \tag{4.42}$$

where $\boldsymbol{U}^{\Omega}$ is the tidal base flow (4.10) and $\boldsymbol{u}$ is a perturbation—which is not necessarily small. As $\boldsymbol{U}^{\Omega}$ is a nonlinear, viscous solution to the flow in the bulk of the fluid, the Navier–Stokes equations reduce to the nondimensional form:

$$\partial_t \boldsymbol{u} + \boldsymbol{U}^{\Omega} \cdot \nabla \boldsymbol{u} + \boldsymbol{u} \cdot \nabla \boldsymbol{U}^{\Omega} + \boldsymbol{u} \cdot \nabla \boldsymbol{u} + 2\boldsymbol{e}_z \times \boldsymbol{u} = -\nabla p + E \nabla^2 \boldsymbol{u} \tag{4.43}$$

$$\nabla \cdot \boldsymbol{u} = 0 \tag{4.44}$$

where lengths are scaled by the mean core radius $R_c = (a+b)/2$, time by $1/\Omega$, and where we have introduced the Ekman number $E = \nu/(R_c^2 \Omega)$ that compares the effects of viscous and Coriolis forces.

According to (4.10), the tidal base flow is proportional to the ellipticity of the tidal deformation β, and may therefore be written as

$$\boldsymbol{U}^{\Omega} = \beta \mathsf{A}(t) \boldsymbol{x}, \tag{4.45}$$

where $\mathsf{A}(t)$ is a linear operator, harmonic in time with frequency 2γ, acting on the position $\boldsymbol{x}$, representing the rotating strain field depicted in Fig. 4.4. We recall that γ is the difference between the planet's rotation rate and the moon's orbital rate, or equivalently the rotation rate of the tidal bulge in the rotating frame. In the linear, inviscid limit, Eq. (4.43) can be recast as

$$\partial_t \boldsymbol{u} + 2\boldsymbol{e}_z \times \boldsymbol{u} + \nabla p = -\beta \left[\mathsf{A}(t) \boldsymbol{u} + \mathsf{A}(t) \boldsymbol{x} \cdot \nabla \boldsymbol{u} \right]. \tag{4.46}$$

This equation is very similar to Eq. (4.25) governing the length-varying pendulum: eigenmodes of the system (i.e. inertial waves here), represented by the left hand side, are excited by a forcing which is harmonic in time, and depends on the amplitude of the eigenmodes. From a qualitative point of view, the periodic stretching of waves by the tidal strain is able to convey energy from tides to waves.

The situation compared to the length-varying pendulum is here slightly enriched by the existence of an infinite number of oscillators: many pairs of inertial modes can cooperate and resonate with the tidal base flow. Let us consider a mode of frequency ω_1: its nonlinear interaction with the base flow corresponding to the RHS of (4.46) bears harmonic terms of frequencies $\pm 2\gamma + \omega_1$ that can match a second mode oscillation, hence reinforcing it, provided its frequency ω_2 matches the following resonance condition:

$$|\omega_1 - \omega_2| = 2\gamma. \tag{4.47}$$

Reciprocally, mode 2 then reinforces mode 1. There is therefore a coherent effect of the tidal base flow and the two resonant waves in building the parametric resonance. Note that this resonance condition includes the single mode resonance for $\omega_1 = -\omega_2 = \pm\gamma$, which is reminiscent of the length-varying pendulum subharmonic resonance. Due to the nature of the base flow, this instability has been coined "elliptical instability". Note also that the resonance condition and the bounded domain of the inertial frequencies implies that the tidally-driven resonance can be excited as long as $|\gamma| \leq 2|\Omega|$.

An illustration of the instability growth is given in Fig. 4.11. In both the experiment and the simulation, the orbital and the spin rate are opposed, that is, $\Omega = -n$, such that $\gamma = 2\Omega$. The resonant modes are therefore at the limit of the inertial modes frequency domain. These modes are composed of horizontal layers of alternating horizontal velocity. They are reminiscent of plane waves in an unbounded domain as, at the frequency 2Ω, the wave vector is purely vertical. Lastly, it is interesting to note that even if the ellipticity is quite small in the experiment shown in Fig. 4.11—$\beta = 0.06$—, the flow is fully turbulent at later times, once the instability has reached saturation.

Quantifying the Growth Rate: A Global Approach

In this section, we detail a method to theoretically determine the growth rate of the parametric instability excited by the tidal base flow. We look for the long-term evolution of the amplitude of the resonant modes, following the general process given in Tilgner (2007b). The perturbation flow $\boldsymbol{u}$ can be written as a superposition of two modes with spatial structures $\boldsymbol{\Psi_1}(\boldsymbol{r})$ and $\boldsymbol{\Psi_2}(\boldsymbol{r})$, that is:

$$\boldsymbol{u}(\boldsymbol{r}, t) \;=\; A_1(t)\boldsymbol{\Psi_1}(\boldsymbol{r}) + A_2(t)\boldsymbol{\Psi_2}(\boldsymbol{r}) \,, \tag{4.48}$$

where the $A_j(t)$ are the amplitudes of the modes. Noting Π_j the pressure field associated to the mode j, $\boldsymbol{\Psi}_j$ and Π_j satisfy the following equation:

$$2\boldsymbol{e}_z \times \boldsymbol{\Psi}_j + \nabla\Pi_j \;=\; i\omega_j\boldsymbol{\Psi}_j \,, \tag{4.49}$$

that is, $\boldsymbol{\Psi}_j$, Π_j are eigenmodes of the linearized, rotating Euler equation. Note that, in general, two modes satisfy an orthogonality relation (Greenspan 1968):

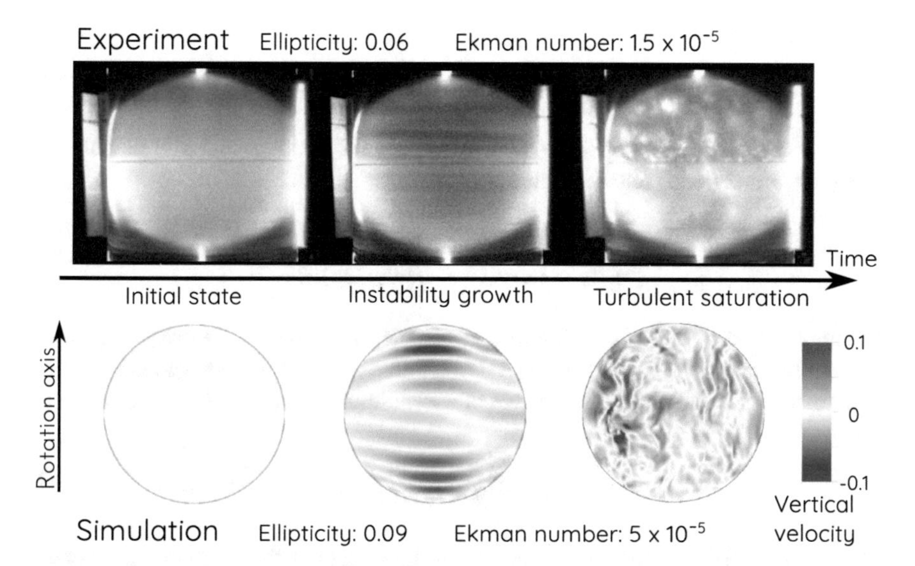

Fig. 4.11 Time evolution of the tidally-driven flow inside an ellipsoid shown in a meridional cross-section. **Top**: experimental visualization using flake-shaped particles materializing shear zones. **Bottom**: numerical simulation showing the vertical velocity. These pictures are adapted from Grannan et al. (2017), Fig. 3

$$\langle \boldsymbol{\Psi}_j | \boldsymbol{\Psi}_k \rangle \equiv \frac{1}{\mathcal{V}} \int_{\mathcal{V}} \boldsymbol{\Psi}_j^* \cdot \boldsymbol{\Psi}_k \mathrm{d}V = \delta_{jk}, \tag{4.50}$$

where $\mathcal{V}$ is the domain volume and δ the Dirac delta function.

Tidal distortion is always a small perturbation of the spherical shape of the container. The modes inside the slightly distorted container are therefore approximated here by the inertial modes of the sphere for simplicity (but see Vidal and Cébron 2017, for a more complete approach). Then we introduce an azimuthal wavenumber m such that the spatial structure of an inertial mode $\boldsymbol{\Psi}$ can be written as

$$\boldsymbol{\Psi}(\boldsymbol{r}) = \boldsymbol{\Phi}(r, z) e^{im\phi}$$

where (r, z, ϕ) are the cylindrical coordinates (Greenspan 1968). Two structures with different m are orthogonal, and the dot product between two modes can be specified as follows:

$$\langle \boldsymbol{\Psi}_j | \boldsymbol{\Psi}_k \rangle = \delta(m_k - m_j) \frac{2\pi}{\mathcal{V}} \int_{r,z} \boldsymbol{\Phi}_j^* \cdot \boldsymbol{\Phi}_k \, r \mathrm{d}r \mathrm{d}z = \delta(m_k - m_j) \left(\boldsymbol{\Phi}_j | \boldsymbol{\Phi}_k \right). \tag{4.51}$$

where we have introduced a reduced dot product $(\cdot|\cdot)$ that acts on the radial and vertical structures of the modes.

We introduce $\mathsf{L}(t)$ the linear operator associated to the RHS of Eq. (4.46), i.e. the linear operator which couples the modes with the tidal base flow. In general, for a field $\boldsymbol{w}$,

$$\mathsf{L}(t)\boldsymbol{w} = \mathsf{A}(t)\boldsymbol{w} + \mathsf{A}(t)\boldsymbol{x}\cdot\nabla\boldsymbol{w}. \tag{4.52}$$

With the orthogonality relation (4.50), the evolution of the amplitudes A_j are inferred from the ansatz (4.48) and the flow Eq. (4.46)

$$\begin{cases} \dot{A}_1 - i\omega_1 A_1 = \beta\left\langle\boldsymbol{\Psi}_1\middle|L(t)\boldsymbol{u}\right\rangle \\ \dot{A}_2 - i\omega_2 A_2 = \beta\left\langle\boldsymbol{\Psi}_2\middle|L(t)\boldsymbol{u}\right\rangle. \end{cases} \tag{4.53}$$

We may get rid of the fast oscillations of the inertial modes in the preceding equations introducing $a_j(t) = \exp(i\omega_j t)A_j(t)$, similarly to what has been done in section "Parametric Sub-harmonic Resonance of a Pendulum". The Eq. (4.53) then reads:

$$\begin{cases} \dot{a}_1 = \beta\left\langle\boldsymbol{\Psi}_1\middle|L(t)\boldsymbol{u}\right\rangle e^{-i\omega_1 t} \\ \dot{a}_2 = \beta\left\langle\boldsymbol{\Psi}_2\middle|L(t)\boldsymbol{u}\right\rangle e^{-i\omega_2 t}. \end{cases} \tag{4.54}$$

The RHS of (4.54) can be specified taking into account the temporal and spatial variations of the modes contained in $\boldsymbol{u}$ and in the coupling operator $\mathsf{L}(t)$. The tidal base flow is transformed into cylindrical coordinates as:

$$\boldsymbol{U}^{\Omega} = -\beta\gamma r\left[\sin(2\gamma t + 2\phi)\boldsymbol{e}_r + \cos(2\gamma t + 2\phi)\boldsymbol{e}_\phi\right] \tag{4.55}$$

showing that it contains the wave numbers $m = 2$ and $m = -2$. L is therefore decomposed as:

$$\mathsf{L}(t) = e^{i(2\gamma t+2\phi)}\mathsf{L}_0 + e^{-i(2\gamma t+2\phi)}\mathsf{L}_0^* \tag{4.56}$$

where L_0 is independent of time and ϕ. Using the orthogonality relation, the RHS of (4.54) is expanded as follows:

$$\begin{aligned} \left\langle\boldsymbol{\Psi}_1\middle|\mathsf{L}(t)\boldsymbol{u}\right\rangle e^{-i\omega_1 t} = & \left(\boldsymbol{\Phi}_1\middle|\mathsf{L}_0\boldsymbol{\Phi}_1\right) a_1\,\delta(2 + m_1 - m_1)e^{i(\omega_1+2\gamma-\omega_1)t} \\ & +\left(\boldsymbol{\Phi}_1\middle|\mathsf{L}_0^*\boldsymbol{\Phi}_1\right) a_1\,\delta(-2 + m_1 - m_1)e^{i(\omega_1-2\gamma-\omega_1)t} \\ & +\left(\boldsymbol{\Phi}_1\middle|\mathsf{L}_0\boldsymbol{\Phi}_2\right) a_2\,\delta(2 + m_2 - m_1)e^{i(\omega_2+2\gamma-\omega_1)t} \\ & +\left(\boldsymbol{\Phi}_1\middle|\mathsf{L}_0^*\boldsymbol{\Phi}_2\right) a_2\,\delta(-2 + m_2 - m_1)e^{i(\omega_2-2\gamma-\omega_1)t} \end{aligned} \tag{4.57}$$

and similarly for $\left\langle\boldsymbol{\Psi}_2\middle|L(t)\boldsymbol{u}\right\rangle e^{-i\omega_2 t}$. When the resonance condition is satisfied, i.e. $\omega_2 - \omega_1 = 2\gamma$, the coupling between the modes 1 and 2 is effective provided the following selection rule applies:

$$m_2 - m_1 = 2. \tag{4.58}$$

All the other coupling terms, whose frequencies do not match the resonance condition, then vanish. A similar derivation for $\left\langle\boldsymbol{\Psi}_2\middle|\mathsf{L}(t)\boldsymbol{u}\right\rangle e^{-i\omega_2 t}$ allows to prove that the system (4.54) reduces to:

$$\begin{cases} \dot{a}_1 = \beta \left(\mathbf{\Phi}_1 \middle| \mathsf{L}_0^* \mathbf{\Phi}_2 \right) a_2 \\ \dot{a}_2 = \beta \left(\mathbf{\Phi}_2 \middle| \mathsf{L}_0 \mathbf{\Phi}_1 \right) a_1. \end{cases} \tag{4.59}$$

The growth rate σ is therefore given by the overlap between the tidal base flow and the two modes:

$$\sigma^2 = \beta^2 \left(\mathbf{\Phi}_1 \middle| \mathsf{L}_0^* \mathbf{\Phi}_2 \right) \left(\mathbf{\Phi}_2 \middle| \mathsf{L}_0 \mathbf{\Phi}_1 \right). \tag{4.60}$$

The amplitudes of the modes grow provided the overlap integrals have the same sign; the growth rate is then proportional to β. This is similar to the length-varying pendulum for which the growth rate was found to be proportional to the amplitude of the perturbation η.

Computing the growth rate is in this case rather difficult as it requires computing the overlap integrals between the modes and the tidal forcing. This is in general nontrivial: although the inertial modes in a sphere, or even a spheroid, are known, there is no analytic formula in the generic case of tri-axial ellipsoids. Computation of the overlap integrals therefore requires numerical solving of the eigenvalue problem of inertial modes, as done for instance in Vidal and Cébron (2017).

Lastly, the amplitude Eq. (4.59) may be refined accounting for the viscous damping of the modes. In the planetary limit of small Ekman number E, viscous dissipation is dominated by friction inside Ekman boundary layers, and yields a correction $\mathcal{O}(\sqrt{E})$; as shown by Le Bars et al. (2010), the correction to the growth rate is then $K\sqrt{E}$ with K a constant typically between 1 and 10 (but see also Lemasquerier et al. 2017, for a discussion on the possible importance of bulk dissipation).

Quantifying the Growth Rate: Short Wavelength Approximation

Although the derivation of the growth rate in the preceding paragraph is rather straightforward, it is quite difficult to extract quantitative information in complex geometries such as tri-axial ellipsoids. This approach requires knowing a priori the spatial structure of the inertial modes that must be computed numerically. Another approach that has proven efficient in past studies to make quantitative prediction (see e.g. Kerswell 2002 and references therein) consists in assuming a scale separation between short wavenumber resonant modes and the large scale tidal flow.

This approach, known as the Wentzel–Kramers–Brillouin (WKB) analysis, assumes that the perturbations take the form of a plane wave packet around a point that is advected by the base flow. The wave packet is affected by tidal distortion as it moves along with the Lagrangian point. This theoretical framework, which resembles the process to infer classical optics from light wave propagation, was formally introduced in the context of hydrodynamic instabilities by Lifschitz and Hameiri (1991). It is particularly suitable for the study of parametric instability of waves interacting with a base flow.

It was applied to the present case of the tidally-driven elliptical instability in rotating flows by Le Dizès (2000). The WKB method allows to retrieve that the short wavelength growing perturbation corresponds to the superposition of two contrapropagating inertial waves of frequency $\pm\gamma$ with an amplitude growth rate

$$\sigma = \frac{\beta\gamma}{16}(2+\gamma)^2 . \tag{4.61}$$

Although this growth rate describes short wavelength perturbations under the form of inviscid plane waves in an infinite domain, it accounts very well for the growth of inertial modes in enclosed containers in the weak tidal distortion and low dissipation regime, i.e. the regime that is relevant for geophysics (Le Bars et al. 2010). A small correction due to boundary friction must then be considered, that is:

$$\sigma_v = \sigma - K\sqrt{E} \tag{4.62}$$

with K between 1 and 10 typically, as explained in the preceding section.

The Elliptical Instability in Planetary Cores

The preceding theoretical results have been used in past studies to evaluate the actual relevance of the tidally-driven elliptical instability in natural systems, for instance by Cébron et al. (2012b). In the case of the Earth, the tidal distortion forced by the Moon—which induces a tidal bulge of ellipticity $\beta \sim 10^{-7}$—is close but a priori not sufficient to overcome the viscous damping of resonant modes. Nevertheless, Cébron et al. (2012b) also considered the early history of the Earth at times where the Moon was closer to its parent planet. Assuming the Earth–Moon distance is reduced by a factor 2, the core of the Earth becomes unstable to the elliptical instability. Mechanical forcings therefore provide an interesting alternative to drive magnetic field before the crystallization of the inner core. Remember that the Earth is known to be surrounded by a magnetic field since at least 3.5 Gy (Tarduno et al. 2010) while the Earth's inner core only started to form around 1 Gy ago (Labrosse 2015).

Following the same process as for the tidally-driven instability, librations have also been shown to drive parametric sub-harmonic instability of inertial waves (Kerswell and Malkus 1998; Cébron et al. 2012a; Grannan et al. 2014; Cébron et al. 2014). In particular, Kerswell and Malkus (1998) have demonstrated that Io's core in unstable, thus providing a possible explanation to the magnetic field measured by the Galileo probe around the Jovian satellite (Kivelson et al. 1996b). Cébron et al. (2012b) have extended the preceding study to show that Europa's core may also be unstable. Note that Ganymede's core remains below the threshold of instability despite being surrounded by a magnetic field.

Finally, precession is similarly capable of driving parametric sub-harmonic instability of inertial waves, but there the coupling flow still remains controversial (see e.g. Kerswell 1993; Lin et al. 2015). This precession-driven instability could be of importance for instance for the magnetic field of the early Earth (Andrault et al. 2016) and Moon (Dwyer et al. 2011).

An Overview of Some Ongoing Research

Present Tools for Investigating Mechanical Forcings and Instabilities in the Laboratory

The investigation of instabilities forced by tides, libration and precession have greatly benefited from experimental studies since the early work of Malkus (1963, 1968, 1989). The typical setups used to study flows driven by tides, libration and precession are given in Fig. 4.12. The interested reader should refer to the review of Le Bars et al. (2015), or research articles such as Grannan et al. (2017) and Noir et al. (2001). Tidal forcing is the most difficult as it requires using a deformable ellipsoid (usually made of silicon); the motion of the tidal deformation is imposed by rollers pressing on the ellipsoid. Today, modern metrology techniques like high-speed, time-resolved particle imaging velocimetry allow describing quantitatively the turbulence in asymptotic regimes of low dissipation and low forcing amplitude bearing some resemblance with planetary ones (Grannan et al. 2017).

Additionally, high-performance computing now allows reaching fully turbulent cases and considering the fully coupled magneto-hydrodynamics problem for dynamo action. Two main strategies have been developed: either consider so-called "local" numerical methods (like finite elements, finite volumes, or spectral elements) to simulate the flow in the tri-axial ellipsoidal geometry relevant for planetary applications (e.g. Cébron et al. 2010a; Ernst-Hullermann et al. 2013; Favier et al. 2015); or use highly efficient numerical methods in simplified geometries (like a sphere or a triply periodic box), then imposing the mechanical forcing as a background force (e.g. Barker and Lithwick 2013; Cébron and Hollerbach 2014; Le Reun et al. 2017).

All methods have pro and cons; only their combination—together with real data analysis—has allowed significant progress towards a better understanding of those flows and their planetary consequences. Some of these advances are presented below.

The Saturation of the Instability

Since the seminal studies by Malkus in the 1960s, many investigations have focused on showing the relevance of mechanical forcing for actual planetary cores. However, only a few works have been devoted to the saturation of the instability, of primary importance for tidal dissipation, the orbital evolution of planets and stars, and dynamo action. In the various experimental and numerical investigations of the saturation flow, a large diversity of behaviors has been observed, depending on the geometry

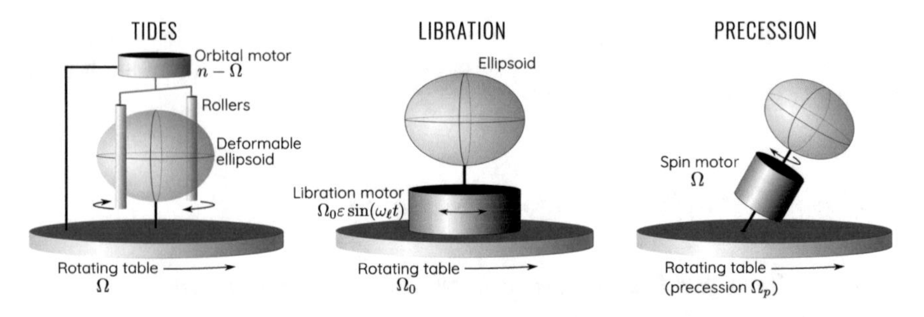

Fig. 4.12 A sketch of the three experimental setups built to study flows driven by tides, libration and precession. The rotation rates of the different motors are associated with geophysical forcings. The different rotation rates and amplitudes are defined in section "Tidal Forcings in Planetary Cores: The Primary Response to Tides"

and the control parameters—be it the tidal distortion, the Ekman number or the libration amplitude and frequency.

In the saturation flow, the emergence of geostrophic vortices, i.e. flows that are invariant along the rotation axis and that have a slow dynamics,[2] seems to play an important role. As they grow, inertial modes feed a geostrophic flow, probably via a secondary instability as shown by Kerswell (1999), which might become of similar amplitude as the excitation flow and lead to the disruption of the instability (Barker and Lithwick 2013). The flow then decays viscously before waves can again be excited and engage in a new cycle of growth and decay. Some experiments, however, such as those carried out by Grannan et al. (2014), have shown that the saturation flow can also be fully turbulent and reach a statistically steady state. Further simulations carried out by Favier et al. (2015) have even shown that the unstable inertial modes are able to excite a priori stable modes, i.e. outside the initial resonance condition, via three modes resonant interactions, a mechanism called "triadic resonance"—see for instance the work of Bordes et al. (2012).

This discrepancy was somewhat clarified by Le Reun et al. (2017). This study used the method introduced by Barker and Lithwick (2013) to study the tidally-driven elliptical instability: instead of simulating the whole planetary interior with complex ellipsoidal outer boundary, only a small parcel is modeled. This parcel can be assumed to be cubic, with periodic boundary conditions, and the tidal excitation is imposed as a background flow. Such a technique allows to study with great details the saturation flow in the regime of low forcing (small ellipticity) and low dissipation (small Ekman number) which is relevant for geophysics. With an artificial and adjustable friction specific to the geostrophic component of the flow, the saturation continuously transitions from a geostrophic-dominated turbulence to a wave-dominated turbulence. In the latter case, it was even shown that the resulting turbulent flow is a superposition of inertial waves with different frequencies, which are all in resonant interaction, a state known as an inertial wave turbulence (Galtier

[2]They correspond to the $\omega \to 0$ limit of inertial waves.

2003; Yarom and Sharon 2014). This regime is expected in the asymptotic limit (low dissipation and low forcing) relevant for planetary flows.

Dynamo Driven By Mechanical Forcing

One of the most striking features of the past few years was to show that the mechanically driven flows are able to drive dynamo action in planetary cores. The few published studies include laminar dynamos from the precession base flow (Ernst-Hullermann et al. 2013), laminar dynamos for tidal instability with *ad hoc* bulk forcing in a spherical domain (Cébron and Hollerbach 2014; Vidal et al. 2017), laminar dynamos in a spheroidal domain for precession and libration instabilities (Wu and Roberts 2009, 2013), and turbulent dynamos in a spherical domain for precession (Tilgner 2005, 2007a; Kida and Shimizu 2011; Lin et al. 2016). Reddy et al. (2018) have carried out simulations of libration, precession, and tidally-driven instabilities in tri-axial ellipsoids leading to sustained turbulence, which is the relevant configuration for planetary applications. They showed that such a flow is able to drive amplification of the magnetic field in the three cases. However, the feedback of the Lorentz force was not taken into account in this work—i.e. the authors addressed kinematic dynamos only. The full problem involving sustained turbulence, magnetic field and the Lorentz force in tri-axial ellipsoids, remains to be explored. In particular, the different features of the magnetic field induced by an inertial wave turbulence or a geostrophic turbulence, as discussed in the preceding paragraph, have only been sparsely investigated (Moffatt 1970; Davidson 2014).

Instabilities in Presence of an Inner Core

The whole set of results discussed above was obtained in the simple context of fully liquid cores. We know that the secular cooling of planets sometimes leads to the progressive crystallization of a solid inner core (like in the Earth). Considering the relevance of mechanical forcing in a shell with inner and outer boundaries is more complicated as the inertial modes become singular; that is, they are locally discontinuous on sheets in the inviscid limit—see for instance discussions in Rieutord et al. (2001). From a mathematical point of view, the existence of such singular solutions is due to the hyperbolic nature of the Poincaré equation describing wave propagation, for which conditions are applied at the boundaries of the domain, which implies ill-posedness.

A few studies have been devoted to tidally-driven instabilities with an inner core, mostly in the case of tidal and libration forcings (Seyed-Mahmoud et al. 2004; Lacaze et al. 2005; Lemasquerier et al. 2017). In these experiments, the elliptical instability mechanism, i.e. the parametric excitation of inertial waves by the base flow, still holds. Nevertheless, the turbulent saturation flow happens to be heterogeneous with

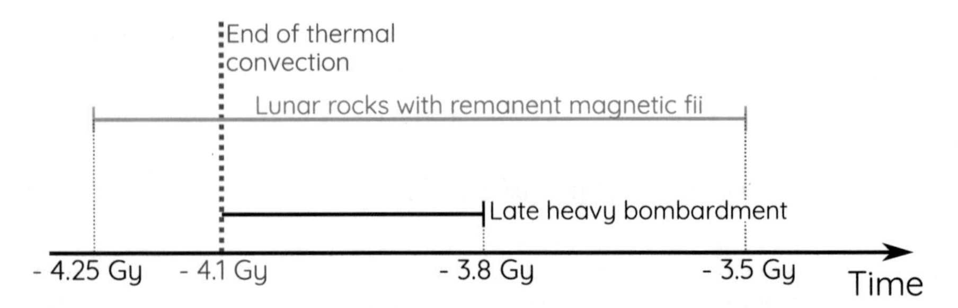

Fig. 4.13 Time line of the events affecting the early Moon with focus on the magnetic era, from Weiss and Tikoo (2014) and Le Bars et al. (2011)

different features inside and outside the cylinder parallel to the axis of rotation and tangent to the inner core (Lemasquerier et al. 2017).

The Mysterious Magnetic Field of the Early Moon

We conclude with a section on how the elliptical instability explains the early history of the Moon, which used to have a magnetic field, even when its core was not convecting.

The early magnetic field of the Moon From measurements of the remanent magnetization of the surface of the Moon, determined on samples collected during the Apollo missions, planetologists show that the Moon used to be surrounded by a strong magnetic field during 700–800 million years atleast, starting around 4 billion years ago (Weiss and Tikoo 2014)—see the schematic in Fig. 4.13. According to Garrick-Bethell et al. (2009), the minimum intensity of the magnetic field, inferred from the abovementioned measurements, was around 1 μT. This finding seems puzzling as thermal evolution models of the Moon state that thermal convection is not possible in its core after -4 Gy.

Satellite data show a strong correlation between the location of the largest values of the magnetic field strength and large impact basins where the collision is thought to have induced a thick melting of the Moon's crust and mantle—see Fig. 4.14. The time line is also striking as the ages of the basins where the strongest remanent magnetization are observed correspond to the late heavy bombardment, an era when several impacts of large bodies occurred approximately 4.1 to 3.8 billion years ago. These data therefore show a strong correlation between large meteoric collisions and the existence of a magnetic field.

Desynchronisation dynamo in the Moon Before an impact, the rotation rate of the Moon is synchronized with its revolution around the Earth. In their work, Le Bars et al. (2011) pointed out that large impacts are able to change the rotation rate of the Moon, and more precisely the rotation rate of its mantle. The fluid outer core of

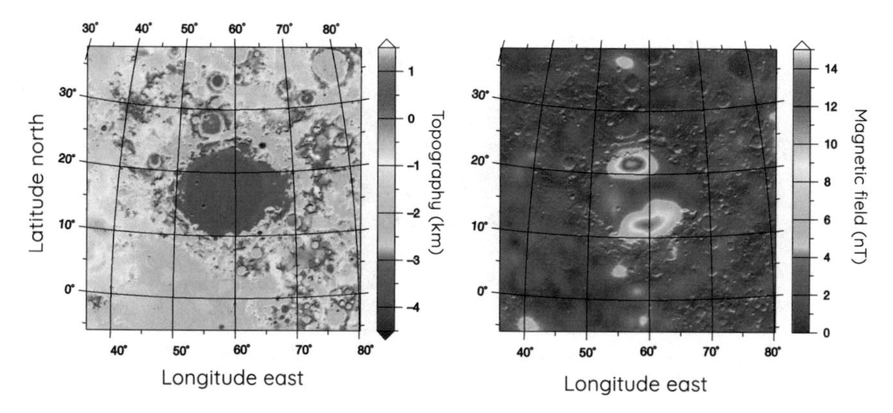

Fig. 4.14 Surface topography and corresponding strength of the magnetic field of the Crisium impact basin, reproduced after Le Bars et al. (2011), Fig. 1. The largest values of the magnetic field are restricted to the impact basin where a thick melt event occurred

the Moon is coupled to the mantle via viscous drag at the boundary. Therefore, after an impact, the core and the mantle rotate at different rates. In addition, the Moon is known to have a frozen, or fossil tidal bulge: it has a deformation that is reminiscent of a tidal bulge, but is not forced by any actual gravitational interaction (it accounts for a rise of 200 m while the present tidal bulge height stands around 10 m). This deformation—or a largest tidal bulge—was thus present in the past. Note that such a frozen distortion is not in hydrostatic equilibrium; the gravitational field of the Moon is not strong enough to force the relaxation of this imbalance.

After an impact, the core thus sees a deformation that rotates at the mantle's rate, a configuration that exactly matches the tidal forcing which is known to excite parametric instability of inertial waves. Le Bars et al. (2011) proved that desynchronization by large impacts is sufficient to drive tidally-driven elliptical instability, and to potentially induce a dynamo. Note that the Moon was closer to the Earth at that time, at about 45 Earth's radii or less, while it is at 60 Earth's radii today. As a consequence, the Moon's orbital rate was larger, thus lowering the Ekman number and the dissipation, which facilitated the growth of the instability. They proposed that the instability takes around 1000 years to develop and saturate. The turbulence driven by the non-synchronous rotation drives enhanced tidal dissipation which tends to resynchronize the spin and the orbit of the Moon, a process that takes around 10 000 years. During that transition period, the magnetic field may have reached up to 2.5 μT, as indicated in Fig. 4.15. The depth of the Curie isotherm, below which rocks acquire remanent magnetization, indicates that the surface of the impact basin could record the magnetic field at its peak—see Fig. 4.15. For a less intense impact, free libration, instead of full desynchronization, might have similarly led to elliptical instability, turbulence, dynamo and recording at the surface while cooling and slowing down.

Note finally that an analog scenario was proposed at the same time by Dwyer et al. (2011), based on the more intense precession forcing that was present at that

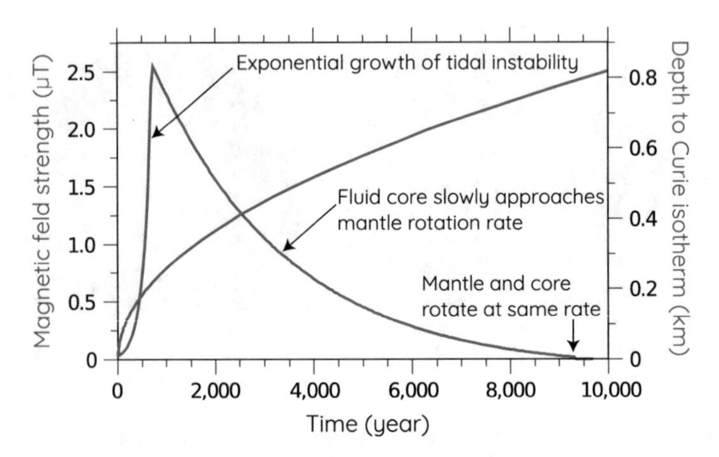

Fig. 4.15 Time evolution of the magnetic field strength and the depth of the Curie isotherm below which rocks acquire a remanent magnetic field. Figure adapted from Le Bars et al. (2011), Fig. 3

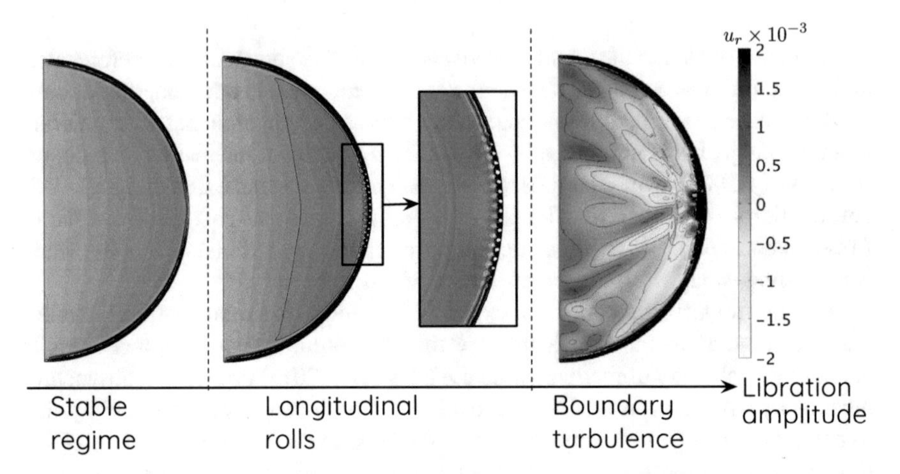

Fig. 4.16 Illustration of the destabilization of the equatorial boundary layer caused by librations. The color scale represents the radial velocity. This figure is adapted from Sauret et al. (2013) Fig. 2

time. It is worth considering that the different scenarios are not exclusive, but might be combined to explain the whole story of the Moon magnetic field evolution.

Conclusion—Beyond the Elliptical Instability in Planetary Cores

In this chapter, we have presented how gravitational interactions are able to create complex, and even turbulent flows, inside planetary cores. We have particularly focused on a specific class of instability related to parametric sub-harmonic resonance, drawing an analogy between rotating fluid cavities and a length-varying pendulum. Such a resonance, also called "elliptical instability", builds on the coupling between two inertial waves and the oscillating strain field induced by tidal interactions. Despite recent advances that have been presented above, many challenges remain to describe the turbulent saturation flows and the induced magnetic field. Also, while tides distortion, libration, precession, and convection have up to now been mainly studied separately, they might be simultaneously present in planetary cores, possibly leading to complex nonlinear interactions that remain to be explored.

A concluding remark: parametric resonance is not the only class of instabilities that can be driven by tides inside planetary cores. For instance, libration leads to centrifugal instability of the equatorial boundary layer (Calkins et al. 2010; Favier et al. 2014). As shown numerically by Sauret et al. (2013), this excited motion in the boundary layer radiates inertial waves in the whole interior—see Fig. 4.16. Lastly, the excitation of inertial modes inside a fluid cavity drives, via interactions with and within the boundary layer, zonal winds, i.e. axisymmetric geostrophic flows, associated with intense shear that may be unstable (Sauret et al. 2010, 2014). All these different excited flows pile up in planetary cores and their respective contributions remain to be evaluated. In any case, we hope to have convinced the reader that tidal interactions lead to complexity in the behavior of fluid interiors, and should not be dismissed for being "small" perturbations.

References

Andrault, Monteux, Le Bars, & Samuel. (2016). The deep earth may not be cooling down. *Earth and Planetary Science Letters*, *443*, 195–203.

Barker. (2016). Nonlinear tides in a homogeneous rotating planet or star: Global simulations of the elliptical instability. *Monthly Notices of the Royal Astronomical Society*, stw702. https://doi.org/10.1093/mnras/stw702.

Barker, & Lithwick. (2013). Non-linear evolution of the tidal elliptical instability in gaseous planets and stars. *Monthly Notices of the Royal Astronomical Society*, *435*(4), 3614–3626.

Barker, Braviner, & Ogilvie. (2016). Nonlinear tides in a homogeneous rotating planet or star: Global modes and elliptical instability. *Monthly Notices of the Royal Astronomical Society*, *459*(1), 924–938. https://doi.org/10.1093/mnras/stw701, arXiv:1603.06839.

Bayly. (1986). Three-dimensional instability of elliptical flow. *Physical Review Letters*, *57*(17), 2160–2163. https://doi.org/10.1103/PhysRevLett.57.2160.

Bordes, Moisy, Dauxois, & Cortet. (2012). Experimental evidence of a triadic resonance of plane inertial waves in a rotating fluid. *Physics of Fluids*, *24*(1), 014105. https://doi.org/10.1063/1.3675627.

Busse. (1968). Steady fluid flow in a precessing spheroidal shell. *Journal of Fluid Mechanics*, *33*(4), 739–751. https://doi.org/10.1017/S0022112068001655.

Busse. (1970). Thermal instabilities in rapidly rotating systems. *Journal of Fluid Mechanics*, *44*(03), 441. https://doi.org/10.1017/S0022112070001921.

Calkins, Noir, Eldredge, & Aurnou. (2010). Axisymmetric simulations of libration-driven fluid dynamics in a spherical shell geometry. *Physics of Fluids*, *22*(8), 086602. https://doi.org/10.1063/1.3475817.
Cébron. (2015). Bistable flows in precessing spheroids. *Fluid Dynamics Research*, *47*(2), 025504.
Cébron, & Hollerbach. (2014). Tidally driven dynamos in a rotating sphere. *The Astrophysical Journal Letters*, *789*(1), L25.
Cébron, Le Bars, Leontini, Maubert, & Le Gal. (2010a). A systematic numerical study of the tidal instability in a rotating triaxial ellipsoid. *Physics of the Earth and Planetary Interiors*, *182*(1–2), 119–128.
Cébron, Le Bars, & Meunier. (2010b). Tilt-over mode in a precessing triaxial ellipsoid. *Physics of Fluids*, *22*(11), 116601.
Cébron, Le Bars, Noir, & Aurnou. (2012a). Libration driven elliptical instability. *Physics of Fluids*, *24*(6), 061703.
Cébron, Le Bars, Moutou, & Le Gal. (2012b). Elliptical instability in terrestrial planets and moons. *Astronomy & Astrophysics*, *539*, A78. https://doi.org/10.1051/0004-6361/201117741.
Cébron, Vantieghem, & Herreman. (2014). Libration-driven multipolar instabilities. *Journal of Fluid Mechanics*, *739*, 502–543. https://doi.org/10.1017/jfm.2013.623.
Davidson. (2014). The dynamics and scaling laws of planetary dynamos driven by inertial waves. *Geophysical Journal International*, *198*(3), 1832–1847. https://doi.org/10.1093/gji/ggu220.
Dwyer, Stevenson, & Nimmo. (2011). A long-lived lunar dynamo driven by continuous mechanical stirring. *Nature*, *479*(7372), 212–214. https://doi.org/10.1038/nature10564.
Ernst-Hullermann, Harder, & Hansen. (2013). Finite volume simulations of dynamos in ellipsoidal planets. *Geophysical Journal International*, *195*(3), 1395–1405.
Favier, Barker, Baruteau, & Ogilvie. (2014). Non-linear evolution of tidally forced inertial waves in rotating fluid bodies. *Monthly Notices of the Royal Astronomical Society*, *439*(1), 845–860. https://doi.org/10.1093/mnras/stu003.
Favier, Grannan, Le Bars, & Aurnou. (2015). Generation and maintenance of bulk turbulence by libration-driven elliptical instability. *Physics of Fluids*, *27*(6), 066601. https://doi.org/10.1063/1.4922085.
Galtier. (2003). Weak inertial-wave turbulence theory. *Physical Review E*, *68*(1), 015301.
Garrick-Bethell, Weiss, Shuster, & Buz. (2009). *Early Lunar Magnetism. Science*, *323*(5912), 356–359. https://doi.org/10.1126/science.1166804.
Glatzmaiers, & Roberts. (1995). A three-dimensional self-consistent computer simulation of a geomagnetic field reversal. *Nature*, *377*(6546), 203–209. https://doi.org/10.1038/377203a0.
Grannan, Le Bars, Cébron, & Aurnou. (2014). Experimental study of global-scale turbulence in a librating ellipsoid. *Physics of Fluids*, *26*(12), 126601. https://doi.org/10.1063/1.4903003.
Grannan, Favier, Le Bars, & Aurnou. (2017). Tidally forced turbulence in planetary interiors. *Geophysical Journal International*, *208*(3), 1690–1703. https://doi.org/10.1093/gji/ggw479.
Greenspan. (1968). *The Theory of Rotating Fluids*. Cambridge: CUP Archive. ISBN 978-0-521-05147-7.
Guéry-Odelin, & Lahaye. (2010). *Classical Mechanics Illustrated by Modern Physics: 42 Problems with Solutions*. Singapore: World Scientific Publishing Company.
Hough. (1895). XII. The oscillations of a rotating ellipsoidal shell containing fluid. *Philosophical Transactions of the Royal Society of London A*, *186*, 469–506. https://doi.org/10.1098/rsta.1895.0012.
Kerswell. (1993). The instability of precessing flow. *Geophysical & Astrophysical Fluid Dynamics*, *72*(1–4), 107–144. https://doi.org/10.1080/03091929308203609.
Kerswell. (1996). Upper bounds on the energy dissipation in turbulent precession. *Journal of Fluid Mechanics*, *321*, 335–370.
Kerswell. (1999). Secondary instabilities in rapidly rotating fluids: Inertial wave breakdown. *Journal of Fluid Mechanics*, *382*, 283–306. https://doi.org/10.1017/S0022112098003954.
Kerswell. (2002). Elliptical instability. *Annual review of fluid mechanics*, *34*(1), 83–113.

Kerswell, & Malkus. (1998). Tidal instability as the source for Io's magnetic signature. *Geophysical Research Letters*, *25*(5), 603–606. https://doi.org/10.1029/98GL00237.
Kida, & Shimizu. (2011). A turbulent ring and dynamo in a precessing sphere. *Journal of Physics: Conference Series*, *318*, 072031. (IOP Publishing).
Kivelson, et al. (1996a). Discovery of Ganymede's magnetic field by the Galileo spacecraft. *Nature*, *384*(6609), 537–541. https://doi.org/10.1038/384537a0.
Kivelson, et al. (1996b). Io's interaction with the plasma torus: Galileo magnetometer report. *Science*, *274*(5286), 396–398. https://doi.org/10.1126/science.274.5286.396.
Kivelson, Khurana, & Volwerk. (2002). The permanent and inductive magnetic moments of Ganymede. *Icarus*, *157*(2), 507–522. https://doi.org/10.1006/icar.2002.6834.
Labrosse. (2015). Thermal evolution of the core with a high thermal conductivity. *Physics of the Earth and Planetary Interiors*, *247*, 36–55. https://doi.org/10.1016/j.pepi.2015.02.002.
Lacaze, Le Gal, & Le Dizès. (2005). Elliptical instability of the flow in a rotating shell. *Physics of the Earth and Planetary Interiors*, *151*(3), 194–205. https://doi.org/10.1016/j.pepi.2005.03.005.
Larmor. (1919). How could a rotating body such as the sun become a magnet? *Report of the British Association for the Advancement of Science*, pp. 159–160.
Le Bars, Le Dizès, & Le Gal. (2007). Coriolis effects on the elliptical instability in cylindrical and spherical rotating containers. *Journal of Fluid Mechanics*, *585*, 323. https://doi.org/10.1017/S0022112007006866.
Le Bars, Lacaze, Le Dizès, Le Gal, & Rieutord. (2010). Tidal instability in stellar and planetary binary systems. *Physics of the Earth and Planetary Interiors*, *178*(1–2), 48–55. https://doi.org/10.1016/j.pepi.2009.07.005.
Le Bars, Wieczorek, Karatekin, Cébron, & Laneuville. (2011). An impact-driven dynamo for the early Moon. *Nature*, *479*(7372), 215–218. https://doi.org/10.1038/nature10565.
Le Bars, Cébron, & Le Gal. (2015). Flows driven by libration, precession, and tides. *Annual Review of Fluid Mechanics*, *47*(1), 163–193. https://doi.org/10.1146/annurev-fluid-010814-014556.
Le Dizès. (2000). Three-dimensional instability of a multipolar vortex in a rotating flow. *Physics of Fluids*, *12*(11), 2762–2774.
Le Reun, Favier, Barker, & Le Bars. (2017). Inertial wave turbulence driven by elliptical instability. *Physical Review Letters*, *119*(3), 034502. https://doi.org/10.1103/PhysRevLett.119.034502.
Lemasquerier, Grannan, Vidal, Cébron, Favier, Le Bars, & Aurnou. (2017). Libration driven flows in ellipsoidal shells. *Journal of Geophysical Research: Planets*, *122*(9), 1926–1950. https://doi.org/10.1002/2017JE005340.
Lifschitz, & Hameiri. (1991). Local stability conditions in fluid dynamics. *Physics of Fluids A: Fluid Dynamics (1989–1993)*, *3*(11), 2644–2651. https://doi.org/10.1063/1.858153.
Lin, Marti, & Noir. (2015). Shear-driven parametric instability in a precessing sphere. *Physics of Fluids*, *27*(4), 046601.
Lin, Marti, Noir, & Jackson. (2016). Precession-driven dynamos in a full sphere and the role of large scale cyclonic vortices. *Physics of Fluids*, *28*(6), 066601.
Malkus. (1963). Precessional torques as the cause of geomagnetism. *Journal of Geophysical Research*, *68*(10), 2871–2886. https://doi.org/10.1029/JZ068i010p02871.
Malkus. (1968). Precession of the Earth as the cause of Geomagnetism: Experiments lend support to the proposal that precessional torques drive the earth's dynamo. *Science*, *160*(3825), 259–264. https://doi.org/10.1126/science.160.3825.259.
Malkus. (1989). An experimental study of global instabilities due to the tidal (elliptical) distortion of a rotating elastic cylinder. *Geophysical & Astrophysical Fluid Dynamics*, *48*(1-3), 123–134. https://doi.org/10.1080/03091928908219529.
Moffatt. (1970). Dynamo action associated with random inertial waves in a rotating conducting fluid. *Journal of Fluid Mechanics*, *44*(4), 705–719. https://doi.org/10.1017/S0022112070002100.
Ness, Behannon, Lepping, & Whang. (1975). Magnetic field of Mercury confirmed. *Nature*, *255*(5505), 204–205. https://doi.org/10.1038/255204a0.

Noir, Brito, Aldridge, & Cardin. (2001). Experimental evidence of inertial waves in a precessing spheroidal cavity. *Geophysical Research Letters*, *28*(19), 3785–3788. https://doi.org/10.1029/2001GL012956.

Olson. (2015). 8.01 - Core Dynamics: An Introduction and Overview. In Schubert (Ed.) *Treatise on Geophysics* (2nd ed.), pp. 1–25. Elsevier, Oxford. ISBN 978-0-444-53803-1. https://doi.org/10.1016/B978-0-444-53802-4.00137-8.

Pierrehumbert. (1986). Universal short-wave instability of two-dimensional Eddies in an inviscid fluid. *Physical Review Letters*, *57*(17), 2157–2159. https://doi.org/10.1103/PhysRevLett.57.2157.

Rambaux, & Castillo-Rogez. (2013). Tides on satellites of giant planets. In *Tides in Astronomy and Astrophysics*, Lecture Notes in Physics, pp. 167–200. Berlin, Heidelberg: Springer. ISBN 978-3-642-32960-9 978-3-642-32961-6. https://doi.org/10.1007/978-3-642-32961-6.

Reddy, Favier, & Le Bars. (2018). Turbulent kinematic dynamos in ellipsoids driven by mechanical forcing. *Geophysical Research Letters*, *45*(4), 1741–1750. https://doi.org/10.1002/2017GL076542.

Rieutord, Georgeot, & Valdettaro. (2001). Inertial waves in a rotating spherical shell: Attractors and asymptotic spectrum. *Journal of Fluid Mechanics*, *435*, 103–144.

Roberts. (1968). On the thermal instability of a rotating-fluid sphere containing heat sources. *Philosophical Transactions of the Royal Society of London Series A*, *263*(1136), 93–117. https://doi.org/10.1098/rsta.1968.0007.

Sarson, Jones, Zhang, & Schubert. (1997). Magnetoconvection dynamos and the magnetic fields of Io and Ganymede. *Science*, *276*(5315), 1106–1108. https://doi.org/10.1126/science.276.5315.1106.

Sauret, Cébron, Morize, & Le Bars. (2010). Experimental and numerical study of mean zonal flows generated by librations of a rotating spherical cavity. *Journal of Fluid Mechanics*, *662*, 260–268. https://doi.org/10.1017/S0022112010004052.

Sauret, Cébron, & Le Bars. (2013). Spontaneous generation of inertial waves from boundary turbulence in a librating sphere. *Journal of Fluid Mechanics*, *728*. https://doi.org/10.1017/jfm.2013.320.

Sauret, Le Bars, & Le Gal. (2014). Tide-driven shear instability in planetary liquid cores. *Geophysical Research Letters*, *41*(17), 6078–6083. https://doi.org/10.1002/2014GL061434.

Seyed-Mahmoud, Aldridge, & Henderson. (2004). Elliptical instability in rotating spherical fluid shells: Application to Earths fluid core. *Physics of the Earth and Planetary Interiors*, *142*(3–4), 257–282. https://doi.org/10.1016/j.pepi.2004.01.001.

Showman, Malhotra, & Renu. (1999). The Galilean satellites. *Science*, *286*(5437), 77–84. https://doi.org/10.1126/science.286.5437.77.

Stevenson. (2001). Mars' core and magnetism. *Nature*, *412*(6843), 214–219.

Strogatz, S.H. (2016). *Nonlinear Dynamics and Chaos* (2nd Ed.). New York: Avalon Publishing.

Tarduno, et al. (2010). Geodynamo, solar wind, and magnetopause 3.4 to 3.45 billion years ago. *Science*, *327*(5970), 1238–1240. https://doi.org/10.1126/science.1183445.

Tilgner. (2005). Precession driven dynamos. *Physics of Fluids*, *17*, 034104.

Tilgner. (2007a). Kinematic dynamos with precession driven flow in a sphere. *Geophysical and Astrophysical Fluid Dynamics*, *101*(1), 1–9.

Tilgner. (2007b). 8.07 - Rotational dynamics of the core. In Schubert (Eds.), *Treatise on Geophysics*, pp. 207–243. Amsterdam: Elsevier. ISBN 978-0-444-52748-6.

Vidal, & Cébron. (2017). Inviscid instabilities in rotating ellipsoids on eccentric Kepler orbits. *Journal of Fluid Mechanics*, *833*, 469–511. https://doi.org/10.1017/jfm.2017.689.

Vidal, Cébron, Schaeffer, & Hollerbach. (2017). Magnetic fields driven by tidal mixing in radiative stars. arXiv:1711.09612.

Waleffe. (1990). On the three dimensional instability of strained vortices. *Physics of Fluids A: Fluid Dynamics*, *2*(1), 76–80. https://doi.org/10.1063/1.857682.

Weiss, & Tikoo. (2014). The lunar dynamo. *Science*, *346*(6214), 1246753. https://doi.org/10.1126/science.1246753.

Wu, & Roberts. (2009). On a dynamo driven by topographic precession. *Geophysical and Astrophysical Fluid Dynamics*, *103*, 467.
Wu, & Roberts. (2013). On a dynamo driven topographically by longitudinal libration. *Geophysical & Astrophysical Fluid Dynamics*, *107*(1–2), 20–44.
Yarom, & Sharon. (2014). Experimental observation of steady inertial wave turbulence in deep rotating flows. *Nature Physics*, *10*(7), 510.

Chapter 5
Fluid Dynamics of Earth's Core: Geodynamo, Inner Core Dynamics, Core Formation

Renaud Deguen and Marine Lasbleis

Abstract This chapter is built from three 1.5 h lectures given in Udine in April 2018 on various aspects of Earth's core dynamics. The chapter starts with a short historical note on the discovery of Earth's magnetic field and core (section "Introduction"). We then turn to an introduction of magnetohydrodynamics (section "A Short Introduction to Magnetohydrodynamics"), introducing and discussing the *induction equation* and the form and effects of the Lorentz force. Section "The Geometry of Earth's Magnetic Field" is devoted to the description of Earth's magnetic field, introducing its spherical harmonics description and showing how it can be used to demonstrate the internal origin of the geomagnetic field. We then move to an introduction of the convection-driven model of the geodynamo (section "Basics of Planetary Core Dynamics"), discussing our current understanding of the dynamics of Earth's core, obtaining heuristically the Ekman dependency of the critical Rayleigh number for natural rotating convection, and introducing the equations and non-dimensional parameters used to model a convectively driven dynamo. The following section deals with the energetics of the geodynamo (section "Energetics of the Geodynamo"). The final two section deal with the dynamics of the inner core, focusing on the effect of the magnetic field (section "Inner Core Dynamics"), and with the formation of the core (section "Core Formation"). Given the wide scope of this chapter and the limited time available, this introduction to Earth's core dynamics is by no means intended to be comprehensive. For more informations, the interested reader may refer to Jones (2011), Olson (2013), or Christensen and Wicht (2015) on the geomagnetic field and the geodynamo, to Sumita and Bergman (2015), Deguen (2012) and Lasbleis and Deguen (2015) on the dynamics of the inner core, and to Rubie et al. (2015) on core formation.

R. Deguen (✉)
Laboratoire de Géologie de Lyon: Terre, Planètes, Environnement, Université de Lyon, UCBL, ENSL, CNRS, Villeurbanne, France
e-mail: renaud.deguen@univ-lyon1.fr

M. Lasbleis
Laboratoire de Planétologie et Géodynamique, UMR-CNRS 6112, Université de Nantes, Nantes, France

M. Le Bars and D. Lecoanet (eds.), *Fluid Mechanics of Planets and Stars*,
CISM International Centre for Mechanical Sciences 595,
https://doi.org/10.1007/978-3-030-22074-7_5

Introduction

The Birth of Geomagnetism

Magnetism, the study of magnetic fields, may be argued to start with the discovery and description of natural magnets or lodestones. Magnets have been known since at least ~600 BC in Greece, and ~300 BC in ancient China. Aristote (384–322 BC) attributes to Thales (~600 BC) the observation that "magnets exert a force on iron". The birth of *geomagnetism* requires the realisation that there is an ambient magnetic field at Earth's scale. The primary observation is that a magnet allowed to rotate freely in a horizontal plane points toward the north (as defined by the direction of the polar star). This seems to have been known since at least the Ist century in China; the compass was used in navigation since at least the Xth century in China, and since the XIIth century in Europe.

In the western world, the first scientific treatise describing the properties of magnets is a letter (*Epistola*) written by Pierre de Maricourt (also known as Petrus Peregrinus) in 1269 to a friend of him, Sygerus de Foucaucourt. Pierre de Maricourt wrote his *epistola* during a military campaign led by Charles d'Anjou, while the French army was besieging the town of Lucera, in southern Italia. In his letter, Pierre de Maricourt describes some of the most important properties of natural magnets: (i) a magnet has two poles (North and South); (ii) the South pole of a magnet attracts the North pole of another magnet, two identical poles repel each other; (iii) the two poles of a magnet cannot be isolated: breaking a magnet into two pieces gives two magnets, each with two poles (in modern terms, this means that there is no magnetic *monopole*); (iv) a magnet free to rotate points toward the North pole. In the second part of his letter, Pierre de Maricourt describes the design of a perpetual motion machine using the properties of magnets.

The exact geometry of Earth's magnetic field turns out to be more complex than what early observations suggested:

1. Compasses do not point exactly toward the geographic north. The *magnetic declination*—the angle between the direction of the true geographic north and the direction given by a compass—has been known since at least the end of XIth's century in China. Christopher Columbus was the first to observe that the magnetic declination varies spatially: while sailing westward from the old world, he observed that the direction given by his compasses changed with longitude along his path. The declination was negative in Europe (i.e. a compass points slightly to the west of the true north), decreased gradually along Columbus' path, reached zero somewhere in the middle of the Atlantic ocean, and became positive on the western side of the Atlantic.
2. Furthermore, the actual direction of the magnetic field is tilted from the horizontal plane: when allowed to rotate around an horizontal axis, a magnetised needle points toward a direction which makes a well-defined angle with the horizontal,

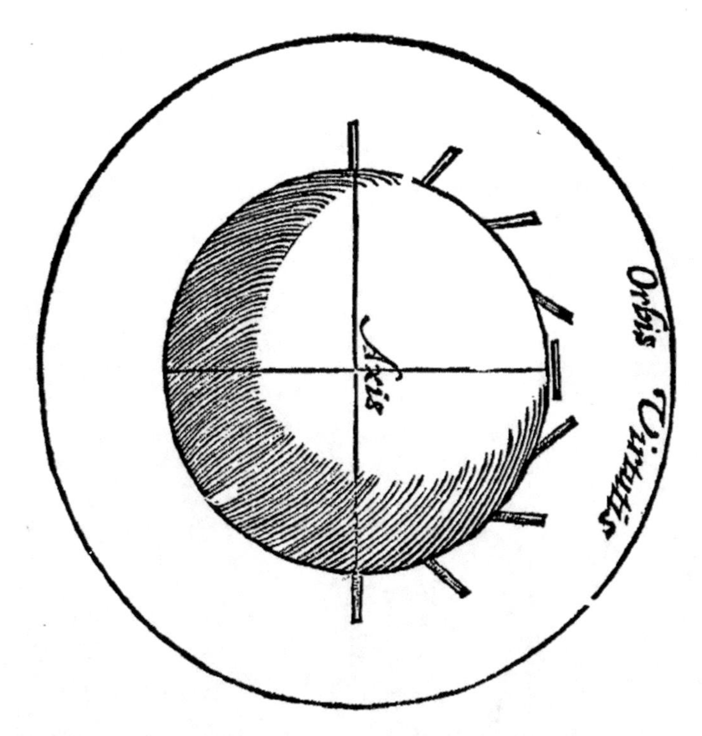

Fig. 5.1 Gilbert's *terella*, a natural magnet carved into a spherical shape (*De Magnete*, 1600). The direction of the magnetic field at the surface of the *terella* is indicated on the right-hand side of the figure: the magnetic field is perpendicular to the surface at the poles, and parallel to the surface at the equator

called the *magnetic inclination*. This has been noticed by Georg Hartmann in 1544 and precisely measured by Robert Norman, who found a declination of 71°50' in London in 1581.

Perhaps the first scientifically grounded theory of the origin of the geomagnetic field is Gilbert's theory published in 1600 in his *De Magnete*, which states that *Earth is a magnet*. Gilbert's claim was based on the similarity between the latitudinal variations of the inclination of Earth's magnetic field and the variations of the magnetic field orientation he measured on a model of Earth, his *terella*, obtained by carving a lodestone into the shape of a sphere (Fig. 5.1). The geometry of Earth's magnetic field indeed strongly resembles that of a magnet, but it has soon been clear that this theory is at best incomplete. Perhaps the strongest argument against Gilbert's theory comes from the observation that the geomagnetic field varies with time. Measurements of the declination in London across the XVIIth century have shown variations of several degrees in a few decades; variations of the declination and inclination on a decadal timescale have been well documented by the mid XVIIIth century. Other evidences of fast magnetic field variations include the westward drift of a line of zero declination inferred to be in the middle of the Atlantic ocean in 1701 by Halley (Fig. 5.2), which had reached the South American continent about 60 years after the

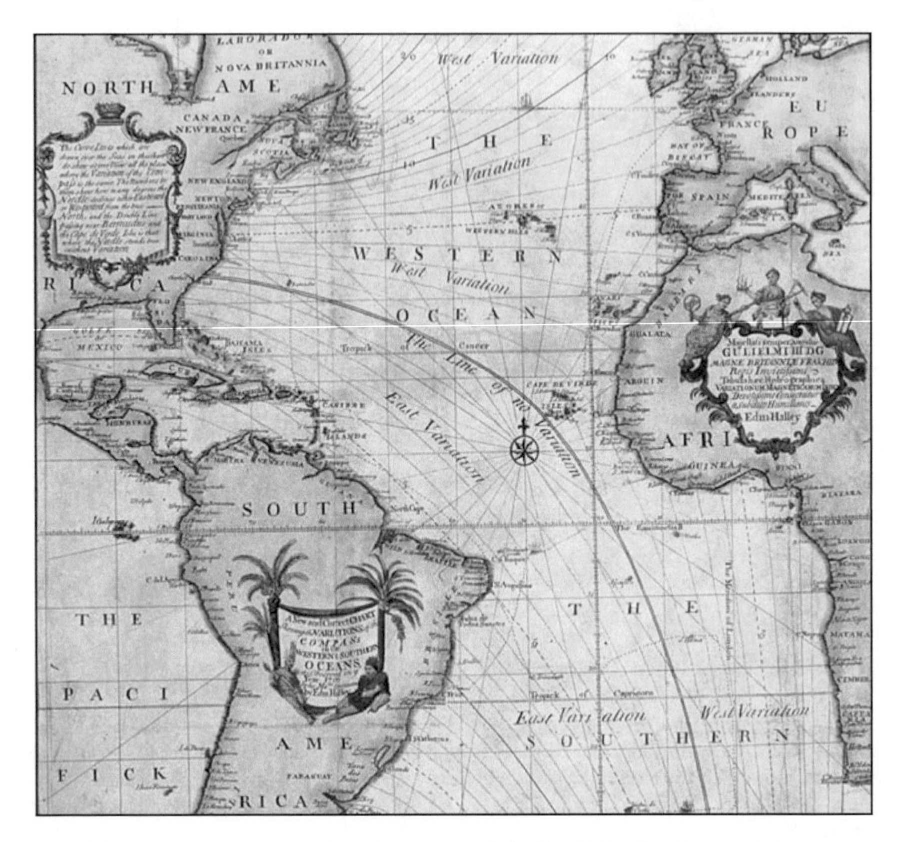

Fig. 5.2 Halley's map of magnetic declination, published in 1701. The thick black line with the annotation "The line of no Variation" is the line of zero declination

publication of Halley's map. Modern measurements and inferences have confirmed that the geomagnetic field varies on timescales spanning a very wide range, including monthly to decadal variations. Such fast time variations are hardly compatible with Gilbert's theory.

The Discovery of the Core

The basic structure of the Earth—a solid, rocky mantle surrounding an iron-alloy core which is molten except for a solid inner core at its center—has not been known before the early XXth century. In the XVIIIth and early XIXth centuries, it was widely believed that most of the Earth is molten, except for a thin crust on which we live. A qualitative argument in favour of this view is the observation of molten lava erupting from volcanoes, which is straightforwardly explained if the interior of the Earth is indeed molten. A more quantitative argument is given by the observation that the temperature increases with depth in mines, at a rate such that the melting temperature of most crustal rocks would be reached at a depth lower than 50 km.[1]

This view was challenged by scientists including Ampère, Poisson, or Kelvin, Ampère arguing that such a thin crust cannot withstand the deviatoric stresses induced by the lunar tides. The fact that the interior of the Earth is solid to a large extent has been demonstrated in the middle of the XIXth century by Lord Kelvin and George Darwin (the son of Charles), by considering the deformation of the interior of the Earth in response to tidal forcing. The oceanic tides that we see at Earth's surface are the relative motion of the oceans with respect to the crust. We would not see them if the Earth was fully molten since it would deform in the same way as the oceans. The fact that we do see oceanic tides means that the interior of the Earth resists deformation, i.e. has a finite rigidity. Calculations based on available data led Kelvin to state that "the Earth is as rigid as steel".

The first quantitative model of the interior of the Earth including a rocky mantle and an iron-rich core was proposed in 1896 by Emil Wiechert based on the following constraints and assumptions. The mean density of the Earth (5.6, obtained from the value of g at Earth's surface) is significantly higher than the density of rocks near the surface ($\sim$3). Wiechert believed that compression alone cannot increase the density of rocks to such extent, and argued that a change of chemical composition at some depth is required. Since only metals were known to have densities higher than 5.6, it was natural for Wiechert to assume the deepest layer (the core) of his model to be metallic. The estimated moment of inertia of the Earth provides further constraints on the repartition of mass within the Earth. By considering a two-layer model with constant density in each layer, the constraints coming from the mean density and moment of inertia of the Earth enable to determine the size of the core and densities of the core and mantle. Wiechert obtained a core radius of about 4970 km, and mantle and core densities of 3.2 and 8.2, respectively. His estimate of the core density being close to that of iron at low pressure (7.8), Wiechert proposed that the core is composed of iron, and attributes the difference of density to compression.

The end of XIXth century and the beginning of the XXth century witnessed the emergence of seismology as a new and growing branch of geophysics. Milne, Wiechert, Oldham and others realised that the study of seismic waves[2] propagating within the Earth offers a way to probe the interior of the Earth. By studying the time of arrival of P-waves as a function of the distance from Earthquakes' epicenters, Oldham (1906) discovered that no P-wave was observed at epicentral distances between $\sim 105^{\circ}$ and $\sim 140^{\circ}$. He explained this *shadow zone* by the presence of a discontinuity (which he interpreted as the boundary between the core and mantle of Wiechert's model), at which the wave propagation velocity decreases, thus refracting the waves downward. Gutenberg (who was a student of Wiechert) estimated the radius of the core to be about 3470 ± 20 km, which is consistent with the modern estimate of

[1]The temperature actually does not reach the melting temperature because (i) the melting temperature increases with pressure; and (ii) the high-temperature gradient observed in mines is limited to depth of $\sim$30 km or less, the temperature gradient becoming much less pronounced at deeper depth because of convective motions in the mantle.

[2]Compression waves, called P-waves, and shear waves, called S-waves.

3480 km. At that time, the core was believed to be solid, mostly because of Kelvin and Darwin's estimates of the rigidity of the Earth. The molten state of the core has been established by Jeffrey in 1926 by comparing the mean rigidity of the Earth to the rigidity of the mantle obtained from estimates of the speed of propagation of seismic waves in the mantle. The mean rigidity of the Earth is significantly smaller than the rigidity of the mantle, which implies that the rigidity of the core must be low.

The final major piece of the basic structure of the Earth has been discovered by Inge Lehmann, a Danish seismologist. In spite of what has been said in the previous paragraph, P-waves are sometimes observed in the core shadow zone. These anomalous arrivals, called P', have been attributed by Lehmann (1936) to P-waves reflected at a discontinuity inside the core at which the seismic waves velocity increases. She thus discovered the *inner core*, a sphere 1221 km in radius (Inge Lehmann's estimate was 1400 km), which was later shown to be solid (Dziewonski and Gilbert 1971).

By comparing seismological estimates of the density and elastic moduli in the core to the results of high pressure experiments, later work established that the liquid and solid parts of the core are both made predominantly of iron, alloyed with lighter elements (Si, O, S, C, H, etc.) (e.g. Birch 1952, 1964; Jephcoat and Olson 1987; Alfè et al. 2000; Badro et al. 2007). These light elements are more abundant in the liquid outer core (about 10 wt.%) than in the solid inner core (less than 5 wt.%), which is consistent with the idea that the inner core is crystallising from the outer core (Birch 1940; Jacobs 1953) since the impurities in an alloy are usually partitioned into the melt upon crystallisation.

A Short Introduction to Magnetohydrodynamics

Magnetohydrodynamics concerns the interactions between the flow of an electrically conducting fluid and a magnetic field. It thus combines hydrodynamics, which is modelled with the Navier–Stokes and continuity equations, and electromagnetism, which is modelled with Maxwell's equations, supplemented with Ohm's law and the conservation of electric charges. Interactions between the flow and magnetic field are twofold: (i) the motion of an electrically conducting material can act as a source term for the evolution of the magnetic field, which can be quantified with the *induction equation*, a prognostic equation for the evolution of the magnetic field; and (ii) the magnetic field generates a force on the electrically conducting material, the *Lorentz force*. The goal of this section is to introduce the induction equation, and discuss the form and effects of the Lorentz force.

Classical Electromagnetism

Maxwell's equations are

$$\nabla \cdot \mathbf{E} = \frac{\rho_c}{\epsilon_0} \qquad \text{Gauss law,} \qquad (5.1)$$

$$\nabla \times \mathbf{E} = -\frac{\partial \mathbf{B}}{\partial t} \qquad \text{Faraday's law of induction,} \tag{5.2}$$

$$\nabla \cdot \mathbf{B} = 0 \qquad \text{No magnetic monopole,} \tag{5.3}$$

$$\nabla \times \mathbf{B} = \mu_0 \mathbf{j} + \mu_0 \epsilon_0 \frac{\partial \mathbf{E}}{\partial t} \qquad \text{Ampere's law + displacement current,} \tag{5.4}$$

where $\mathbf{E}$ is the electric field, $\mathbf{B}$ is the magnetic field, ρ_c is the electric charge density, $\mathbf{j}$ is the electric current density, μ_0 is the permeability of free space, and ϵ_0 is the permittivity of free space. Note that $\mu_0\epsilon_0 = 1/c^2$, where c is the speed of light. Maxwell's equations are supplemented by an equation expressing that electric charges are conserved, which we write

$$\frac{\partial \rho_c}{\partial t} + \nabla \cdot \mathbf{j} = 0, \tag{5.5}$$

and by Ohm's law, which can be written in a moving electrically conducting material as

$$\mathbf{j} = \rho_c \mathbf{u} + \sigma \left(\mathbf{E} + \mathbf{u} \times \mathbf{B} \right), \tag{5.6}$$

where σ is the electrical conductivity, and $\mathbf{u}$ the velocity of the conducting material.

The Charge Density in an Electrically Conducting Material

Replacing $\mathbf{j}$ in Eq. (5.5) by its expression from Ohm's law and then using Gauss law [Eq. (5.1)], we obtain the following equation for ρ_c,

$$\frac{\epsilon_0}{\sigma}\frac{D\rho_c}{Dt} + \rho_c = -\epsilon_0 \nabla \cdot (\mathbf{u} \times \mathbf{B}), \tag{5.7}$$

which means that the charge density follows time variations of $\epsilon_0 \nabla \cdot (\mathbf{u} \times \mathbf{B})$ with a response time, or time lag, ϵ_0/σ. In most liquid metals, ϵ_0/σ is between 10^{-16} and 10^{-18} s and the charge density responds almost instantaneously to changes in $\mathbf{u}$ and $\mathbf{B}$. The charge density must therefore be very close to

$$\rho_c = -\epsilon_0 \nabla \cdot (\mathbf{u} \times \mathbf{B}). \tag{5.8}$$

This estimate can be used to discuss the importance of the term $\rho_c\mathbf{u}$ in Ohm's law. Denoting by L and U typical length and velocity scales of a given problem, comparing $\rho_c\mathbf{u}$ and $\sigma\mathbf{u} \times \mathbf{B}$ shows that

$$\frac{|\rho_c \mathbf{u}|}{|\sigma \mathbf{u} \times \mathbf{B}|} \sim \frac{|\epsilon_0 \nabla \cdot (\mathbf{u} \times \mathbf{B})\, \mathbf{u}|}{|\sigma \mathbf{u} \times \mathbf{B}|} \sim \frac{\epsilon_0/\sigma}{L/U}. \tag{5.9}$$

The term $\rho_c \mathbf{u}$ can therefore be neglected if the macroscopic timescale L/U is large compared to the charge response time ϵ_0/σ. Since, again, $\epsilon_0/\sigma \sim 10^{-18} - 10^{-16}$ s, $\rho_c \mathbf{u}$ can be neglected in any situation typically encountered in MHD problems, including planetary core dynamics. Ohm's law can therefore be written as

$$\mathbf{j} = \sigma \left(\mathbf{E} + \mathbf{u} \times \mathbf{B} \right). \tag{5.10}$$

The Non-relativistic Limit

When applied to the geodynamo problem (as well as most MHD problems), Ampère's law [Eq. (5.4)] can be simplified as follows in the non-relativistic limit. Consider a problem with $\mathbf{B}$ and $\mathbf{E}$ having magnitude B and E varying on typical lengthscale L and timescale T. The ratio of the displacement current $\mu_0 \epsilon_0 \partial_t \mathbf{E}$ to the curl of $\mathbf{B}$ is on the order of

$$\frac{\left| \dfrac{1}{c^2} \dfrac{\partial \mathbf{E}}{\partial t} \right|}{|\nabla \times \mathbf{B}|} \sim \frac{L}{c^2 T} \frac{E}{B}. \tag{5.11}$$

Using Faraday's law of induction [Eq. (5.2)], which implies that $E/L \sim B/T$, we obtain

$$\frac{\left| \dfrac{1}{c^2} \dfrac{\partial \mathbf{E}}{\partial t} \right|}{|\nabla \times \mathbf{B}|} \sim \left(\frac{L/T}{c} \right)^2. \tag{5.12}$$

The ratio L/T is an estimate of typical velocities, which in the non-relativistic limit is $\ll c$. This implies that

$$\left| \frac{1}{c^2} \frac{\partial \mathbf{E}}{\partial t} \right| \ll |\nabla \times \mathbf{B}|. \tag{5.13}$$

This will always be assumed here, and we will use the non-relativistic Maxwell's equations:

$$\nabla \cdot \mathbf{E} = \frac{\rho_c}{\epsilon_0} \qquad \text{Gauss law,} \tag{5.14}$$

$$\nabla \times \mathbf{E} = -\frac{\partial \mathbf{B}}{\partial t} \qquad \text{Faraday's law of induction,} \tag{5.15}$$

$$\nabla \cdot \mathbf{B} = 0 \qquad \text{No magnetic monopole,} \tag{5.16}$$

$$\nabla \times \mathbf{B} = \mu_0 \mathbf{j} \qquad \text{Ampere's law.} \tag{5.17}$$

The Induction Equation

A prognostic equation for **B** can be obtained from Ohm's law and Maxwell's equations. Taking the curl of Ohm's law and using Faraday and Ampère's laws, one can obtain the so-called *induction equation*, which writes

$$\frac{\partial \mathbf{B}}{\partial t} = \nabla \times (\mathbf{u} \times \mathbf{B}) + \eta \nabla^2 \mathbf{B}, \tag{5.18}$$

where η, the *magnetic diffusivity*, is defined as

$$\eta = \frac{1}{\mu_0 \sigma}. \tag{5.19}$$

An alternative—and sometimes more useful—form of the induction equation can be obtained by noting that $\nabla \times (\mathbf{u} \times \mathbf{B}) = (\mathbf{B} \cdot \nabla)\,\mathbf{u} - (\mathbf{u} \cdot \nabla)\,\mathbf{B}$, which allows to transform Eq. (5.18) into

$$\frac{\partial \mathbf{B}}{\partial t} + \underbrace{(\mathbf{u} \cdot \nabla)\,\mathbf{B}}_{\text{Advection}} = \underbrace{(\mathbf{B} \cdot \nabla)\,\mathbf{u}}_{\text{Stretching}} + \eta \nabla^2 \mathbf{B}, \tag{5.20}$$

or

$$\frac{D\mathbf{B}}{Dt} = (\mathbf{B} \cdot \nabla)\,\mathbf{u} + \eta \nabla^2 \mathbf{B}. \tag{5.21}$$

This shows that the Lagrangian derivative of **B** depends on a competition between two terms: a *stretching* term $(\mathbf{B} \cdot \nabla)\,\mathbf{u}$, and a diffusion term $\eta \nabla^2 \mathbf{B}$. Diffusion will always tend to smooth out spatial variations of **B**. In contrast, the stretching term can increase the magnitude of **B**.

To estimate the relative importance of these two terms, we denote by B and U the magnitudes of the magnetic and velocity fields, which are both assumed to vary over the same length scale L. Forming the ratio of the stretching and diffusion terms, we obtain

$$\frac{|\nabla \times (\mathbf{u} \times \mathbf{B})|}{|\eta \nabla^2 \mathbf{B}|} \sim \frac{UB/L}{\eta B/L^2} = \frac{UL}{\eta} = R_m, \tag{5.22}$$

which defines the magnetic Reynolds number R_m. R_m is a measure of the relative importance of stretching and advection of the magnetic field to its diffusion (it could also have been called a "magnetic Péclet number").

To understand the induction equation, we will first consider the limits of small and large R_m, before considering a more general case.

The $R_m \ll 1$ Limit

If $R_m \ll 1$, the advection and stretching terms are both negligible compared to the diffusion term, and the induction equation becomes a simple diffusion equation:

$$\frac{\partial \mathbf{B}}{\partial t} = \eta \nabla^2 \mathbf{B}. \tag{5.23}$$

In this limit, a magnetic field varying over a length scale L would smooth out by diffusion in a timescale

$$\tau_\eta = \frac{L^2}{\eta} = \mu_0 \sigma L^2. \tag{5.24}$$

In the Earth's core ($L \sim 3000$ km, $\eta \sim 1$ m^2.s^{-1}), the diffusion timescale is $\tau_\eta \sim$ 300 000 years. Properly taking into account the spherical geometry of the core yields a smaller timescale, $\tau_\eta = L^2/(\pi^2 \eta) \sim 30\,000$ years, which implies that a spatially varying magnetic field can perdure only during a few tenths of thousand years. Since this is small compared to the age of the Earth (4.56 Gy), this implies that the geomagnetic field cannot be of primordial origin.

In contrast, the magnetic field would diffuse very rapidly in any reasonable size laboratory experiment: in a one meter size experiment using liquid sodium (which has a very low magnetic diffusivity, $\eta \sim 0.1$ m^2), the diffusion time is $\sim$10 s.

The $R_m \gg 1$ Limit

At $R_m \gg 1$, the induction equation becomes:

$$\frac{D\mathbf{B}}{Dt} = \frac{\partial \mathbf{B}}{\partial t} + \underbrace{(\mathbf{u} \cdot \nabla)\,\mathbf{B}}_{\text{Advection}} = \underbrace{(\mathbf{B} \cdot \nabla)\,\mathbf{u}}_{\text{Stretching}}. \tag{5.25}$$

Let us consider the evolution of an initially uniform magnetic field $\mathbf{B} = B_0 \mathbf{e}_z$ in response to a flow with velocity $\mathbf{u} = (u_x, u_z)$. At time $t = 0$, at which $\mathbf{B} = B_0 \mathbf{e}_z$, the term $(B \cdot \nabla)\,\mathbf{u}$ writes

$$B_0 \begin{bmatrix} \partial_z u_x \\ \partial_z u_z \end{bmatrix} \tag{5.26}$$

From this one can see that:

1. shearing the magnetic field ($\mathbf{u} = u_x(z)\mathbf{e}_x$) produces a magnetic field component perpendicular to the initial direction of the magnetic field. Solving the induction equation with a simple shear flow ($u_x = \dot{\gamma} z$, where $\dot{\gamma}$ is the shear rate) gives

$$B_x = \dot{\gamma} B_0 t, \tag{5.27}$$

$$B_z = B_0. \tag{5.28}$$

The magnetic lines are tilted by the shear, and tends to be aligned with the shear direction. Note also that $|\mathbf{B}|$ increases as $|\mathbf{B}| = B_0\sqrt{1 + (\dot{\gamma}t)^2}$. There is therefore a net production—and not just a re-distribution—of magnetic energy.

2. stretching (resp. compressing) the magnetic field increases (resp. decreases) its magnitude. Solving the induction equation with $\mathbf{u} = u_z(z)\mathbf{e}_z$ gives

$$B_x = 0, \tag{5.29}$$

$$B_z = B_0 \exp\left(\int_0^t \frac{\partial u_z}{\partial z} dt\right). \tag{5.30}$$

A constant $\partial u_z/\partial z$ results in exponential growth (if it is positive) or decrease (if negative) of B_z.

In the limit of infinite R_m, one can obtain two useful theorems:

Helmholtz theorem: *magnetic lines are material lines.*[3]

Alfven's frozen flux theorem: *a magnetic tube is material and its magnetic flux is conserved: denoting by S any cross-section of a magnetic tube, then the magnetic flux* $\int_S \mathbf{B} \cdot d\mathbf{S}$ *(which is constant along the tube because* $\nabla \cdot \mathbf{B} = 0$*) does not vary with time.*[4]

---

[3]**Proof**: consider a small vector $\boldsymbol{\delta}$ having material end-points, advected by the flow. At time $t + dt$ this small vector will be equal to

$$\boldsymbol{\delta}(t + dt) = -\mathbf{u}(\mathbf{x}, t)dt + \boldsymbol{\delta}(t) + \mathbf{u}(\mathbf{x} + \boldsymbol{\delta}, t)dt, \tag{5.31}$$

$$= \boldsymbol{\delta}(t) + \boldsymbol{\delta} \cdot \nabla \mathbf{u}\, dt + \mathcal{O}(\delta^2). \tag{5.32}$$

Taking the $dt \to 0$, $\boldsymbol{\delta} \to 0$ limit gives

$$\frac{D\boldsymbol{\delta}}{Dt} = (\boldsymbol{\delta} \cdot \nabla)\, \mathbf{u}. \tag{5.33}$$

$\boldsymbol{\delta}$ therefore evolves according to the same equation as $\mathbf{B}$ [Eq. (5.25)].

[4]**Proof**: The fact that a magnetic tube is material follows directly from Helmholtz theorem. To show that its magnetic flux does not vary with time, write its time derivative as

$$\frac{d}{dt}\int_S \mathbf{B} \cdot d\mathbf{S} = \int_S \frac{\partial \mathbf{B}}{\partial t} \cdot d\mathbf{S} + \oint_C \mathbf{B} \cdot \mathbf{u} \times d\mathbf{l}, \tag{5.34}$$

use the diffusion-free induction equation to write the first term on the RHS as

$$\int_S \frac{\partial \mathbf{B}}{\partial t} \cdot d\mathbf{S} = \int_S \nabla \times (\mathbf{u} \times \mathbf{B}) \cdot d\mathbf{S}, \tag{5.35}$$

and the identity $(\mathbf{A} \times \mathbf{B}) \cdot \mathbf{C} = -\mathbf{B} \cdot (\mathbf{A} \times \mathbf{C})$ plus Stokes' theorem to write the second term as

An Example of a Kinematic Solution of the Full Induction Equation

Let us now consider one simple example of solution of the full induction equation. We consider a velocity field of the form $\mathbf{u} = U \sin(2\pi z/L)\mathbf{e}_x$ (i.e. a pure shear flow with a shear rate varying periodically with z). With a velocity field of this form, the induction equation writes

$$\frac{\partial B_x}{\partial t} = 2\pi \frac{U}{L} \cos(2\pi z/L) B_z + \eta \frac{\partial^2 B_x}{\partial z^2}, \tag{5.38}$$

$$\frac{\partial B_z}{\partial t} = \eta \frac{\partial^2 B_z}{\partial z^2}. \tag{5.39}$$

The magnetic field is assumed to be initially uniform, $\mathbf{B}(t=0) = B_0 \mathbf{e}_z$. Solving these equations (by looking for a solution of the form $B_x = f(t) \cos(2\pi z/L)$) gives

$$B_x = B_0 R_m \left(1 - \mathrm{e}^{-t/\tau}\right) \cos(2\pi z/L), \tag{5.40}$$

$$B_z = B_0, \tag{5.41}$$

where

$$R_m = \frac{UL}{\eta}, \quad \tau = \frac{1}{4\pi^2} \frac{L^2}{\eta}. \tag{5.42}$$

The magnitude of B_x increases linearly with t when $t \ll \tau$, and saturates at a value $R_m B_0$ when vertical diffusion balances magnetic field production, which happens at $t \sim \tau$. Compared to the infinite R_m limit, there is still a tendency to align $\mathbf{B}$ with the shear, but this is now mitigated by diffusion.

The Lorentz Force

Experiments show that a charged particle (q) moving with velocity $\mathbf{u}$ in an electric field $\mathbf{E}$ and magnetic field $\mathbf{B}$ experiences a force

$$\mathbf{F}_L = q\,(\mathbf{E} + \mathbf{u} \times \mathbf{B}) \tag{5.43}$$

$$\oint_C \mathbf{B} \cdot \mathbf{u} \times d\mathbf{l} = -\oint_C (\mathbf{u} \times \mathbf{B}) \cdot d\mathbf{l}, \tag{5.36}$$

$$= -\int_S \nabla \times (\mathbf{u} \times \mathbf{B}) \cdot d\mathbf{S}, \tag{5.37}$$

which is equal to the opposite of the first term on the RHS of Eq. (5.34).

called the Lorentz force. This can be generalised to the case of a charged continuous medium: electric and magnetic fields produce a volumetric force given by

$$\mathbf{f}_L = \rho_c \mathbf{E} + \mathbf{j} \times \mathbf{B}. \tag{5.44}$$

The Non-relativistic Limit

In the non-relativistic limit, one can first use Faraday's law (without the current displacement) to write

$$\mathbf{f}_L = \rho_c \mathbf{E} + \frac{1}{\mu_0}(\nabla \times \mathbf{B}) \times \mathbf{B}. \tag{5.45}$$

Using the estimate of ρ_c for electrical conductors [Eq. (5.8)], the relative importance of the two terms of the Lorentz force can be estimated as

$$\frac{|\rho_c \mathbf{E}|}{\left|\frac{1}{\mu_0}(\nabla \times \mathbf{B}) \times \mathbf{B}\right|} \sim \frac{|\epsilon_0 \nabla \cdot (\mathbf{u} \times \mathbf{B})\, \mathbf{E}|}{\left|\frac{1}{\mu_0}(\nabla \times \mathbf{B}) \times \mathbf{B}\right|} \sim \frac{UE}{B/(\mu_0 \epsilon_0)} \sim \left(\frac{U}{c}\right)^2, \tag{5.46}$$

where we have used the scaling relation $E/B \sim L/T \sim U$ obtained from Faraday's law. In a non-relativistic problem, we can therefore safely neglect the $\rho_c \mathbf{E}$ term and write the Lorentz force as

$$\mathbf{f}_L = \frac{1}{\mu_0}(\nabla \times \mathbf{B}) \times \mathbf{B}. \tag{5.47}$$

Magnetic Pressure and Tension

A bit of algebra on $(\nabla \times \mathbf{B}) \times \mathbf{B}$ allows to decompose the Lorentz force into two terms:

$$\mathbf{f}_L = \underbrace{-\nabla\left(\frac{B^2}{2\mu_0}\right)}_{\text{magnetic pressure}} + \underbrace{\frac{1}{\mu_0}(\mathbf{B} \cdot \nabla)\, \mathbf{B}}_{\text{magnetic tension}}. \tag{5.48}$$

The first term acts as a pressure term : a force equal to the gradient of the *magnetic pressure* $B^2/(2\mu_0)$. The second term is called *magnetic tension*, for reasons which will be explained below. This decomposition is often useful because in many cases the gradient of the magnetic pressure can be balanced to a large extent by pressure gradients, so that only the magnetic tension has a strong dynamic effect (see section "Deformation Induced by the Lorentz Force" for example). Only the magnetic tension can produce vorticity.

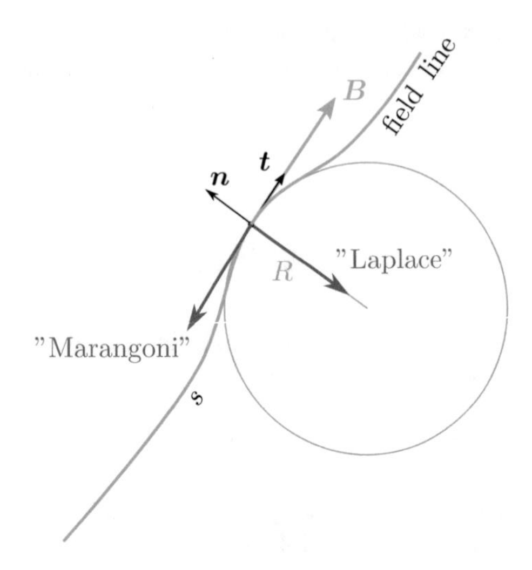

Fig. 5.3 The magnetic tension induced by bending a field line can be decomposed as the sum of a "Laplace" term oriented perpendicularly to the field line, with a magnitude proportional to the field line curvature, and of a "Marangoni" term, which acts parallel to the field line

We restrict here our analysis to 2D for the sake of simplicity, but the following reasoning can be generalised to 3D. Let us consider a curvilinear Frenet coordinate system $(\mathbf{t}, \mathbf{n})$ attached to a given magnetic field line (Fig. 5.3). We denote by s the coordinate along this field line. In this coordinate system, we can write

$$\frac{1}{\mu_0}(\mathbf{B}\cdot\nabla)\,\mathbf{B} = \frac{B}{\mu_0}\frac{\partial}{\partial s}(B\mathbf{t}), \tag{5.49}$$

$$= \frac{B^2}{\mu_0}\frac{\partial \mathbf{t}}{\partial s} + \frac{\partial}{\partial s}\left(\frac{B^2}{2\mu_0}\right)\mathbf{t}, \tag{5.50}$$

$$= -2\frac{B^2}{2\mu_0}\mathcal{K}\mathbf{n} + \nabla_{\mathbf{t}}\left(\frac{B^2}{2\mu_0}\right), \tag{5.51}$$

where $\mathcal{K} = \nabla\cdot\mathbf{n}$ is the (signed) curvature of the field line, the derivative of $\mathbf{t}$ according to s being equal to $-\mathcal{K}\mathbf{n}$. The operator $\nabla_{\mathbf{t}}(...) = \nabla(...) - \mathbf{n}(\mathbf{n}\cdot\nabla)(...)$ is the component of the gradient parallel to the field line.

This equation bears a strong resemblance with the expression of the stress jump induced by interfacial tension across the interface between two immiscible fluids, which is equal to

$$-\gamma\mathcal{K}\mathbf{n} + \nabla_{\mathbf{t}}\gamma, \tag{5.52}$$

where γ is the interfacial tension. The first term is the *Laplace pressure* term, which expresses the fact that interfacial tension induces a pressure jump across a curved interface. This pressure jump is proportional to the interfacial tension and to the curvature of the interface. The second term is a stress tangential to the interface associated with gradients of interfacial tension, which is responsible of *Marangoni* effects.

Equation (5.51) has a similar mathematical form with $B^2/(2\mu_0)$ taking the role of interfacial tension. The analog of the *Laplace pressure* implies that deforming a magnetic line induces a volumetric force proportional to B^2 normal to the line, and directed toward the center of curvature of the field lines. This force thus acts against deformation of the magnetic lines and tends to straighten them. The analog of the *Marangoni* tension is a force acting parallel to the magnetic lines toward regions of higher magnetic field intensity. It produces a tension parallel to the fields lines if the magnetic field intensity varies along field lines.

Putting back the magnetic pression and magnetic tension terms together, the Lorentz force writes

$$\mathbf{j} \times \mathbf{B} = -\nabla\left(\frac{B^2}{2\mu_0}\right) + \frac{1}{\mu_0}\left(\mathbf{B}\cdot\nabla\right)\mathbf{B}, \tag{5.53}$$

$$= -\nabla\left(\frac{B^2}{2\mu_0}\right) + \nabla_{\mathbf{t}}\left(\frac{B^2}{2\mu_0}\right) - \frac{B^2}{\mu_0}\mathcal{K}\mathbf{n}, \tag{5.54}$$

$$= -\nabla_{\mathbf{n}}\left(\frac{B^2}{2\mu_0}\right) - \frac{B^2}{\mu_0}\mathcal{K}\mathbf{n}, \tag{5.55}$$

where $\nabla_{\mathbf{n}}$ is the component of the gradient normal to the field line. Both terms acts perpendicularly to the field lines (consistently with the $\frac{1}{\mu_0}(\nabla \times \mathbf{B}) \times \mathbf{B}$ expression of the Lorentz force).

Alfvèn Waves

The relationship between the Lorentz force and the magnetic field lines curvature suggests that a magnetised fluid can carry waves: if magnetic lines are deformed from an initial equilibrium state by a flow with velocity $\mathbf{u}$, deformation of the lines will produce a restoring force (the Lorentz force) normal to the line and of magnitude proportional to the curvature of the lines (Fig. 5.4). Only a velocity field perpendicular

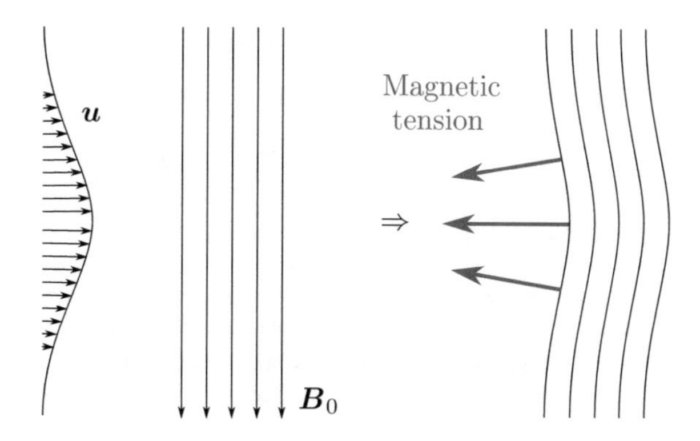

Fig. 5.4 The bending of magnetic field lines produces a Lorentz force acting against the flow that deformed the field lines

to the magnetic lines would curve the lines and thus produce a restoring force. Taking into account that *a velocity perturbation propagating as a plane wave in an incompressible fluid must be a transverse wave*,[5] We can thus expect velocity perturbations perpendicular to the field lines to propagate in the form of a transverse wave travelling in the direction of magnetic lines. These are called *Alfvèn waves*. The celerity of these waves can be obtained from dimensional analysis: the propagation velocity v_A can be a function of the magnetic field intensity B_0 [T = kg s^{-2} A^{-1}], of the magnetic permeability μ_0 [m kg s^{-2} A^{-2}], of the density ρ [kg m^{-3}] of the fluid, and of the wave number k [m^{-1}] of the wave, which are the only parameters of the problem. There is only one way to build a group of parameters with the dimension of velocity from this list of parameters (this is proved with the Vaschy-Buckingham theorem), which is

$$v_A \sim \frac{B_0}{\sqrt{\rho\mu_0}}. \tag{5.58}$$

An important point is that it is not possible to build a velocity involving the wave number k, which implies that Alfvèn waves must be non-dispersive.

A classical way of deriving the dispersion equation of these waves is to work directly from the Navier–Stokes (including the Lorentz force) and induction equations.[6] As an alternative, we will here obtain the wave equation from an analysis based on magnetic tension. We consider a conducting, incompressible fluid of uniform density permeated by a magnetic field, and neglect magnetic diffusion in the induction equation (thus taking the infinite R_m limit). Noting that $\mathbf{u}\cdot\boldsymbol{\nabla}\mathbf{u} = 0$ (since we are considering plane transverse waves), the Navier–Stokes and induction equations then write

$$\rho\frac{\partial \mathbf{u}}{\partial t} = -\boldsymbol{\nabla} p - \boldsymbol{\nabla}_{\mathbf{n}}\left(\frac{B^2}{2\mu_0}\right) - \frac{B^2}{\mu_0}\mathcal{K}\mathbf{n}, \tag{5.59}$$

$$\frac{\partial \mathbf{B}}{\partial t} = (\mathbf{B}\cdot\boldsymbol{\nabla})\,\mathbf{u} - (\mathbf{u}\cdot\boldsymbol{\nabla})\,\mathbf{B}. \tag{5.60}$$

---

[5]To show this, consider a plane wave propagating in the z-direction, with fluid velocity

$$\mathbf{u} = \mathcal{R}\{\mathbf{u}_0 \exp[i(k_z z - \omega t)]\}, \tag{5.56}$$

where $\mathcal{R}$ indicates the real part. If the fluid is incompressible, $\boldsymbol{\nabla}\cdot\mathbf{u} = 0$, and hence

$$\frac{\partial u_z}{\partial z} = -\frac{\partial u_x}{\partial x} - \frac{\partial u_y}{\partial y}, \tag{5.57}$$

which is equal to 0 since the plane wave assumption implies that the velocity field is a function of z and t only. This implies that the component u_z of the velocity field must be spatially uniform. In other words, only the component of the velocity field perpendicular to the propagation direction can oscillate spatially: the wave must be transverse.

[6]Consider small perturbations of the velocity and magnetic fields, linearise the Navier–Stokes and induction equations, take the curl of these two equations, and combine them after taking the time derivative of the curled Navier–Stokes equation (vorticity equation).

We consider a uniform background magnetic field $\mathbf{B}_0 = B_0\mathbf{e}_z$, which is slightly perturbed by a small velocity field perturbation. By "slightly pertubed", we mean here that the curvature $\mathcal{K}$ of the magnetic lines is assumed to remain small. The restoring Lorentz force being perpendicular to **B**, we expect oscillations of the fluid perpendicular to **B**, in the x direction, and, since only transverse waves can be carried by an incompressible fluid, propagation parallel to the magnetic fields lines, in the z direction. A transverse wave propagating along a field line will shear the magnetic line and produce by induction a magnetic field in the x direction. We thus consider small velocity and magnetic fields perturbations of the form $\mathbf{u} = u_x(z,t)\mathbf{e}_x$ and $\mathbf{b} = b_x(z,t)\mathbf{e}_x$, small curvature $\mathcal{K}(z)$, and linearize Eqs. (5.59) and (5.60) to obtain

$$\rho\frac{\partial u_x}{\partial t} = -\frac{B_0^2}{\mu_0}\mathcal{K}, \tag{5.61}$$

$$\frac{\partial b_x}{\partial t} = B_0\frac{\partial u_x}{\partial z}. \tag{5.62}$$

We denote by δ_x the x-displacement of a magnetic field line. Since in the infinite R_m limit magnetic field lines are material lines, the horizontal displacement of the field lines is linked to the velocity field by

$$u_x = \frac{\partial \delta_x}{\partial t}. \tag{5.63}$$

With $\mathbf{n} \simeq \mathbf{e}_x - \frac{\partial \delta_x}{\partial z}\mathbf{e}_z$, the curvature is

$$\mathcal{K} = \nabla \cdot \mathbf{n} \simeq -\frac{\partial^2 \delta_x}{\partial z^2}. \tag{5.64}$$

The x-component of Navier–Stokes then writes

$$\frac{\partial^2 \delta_x}{\partial t^2} - \frac{B_0^2}{\rho\mu_0}\frac{\partial^2 \delta_x}{\partial z^2} = 0, \tag{5.65}$$

which is a non-dispersive wave equation, with celerity

$$v_A = \frac{B_0}{\sqrt{\rho\mu_0}}. \tag{5.66}$$

The period T of theses waves is related to their wavelength λ by

$$T = \frac{\sqrt{\rho\mu_0}\lambda}{B_0}. \tag{5.67}$$

The Geometry of Earth's Magnetic Field

Spherical Harmonics Decomposition

Figure 5.5a shows a map of the intensity of the radial component B_r of the geomagnetic field at the surface of the Earth. As already discussed in the introduction, and first noted by Gilbert in 1600, Earth's magnetic field strongly resembles the magnetic field which would be produced by a magnet inside the Earth and aligned with the rotation axis: the field is strongly dipolar, close to axisymmetric, with negative B_r in the northern hemisphere and positive B_r in the southern hemisphere. However, closer inspection of Fig. 5.5a reveals deviations from a NS-oriented dipolar field and smaller scales details. For example, the position of the magnetic equator (the line at the surface where **B** is horizontal) wanders quite far from the geographic equator. Figure 5.5a also shows patches of stronger field intensity under North America, Siberia, and the south Indian ocean.

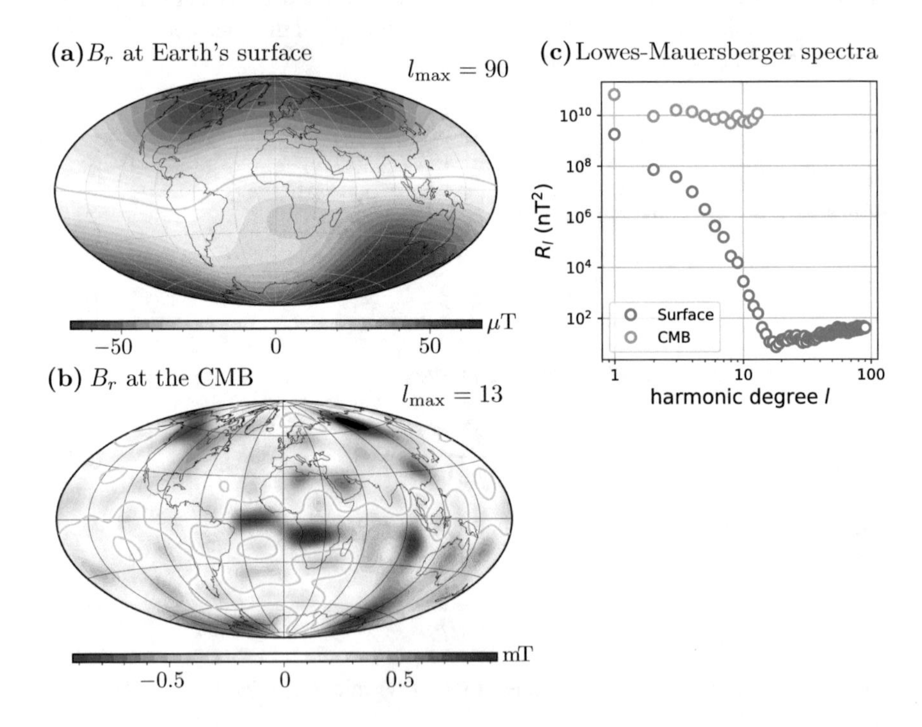

Fig. 5.5 **a** Radial component B_r of the geomagnetic field at the surface of the Earth in 2003 up to spherical harmonic degree $l = 90$, according to the Potsdam Magnetic Model of the Earth (POMME) (Maus et al. 2006); **b** radial component B_r of the geomagnetic field at the core-mantle boundary, obtained from the model POMME; **c** Lowes-Mauersberger spectra for the magnetic field at the surface (blue circles) and at the core-mantle boundary (orange circles)

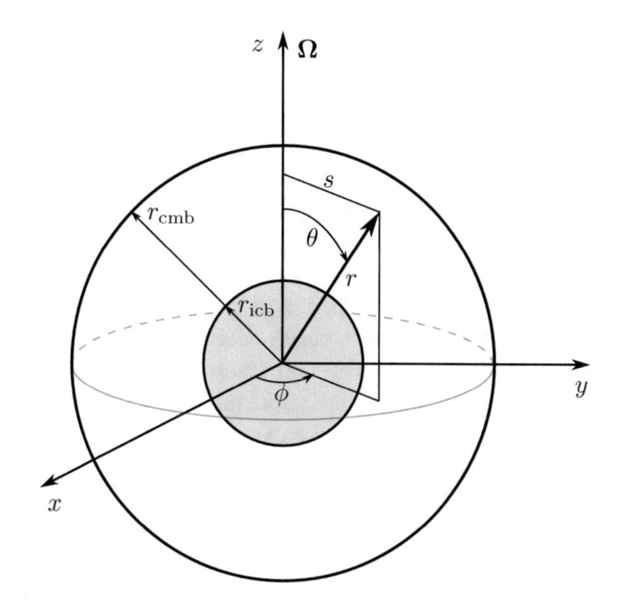

Fig. 5.6 Spherical (r, θ, ϕ) and cylindrical (s, ϕ, z) coordinate systems

The fact that the mantle, crust, and lower atmosphere of the Earth can be considered as current-free regions (they are good electric insulators) greatly simplifies the mathematical description of the geomagnetic field: Ampère's law [Eq. (5.17)] with $\mathbf{j} = 0$ writes $\nabla \times \mathbf{B} = 0$, which implies that $\mathbf{B}$ can be written as the gradient of a scalar field, the geomagnetic potential V:

$$\mathbf{B} = -\nabla V. \tag{5.68}$$

Since $\mathbf{B}$ is a divergence-free vector field and $\nabla \cdot \nabla(...) = \Delta(...)$, the geomagnetic potential obeys Laplace equation,

$$\Delta V = 0. \tag{5.69}$$

The general solution of this equation can be written as

$$V(r, \theta, \phi) = \sum_{l=1}^{\infty} \sum_{m=-l}^{l} \left(\frac{A_{lm}}{r^{l+1}} + B_{lm} r^l \right) Y_{lm}(\theta, \phi), \tag{5.70}$$

where (r, θ, ϕ) are the usual spherical coordinate systems (radius, colatitude, longitude) as defined on Fig. 5.6, and $Y_{lm}(\theta, \phi)$ are the spherical harmonics, l and m being the degree and order, which form a complete set of orthogonal functions on the sphere.

Demonstrating the Internal Origin of the Geomagnetic Field

Written this way, the geomagnetic potential can be seen as the sum of terms of internal origins, which decrease with increasing r (the A_{lm} terms), and terms of external origins, which increase with increasing r (the B_{lm} terms). By determining the A_{lm} and B_{lm} coefficients, it is therefore possible to determine whether the magnetic field observed at the surface of the Earth is predominantly of internal or external origin. In his *General Theory of Terrestrial Magnetism* published in 1839, Gauss introduced the geomagnetic potential and its spherical harmonics expansion. Considering only terms of internal origin, Gauss determined the coefficients of the expansion up to degree $l = 4$ using least-square inversion from intensity, inclination, and declination maps available at that time. From the excellent agreement between the observations and his spherical harmonics expansion, he concluded that the geomagnetic field must be of internal origin.

The internal and external terms only differ by the way they vary with r, and it may not be obvious at first sight how it is possible to differentiate between them from measurements all made at the same radius. But the actual data are measurements of the magnetic field $\mathbf{B} = -\nabla V$, and not V. The latitudinal variations of the direction of $\mathbf{B}$ at a given r happen to be sensitive to the origin of the magnetic field. To get a feeling of how this works, let us consider the $l = 1$ terms only, and write the potential as

$$V = \left[\frac{A_{10}}{r^2} + B_{10}r\right] Y_{10}(\theta) = \left[\frac{A_{10}}{r^2} + B_{10}r\right] \cos\theta. \tag{5.71}$$

From this, we obtain the corresponding magnetic field components:

$$B_r = -\frac{\partial V}{\partial r} = -\left[-2A_{10}/r^3 + B_{10}\right] Y_{10} = \left[2A_{10}/r^3 - B_{10}\right] \cos\theta, \tag{5.72}$$

$$B_\theta = -\frac{1}{r}\frac{\partial V}{\partial \theta} = -\left[A_{10}/r^3 + B_{10}\right] \frac{\partial Y_{10}}{\partial \theta} = \left[A_{10}/r^3 + B_{10}\right] \sin\theta, \tag{5.73}$$

$$B_\phi = -\frac{1}{r\sin\theta}\frac{\partial V}{\partial \phi} = 0, \tag{5.74}$$

and calculate the *inclination* I (the angle between $\mathbf{B}$ and the horizontal plane) as

$$\tan I = \frac{B_r}{B_\theta} = \frac{2A_{10}/r^3 - B_{10}}{A_{10}/r^3 + B_{10}} \tan\lambda, \tag{5.75}$$

where $\lambda = \pi/2 - \theta$ is the latitude. If the geomagnetic field is either of internal origin only ($B_{10} = 0$), or of external origin only ($A_{10} = 0$), then

$$\tan I = 2\tan\lambda \qquad \text{if } \mathbf{B} \text{ is of internal origin,} \tag{5.76}$$

$$I = -\lambda \qquad \text{if } \mathbf{B} \text{ is of external origin.} \tag{5.77}$$

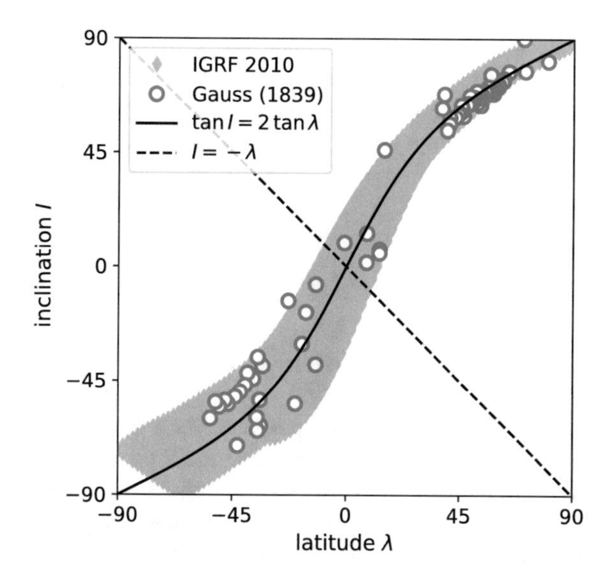

Fig. 5.7 Inclination I as a function of latitude λ for IGRF 2010 geomagnetic model (grey) and for magnetic field measurements used by Gauss (blue circles). Also shown are the predictions for a $l = 1$, $m = 0$ magnetic field of internal (solid black line) and external (dashed black line) origins

Figure 5.7 shows the inclination I as a function of λ for the IGRF 2010 geomagnetic model (grey diamonds), and for the data used by Gauss (blue diamonds). The data points are quite dispersed around the $\tan I = 2\tan\lambda$ curve, but this is mostly due to the fact that the geomagnetic dipole is actually tilted from the rotation axis, and also because Earth's magnetic field includes non-negligible higher degree terms.

From now on we will thus keep only the contributions of internal origin. The spherical harmonics can be written as $Y_{lm} = \mathrm{e}^{im\phi} P_{lm}(\cos\theta)$, where P_{lm} are the associated Legendre polynomials. Keeping only the terms in r^{-l-1} in Eq. (5.70) and using real coefficients, one can write the magnetic potential as

$$V(r,\theta,\phi) = r_\oplus \sum_{l=1}^{\infty} \sum_{m=-l}^{l} \left(\frac{r_\oplus}{r}\right)^{l+1} \left[g_l^m \cos m\phi + h_l^m \sin m\phi\right] P_{lm}(\cos\theta), \quad (5.78)$$

where g_l^m and h_l^m are the *Gauss coefficients*, and $r_\oplus$ is the radius of the Earth.

It is useful to look at the spectrum of magnetic energy as a function of the spherical harmonic degree l. Using the fact that spherical harmonics form an orthogonal basis, one can show that the magnetic energy at a radius r corresponding to all components of degree l is given by

$$R_l(r) = \left(\frac{r_\oplus}{r}\right)^{2l+4} (l+1) \sum_{m=0}^{l} \left(g_{lm}^2 + h_{lm}^2\right). \quad (5.79)$$

The resulting spectrum is called the *Lowes-Mauersberger* spatial power spectrum. Figure 5.5c shows this spectrum at Earth's surface in blue circles. The energy of the dipolar components ($l = 1$) is more than an order of magnitude larger than the

$l = 2$ component. R_l then decreases with l up to l equal 13 or 14 where it reaches a plateau. At $l > 13$, the magnetic energy is dominated by contributions of the crustal magnetisation, which obscures the field from deeper origin. Only the $l \leq 13$ spherical harmonics components can be attributed to the magnetic field originating from the core.

The Magnetic Field at the Core-Mantle Boundary

One very interesting property of a curl-free magnetic field is that it is possible to extrapolate to other radii the magnetic potential once we know the coefficients of the spherical harmonics decomposition. Since the mantle of the Earth can be considered to be current-free, it is in particular possible to extrapolate the surface magnetic field down to the base of the mantle, though this can be done only for harmonic degrees $l \leq 13$ since the higher degrees components of the core field are obscured by the crustal field. The potential for $l \leq 13$ at the core-mantle boundary can be obtained by writing V at $r = r_{\text{cmb}}$ from Eq. (5.78). From this one can obtain the associated magnetic field map and spectrum.

Figure 5.5b shows the radial component of the field at the core-mantle boundary (CMB). Though the field is still quite dipolar, it exhibits much more smaller scale variations than the surface field (Fig. 5.5a). This is confirmed by inspection of the energy spectrum (Fig. 5.5c, orange circles): the spectrum is about flat for $l \geq 2$, with the energy of the dipolar components still about an order of magnitude larger than the higher l components. The fact the dipolar components still stand out at the CMB is an important result. Since the spherical components decrease with r as r^{-l-1}, any locally produced magnetic field would appear dipolar when seen from far enough. The fact that the field at the CMB is still dominated by the $l = 1$ terms suggests that the dipolar nature of the geomagnetic field is a robust feature of the geodynamo.

The Field Within the Core: Poloidal–Toroidal Decomposition

In planetary cores, $\mathbf{B}$ cannot be considered to be curl-free anymore ($\mathbf{j} \neq 0$), and thus cannot be written as the gradient of a potential. However, the fact that $\mathbf{B}$ is a divergence-free vector field allows to write it as

$$\mathbf{B} = \nabla \times (Tr\mathbf{e}_r) + \nabla \times \nabla(Pr\mathbf{e}_r) \tag{5.80}$$

where T and P are the *poloidal* and *toroidal* potentials. Note that the toroidal part has no radial component, which means that the magnetic field reconstructed at the CMB only corresponds to the poloidal part. We have no direct constraints on the toroidal part of the magnetic field in the core.

Core Flow Inversion

The magnetic field at the CMB evolves with time in a measurable way. These variations can be interpreted with the help of the induction equation, remembering that at the CMB only the radial component of the magnetic field is known. Taking the dot product of the induction equation with $\mathbf{e}_r$ gives

$$\frac{\partial B_r}{\partial t} = -\nabla_h \cdot (B_r \mathbf{u}_h) + \eta \frac{1}{r} \nabla^2 (r B_r), \tag{5.81}$$

where $\nabla_h \cdot (...)$ is the horizontal part of the divergence operator, and $\mathbf{u}_h = (0, u_\theta, u_\phi)$ is the horizontal part of the velocity field. The diffusion term is usually neglected, which can be justified *a posteriori* on the basis that the magnetic Reynolds number based on the smallest spatial scale considered ($\sim$1000 km) and estimated velocity is $\sim$500. Dropping the diffusion term, Eq. (5.81) writes

$$\frac{\partial B_r}{\partial t} = -\nabla_h \cdot (B_r \mathbf{u}_h). \tag{5.82}$$

Knowing B_r and its time derivative, one can in principle invert Eq. (5.82) to obtain the horizontal velocity field just below the CMB. This happens to be a severely ill-posed inverse problem, the most obvious reason being that we are trying to estimate a two-components vector field ($\mathbf{u}_h$) from a scalar field (B_r). In other words, we have only one equation for two unknowns. Inverting Eq. (5.82) thus requires a second equation for $\mathbf{u}_h$. Various assumptions have been made (e.g. steady flow, toroidal flow, tangentially geostrophic flow, columnar flow, quasi-geostrophic flow, ...), and the choice does impact the resulting flow pattern (see Holme 2015 for a review). Robust features of the inverted flows include a strong westward flow under the Atlantic, a much weaker flow under the Pacific, and some degree of symmetry about the equatorial plane. The RMS flow velocity is around 12–14 km per year, or about 4×10^{-4} m.s^{-1}.

Basics of Planetary Core Dynamics

The Geodynamo Hypothesis

As discussed in section "Introduction", Gilbert's claim that *Earth is a magnet* has been dismissed by the observation of fast temporal changes of the geomagnetic field. In addition, we now know that permanent magnets (i.e. ferromagnetic or ferrimagnetic materials) lose their permanent magnetic properties (by becoming paramagnetic) above a critical temperature called the *Curie temperature*. Magnetite and iron have Curie temperatures of 858 K and 1043 K, respectively. The temperature in the Earth exceeds these temperatures at depths larger than around 100–150 km (Jaupart and Mareschal 2010), which confines ferromagnetism to rather shallow depths. Magnetisation of crustal material can, in fact, be quite strong, but cannot explain the large scale part of Earth' s magnetic field.

Leaving aside *ad-hoc* theories, we are thus left with the MHD equations introduced in section "A Short Introduction to Magnetohydrodynamics" to understand the origin of Earth's magnetic field. We have already shown (section "The Induction Equation") that the geomagnetic field cannot be of primordial origin, and must be sustained in some way against ohmic dissipation: solving the induction equation with no velocity field indeed shows that spatial variations of the magnetic field would be smoothed out by diffusion on a timescale of $\sim$30 000 years, while we know from paleomagnetism that the geomagnetic field has been sustained for (at least) the last $\sim$3.5 Gy.

Self-exciting Dynamos

The current theory for the origin of the Sun and Earth's magnetic fields has been proposed in 1919 by Sir Joseph Larmor in a meeting of the *British Association for the Advancement of Science*. In the short report of his presentation (Larmor 1919), he wrote:

> In the case of the Sun, surface phenomena point to the existence of a residual internal circulation mainly in meridian planes. Such internal motion induces an electric field acting on the moving matter; and if any conducting path around the solar axis happens to be open, an electric current will flow round it, which may in turn increase the inducing magnetic field. In this way it is possible for the internal cyclic motion to act after the manner of the cycle of a self-exciting dynamo, and maintain a significant magnetic field from insignificant beginnings, at the expense of some of the energy of the internal circulation.
>
> [...]
>
> The very extraordinary feature of the Earth magnetic field is its great and rapid changes, comparable with its whole amount. [...] [*a self-exciting dynamo*] would account for magnetic change, sudden or gradual, on the Earth merely by change of internal conducting channels: though, on the other hand, it would require fluidity and residual circulation in deep-seated regions.

Larmor's proposition was inspired by the *self-exciting dynamos* developed in the second half of 19th century, which were the first efficient electric generators. The underlying mechanism can be understood as a positive feedback loop involving Faraday's law of induction and Ampère's law. Consider an electrically conducting material moving into an arbitrarily small seed magnetic field. Faraday's law of induction tells that time variations of a magnetic field in an electrically conducting material produces electric currents. Equivalently, the motion of the conducting material into the seed magnetic field produces electric currents. According to Ampère's law, these electric currents would themselves produce a magnetic field. If the orientation of the electric current is such that the induced magnetic field reinforces the initial seed magnetic field, then the intensity of the magnetic field can grow.

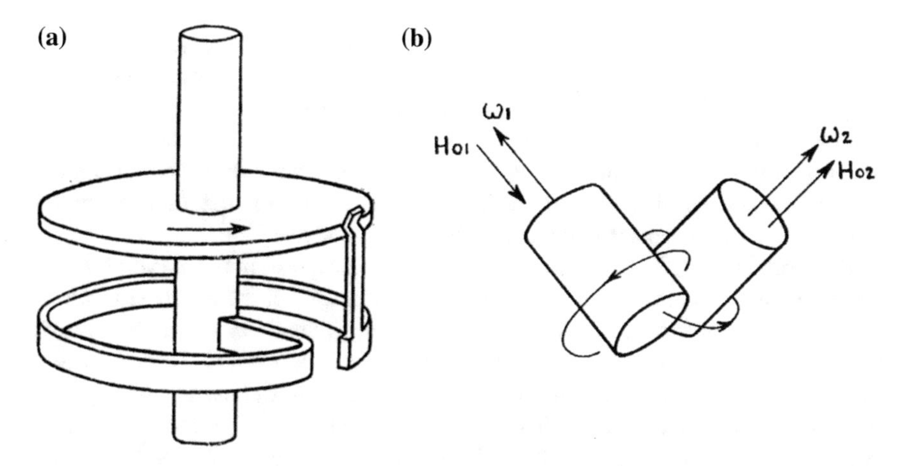

Fig. 5.8 **a** Bullard's disk dynamo (Bullard and Gellman 1954). It consists in a Faraday disk whose center and periphery are connected by a conducting wire forming a loop around the axle of the disk. In the presence of a 'seed' magnetic field oriented parallel to the axis, rotating the copper disk produces an electric current, which intensity depends on the rate of rotation of the disk and of the electric resistance R of the wire. The arrangement of the electric wire loop is such that the induced magnetic field reinforces the seed magnetic field. **b** A schematic of Lowes and Wilkinson's experimental dynamo (Lowes and Wilkinson 1963), based on Herzenberg (1958)'s kinematic dynamos

The word dynamo was coined by Faraday to name the electric generator he invented, now called the *Faraday disk*. Faraday's disk consists in a copper disk rotating within a magnetic field produced by a horseshoe magnet. The rotating motion of the copper disk produces a difference of electric potential between its center and periphery by virtue of Faraday's law of induction, and hence an electrical current if the center and periphery of the disk are linked through an electric circuit. Self-exciting dynamos are based on a similar principle except that the permanent magnet is replaced by electromagnets fed by the induced electric current. This creates a positive feedback loop which increases the intensity of the magnetic field, thus increasing the currents produced by induction. The concept has been formulated by Anyos Jedlik around 1856; practical designs of working self-exciting dynamos have been patented by Varley in 1866, and presented by Wheatstone and Siemens in 1867. These self-exciting dynamos were able to produce a much higher power output than permanent magnet dynamos, which opened the way to the industrial use of electricity. A simple conceptual model of a self-exciting dynamo was proposed by Bullard and Gellman (1954) as a toy model of the geodynamo (Fig. 5.8).

Compared to self-exciting dynamos such as Bullard's, the concept of *geodynamo* faces several additional difficulties. To quote Bullard and Gellman (1954): "A central problem [...] is to determine whether there exist motions of a simply connected, symmetrical fluid body which is homogeneous and isotropic that will cause it to act as a self-exciting dynamo [...]. We call such dynamos 'homogeneous' to distinguish them from the dynamos of the electrical engineer, which are multiply connected and of low symmetry." In industrial dynamos, the electric currents produced by induction are fed into electric wires which are arranged in such a way that the induced magnetic field indeed reinforces the seed field. In Earth's core, no such wiring exists and the electric currents path is determined from Ohm's law by the direction of the electromotive force (and thus ultimately by the geometry of the velocity and magnetic fields). The motion of molten iron in the core has to be such that, on average, the induced electric currents can maintain the geomagnetic field.

As formulated by Bullard and Gellman (1954), the question is not (yet) whether a free-flowing fluid can maintain a magnetic field in spite of the expected feedback of the Lorentz force onto the flow. At first no such feedback was considered: early proponents of the geodynamo hypothesis first tried to find *kinematic dynamos*, where only the induction equation is solved, the velocity field being chosen in the hope that it could produce dynamo action. This is a difficult problem, even if the velocity field is prescribed, but working kinematic dynamos have indeed been found. Classical examples include the dynamos found by Herzenberg (1958), Roberts (1972), and Ponomarenko (1973).

The first laboratory demonstration of homogeneous dynamo action was given by Lowes and Wilkinson (1963) with an apparatus inspired by Herzenberg's dynamo. Lowes and Wilkinson's apparatus consists in two rotating solid cylinders made of a high magnetic permeability iron-alloy ("Perminvar") with their axes at right angles, embedded in a solid block of the same material (Fig. 5.8). The electrical contact between the cylinders and the block is ensured by a thin layer of liquid mercury. Denoting by ω the rotation rate of the cylinders, a their radii, and $\lambda = 1/(\mu\sigma)$ the magnetic diffusivity of the "Perminvar" alloy, Lowes and Wilkinson (1963) found that their apparatus operates as a self-exciting dynamo when the non-dimensional number $\omega a^2/\lambda$ exceeds a critical value. This number is a magnetic Reynolds number (with velocity scale ωa), so the fact that it must exceed a critical value for dynamo action is consistent with our analysis of the induction equation (section "The Induction Equation"). In spite of being made of two materials (Perminvar plus a thin layer of mercury), this is effectively a homogeneous dynamo in the sense that the electric currents are not forced into specific paths.

A next critical step for establishing the viability of the geodynamo hypothesis has been to obtain dynamos in which the velocity field is not imposed *a priori*, and in which the feedback of the magnetic field on the flow is taken into account (*via* the Lorentz force). Fluid dynamos obtained numerically by self-consistently solving the coupled Navier–Stokes, induction, and heat transfer equations have started to appear in the eighties in the contexts of the solar dynamo (Gilman and Miller 1981; Glatzmaier 1984, 1985a, b) and geodynamo (Zhang and Busse 1988; Glatzmaier and Roberts 1995; Kageyama et al. 1995). In spite of a number of shortcomings, these numerical dynamos driven by convective motions have been successful in reproducing some of the most salient features of Earth's magnetic field (most on this in section "Convective Dynamos").

Experimental liquid homogeneous self-exciting dynamos have been obtained more recently, in two apparatus developed in Riga (Latvia) (Gailitis et al. 2001) and Karlsruhe (Germany) (Stieglitz and Müller 2001). In both experiments, liquid sodium was forced by propellers into a system of pipes, which effectively imposed the velocity field. Liquid sodium is the best electrically conducting liquid available in large quantities (it has a magnetic diffusivity $\eta = 0.1\ m^2.s^{-1}$, about ten times lower than molten iron), and has been systematically used as working fluid in dynamo experiments. In both apparatus the imposed flow is inspired by known kinematic dynamos: the Riga experiment is based upon the Ponomarenko dynamo, and the Karlsruhe experiment is based upon the G.O. Roberts dynamo. The Riga experiment used 2 m^3 of sodium and required a power input of 200 kW; the Karlsruhe experiment used 1.6 m^3 of sodium and required a power input of 630 kW.

Obtaining experimentally a dynamo with a much less constrained flow proved to be even more arduous. To date the only successful free-flowing experiment is the VKS experiment developed in Cadarache (France) (Monchaux et al. 2007; Berhanu et al. 2007). Though numerical dynamo calculations are now well established, liquid metal experiments are still very valuable tools for understanding planetary core dynamics (e.g. Nataf and Gagnière 2008; Cabanes et al. 2014). On certain aspects, liquid metal experiments are dynamically closer to Earth's core conditions than numerical simulations. In particular, liquid sodium has a magnetic Prandtl number (the ratio of kinematic viscosity to the magnetic diffusivity) of about 10^{-6}, similar to that of Earth's core, while working numerical dynamos have been so far limited to magnetic Prandtl number values above ~ 0.1.

What Drives the Geodynamo?

In parallel to the question of the feasibility of a liquid homogeneous dynamo, a central question has been the source of power of the geodynamo. What drives the motion and provides the energy that is being lost by ohmic dissipation?

In industrial devices or laboratory experiences, the power is provided by an external mean and therefore is not, theoretically speaking, an issue (though it can be a practical issue, since driving a liquid dynamo in the laboratory does require a quite large power input).

In Earth's core, the possible sources of motion and power fall into two broad categories: (i) natural convection, either of thermal origin (Bullard 1949, 1950; Verhoogen 1961) or compositional origin (Braginsky 1963; Gubbins 1977; Loper 1978; O'Rourke and Stevenson 2016); and (ii) stirring produced by astronomical forcing, i.e. motions forced in the core by either tidal forcing or changes of the mantle rotation vector (in direction—precession or nutation—or intensity—libration) (Bondi and Lyttleton 1948; Bullard 1949; Malkus 1963, 1968). The second class of models is the subject of the chapter by Le Reun and Le Bars; we will focus here on natural convection.

Core convection, whether it be thermal or compositional, is controlled by the rate at which the core loses heat to the mantle. From a thermal point of view, the core is the slave of the mantle: The heat flux from the core to the mantle is dictated by the efficiency of heat transport by mantle convection, and core convection is thus tied to mantle convection. Cooling of the core can drive convective motions is several ways:

1. Cooling the core from above can potentially produce thermal convection, if the imposed heat flux is larger than the flux which can be conducted along an adiabat in the core (see section "Rotating Convection").
2. The secular cooling of the core is responsible for its slow solidification and the formation of the solid inner core. In spite of the core being colder at the CMB than at deeper depth, the core started to solidify at its center and the inner core is now growing outward. The reason for this is that the solidification temperature of the core material increases with pressure faster than the actual temperature (Jacobs 1953). If the core started hot and fully molten and then gradually cooled down, the temperature first reached the solidification temperature at the center of the core, which allowed the inner core to nucleate. Further cooling results in the slow growth of the inner core.
   One key point here is that the core is made of an iron-rich alloy rather than pure iron. Since the density of the core is lower than that of pure iron, we know that the main impurities are "light elements" (i.e. lighter than iron), likely a mixture of mainly oxygen, sulfur and silicium. Upon solidification, these elements are partitioned preferentially into the liquid outer core: solidification of the inner core thus results in an outward flux of light elements at the inner core boundary (ICB), which can drive *compositional* convection. In addition, the latent heat of solidification helps thermal convection by slowing down the cooling of the inner core boundary.
3. Finally cooling of the core may also be at the origin of a compositional flux across the core-mantle boundary.
   Light elements like oxygen, silicium or magnesium, are present both in the core and mantle (composed predominantly of silicates and oxides of iron and magnesium). Since the core and mantle materials in contact at the core-mantle boundary should be very close to thermodynamic equilibrium, the relative abundance of these elements in the core and silicates at the CMB is set by the partitioning coefficient of these elements between silicates and the core alloy. This in general depends on temperature: cooling of the core thus modifies this chemical equilibrium and results in a flux of elements between mantle and core. If light elements are transferred from the mantle to the core, then a stably stratified layer may form below the CMB. If in the other way, the flux of light elements would drive compositional convection in the core. The actual flux of element is limited by the rate at which convection in the mantle provides "fresh" material which can react with the molten iron of the core; again, mantle convection controls to a large part the buoyancy flux available to drive core convection.
   Another possible mechanism is exsolution of light elements from the core. It has been proposed that MgO (O'Rourke and Stevenson 2016; Badro et al. 2016) and SiO_2 (Hirose et al. 2017) can precipitate (or exsolve) from the core alloy if initially abundant enough. The saturation concentration of these species happens to decrease with decreasing temperature, so a gradual cooling of the core results in the progressive removal of magnesium, oxygen and silicon from the core. If exsolution happens predominantly in the vicinity of the core-mantle boundary, then removing MgO and SiO_2 from the core induces a buoyancy flux (by leaving behind an iron-rich, dense melt) which can drive convection in the core.

Rotating Convection

Governing Equations and Non-dimensional Parameters

The liquid outer core is modelled as a spherical shell of outer and inner radii $r_{\rm cmb}$ and $r_{\rm icb}$. The shell thickness is denoted by $H = r_{\rm cmb} - r_{\rm icb}$. In what follows, we will use either a spherical coordinate system (r, θ, ϕ) or a cylindrical coordinate system (s, ϕ, z) (Fig. 5.6). The core is rotating at a rate Ω, and the gravity field is $\mathbf{g} = -g(r)\mathbf{e}_r$. We assume that there is a positive temperature difference ΔT between the inner and outer boundaries. We denote by ν the kinematic viscosity of the liquid core, by ρ its density, by κ its thermal diffusivity, by α its thermal expansion coefficient, and by c_p its specific heat capacity. Though this is a somewhat dubious assumption, the outer core is often modelled as a Boussinesq fluid (which in particular includes the assumption that the fluid is incompressible). Molten iron in the outer core is also typically assumed to behave as a Newtonian fluid with temperature-independent kinematic viscosity.

Rotating thermal convection is governed by the Navier–Stokes equation in a rotating frame of reference, mass conservation, and a transport equation for temperature T. Denoting by $\mathbf{u}$ the velocity field, by P the pressure, and by T the temperature, this set of equations can be written (under the Boussinesq approximation), as

$$\rho\frac{D\mathbf{u}}{Dt} + \underbrace{2\rho\Omega \times \mathbf{u}}_{\text{Coriolis}} = -\nabla P + \underbrace{\alpha\rho T g\mathbf{e}_r}_{\text{buoyancy}} + \rho\nu\nabla^2\mathbf{u}, \tag{5.83}$$

$$\nabla \cdot \mathbf{u} = 0, \tag{5.84}$$

$$\frac{DT}{Dt} = \kappa\nabla^2 T. \tag{5.85}$$

These equations can be made dimensionless with a variety of different choices of scales. For exemple, scaling lengths by the outer core thickness H, time by H^2/ν, velocity by ν/H, temperature by the difference of temperature ΔT across the shell, and pressure by $\rho\Omega\nu$, gives

$$E\frac{D\mathbf{u}}{Dt} + 2\mathbf{e}_z \times \mathbf{u} = -\nabla P + \frac{E}{Pr} Ra\, T\mathbf{e}_r + E\nabla^2\mathbf{u}, \tag{5.86}$$

$$\nabla \cdot \mathbf{u} = 0, \tag{5.87}$$

$$\frac{DT}{Dt} = \frac{1}{Pr}\nabla^2 T, \tag{5.88}$$

where the Ekman, Rayleigh and Prandtl numbers are defined as

$$\text{Ekman number } E = \frac{\nu}{\Omega H^2}, \tag{5.89}$$

$$\text{Rayleigh number } Ra = \frac{\alpha g H^3 \Delta T}{\kappa\nu}, \tag{5.90}$$

$$\text{Prandtl number } Pr = \frac{\nu}{\kappa}. \tag{5.91}$$

The quantity RaE/Pr is sometimes taken as a "modified Rayleigh number" defined as

$$Ra^* = \frac{E}{Pr} Ra = \frac{\alpha g \Delta T H}{\Omega \nu}. \tag{5.92}$$

If a heat flux q is imposed at the boundaries rather than a temperature difference, Ra and Ra^* can be modified to give flux-based Rayleigh numbers by replacing ΔT by qH/k, where k is thermal conductivity.

The Ekman number is a measure of the ratio of the viscous forces and Coriolis acceleration, if the viscous forces are estimated assuming a flow varying spatially on a length scale $\sim H$:

$$\frac{\text{viscous forces}}{\text{Coriolis}} \sim \frac{|\rho\nu\nabla^2\mathbf{u}|}{|2\rho\Omega \times \mathbf{u}|} \sim \frac{\nu V/H^2}{\Omega V} \sim \frac{\nu}{\Omega H^2}. \tag{5.93}$$

It is about 10^{-15} for the outer core, which means that viscous forces would become of importance only at relatively small length scales (more about this later).

It can also be useful to consider the vorticity equation, obtained by taking the curl of Eq. (5.86),

$$\frac{D\zeta}{Dt} = (2\Omega + \zeta) \cdot \boldsymbol{\nabla}\mathbf{u} + \alpha g \boldsymbol{\nabla} T \times \mathbf{e}_r + \nu \boldsymbol{\nabla}^2 \zeta, \tag{5.94}$$

which can usually be simplified to

$$\frac{D\zeta}{Dt} = 2\Omega \frac{\partial \mathbf{u}}{\partial z} + \alpha g \boldsymbol{\nabla} T \times \mathbf{e}_r + \nu \boldsymbol{\nabla}^2 \zeta, \tag{5.95}$$

by assuming that the vorticity ζ is small compared to the rotation rate Ω. Here the term $2\Omega\frac{\partial \mathbf{u}}{\partial z}$ is the production of vorticity through stretching of planetary vorticity, $\alpha g \boldsymbol{\nabla} T \times \mathbf{e}_r$ is the *baroclinic* production of vorticity, and $\nu \boldsymbol{\nabla}^2 \zeta$ is the viscous diffusion of vorticity.

Geostrophy

The very small value of the Ekman number of the core suggests that viscous forces may be neglected, at least as long as only large-scale motions are considered. Keeping only the Coriolis acceleration and the pressure gradient (thus neglecting also inertia and buoyancy), the Navier–Stokes equation reduces to

$$2\rho\Omega \times \mathbf{u} = -\boldsymbol{\nabla} p, \tag{5.96}$$

which is called the geostrophic balance. Taking the curl of Eq. (5.96), which gives

$$\Omega \cdot \nabla \mathbf{u} = \Omega \frac{\partial \mathbf{u}}{\partial z} = 0,$$

shows that the velocity field is invariant along the rotation axis if this balance holds. This is known as the *Taylor–Proudman theorem.*

In a rotating container with sloped boundaries, such as planetary cores, the motion is further restricted by the condition that no fluid can cross the boundary. In a spherical container, this condition writes $\mathbf{u} \cdot \mathbf{e}_r = 0$. In other words, the velocity field at the boundary can only have θ and ϕ components. Of these, only the ϕ component is allowed by the Taylor–Proudman constraint. This means that the only possible motions are longitudinal (zonal) motions of the form

$$\mathbf{u} = u_\phi(s, \phi, t)\mathbf{e}_\phi.$$

Mass conservation ($\nabla \cdot \mathbf{u} = (r \sin\theta)^{-1} \partial u_\phi / \partial \phi = 0$) further implies that $\mathbf{u}$ cannot depend on ϕ. The only possible geostrophic flows in a spherical container thus consist in rigid cylinders rotating about Earth's axis of rotation, called *geostrophic cylinders*:

$$\mathbf{u} = u_\phi(s, t)\mathbf{e}_\phi.$$

Torsional Oscillations

Though it is somewhat of a side note in the context of a thermal convection-oriented section, it is worth discussing quickly the effect of a magnetic field on geostrophic motions. The addition of a magnetic field permeating the geostrophic cylinders allows the propagation of a subclass of Alfvèn waves called *torsional waves*, or *torsional oscillations* (Braginsky 1970).

The basic mechanism behind these waves can be easily understood using the concept of magnetic tension introduced in section "The Lorentz Force". Assume one geostrophic cylinder rotates at a different rate (Fig. 5.9a). The differential rotation of this cylinder with respect to its neighbours will bend the magnetic lines as shown in Fig. 5.9b. The resulting magnetic tension will produce a restoring force of magnitude inversely proportional to the azimuthal displacement of the cylinder. The net effect on the geostrophic cylinder is a restoring torque. As shown in section "The Lorentz Force" Alfvèn waves are longitudinal waves; torsional waves therefore must propagate in the s-direction.

A dispersion equation for these waves can be obtained by applying conservation of angular momentum to a geostrophic cylinder of inner radius s and thickness ds. By restricting the analysis to small relative motions of the cylinders, one find that the angular velocity $\omega(s)$ obeys the following equation:

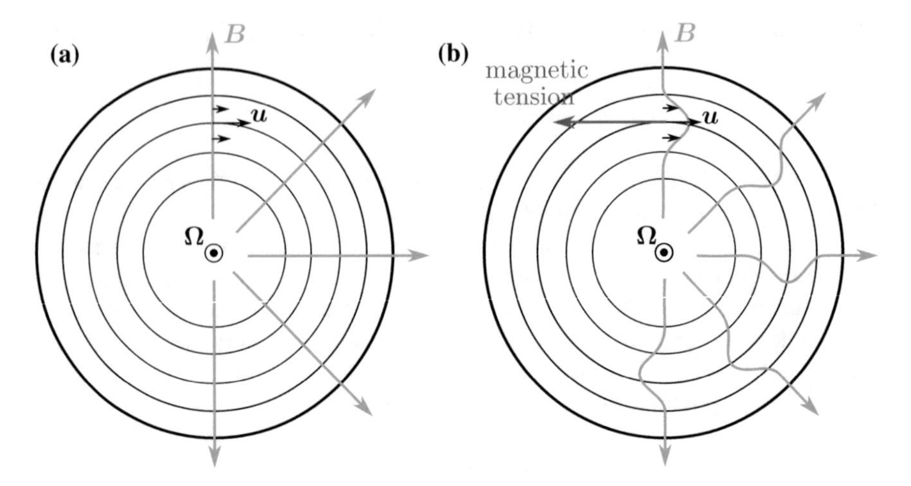

Fig. 5.9 Torsional waves mechanism

$$\frac{\partial^2 \omega}{\partial t^2} = \frac{1}{\mu_0 \rho s^2 H(s)} \frac{\partial}{\partial s} \left(s^3 H(s) \bar{B}_s \frac{\partial \omega}{\partial s} \right), \tag{5.97}$$

where

$$\bar{B}_s = \sqrt{\frac{1}{4\pi H} \int_{-H}^{H} \int_0^{2\pi} B_s^2 d\phi dz}. \tag{5.98}$$

is the r.m.s. value of B_s on the geostrophic cylinder of radius s (Braginsky 1970).

The propagation velocity of these oscillations being proportional to $\bar{B}_s$, identifying torsional oscillations in core flow inversions can provide an estimate of the magnetic fields intensity *within* the core. Gillet et al. (2010) found a 6-year period signal in their core flow inversions which they interpreted as a torsional wave. It takes about 4 years for this wave to propagate through the outer core; the resulting propagation velocity gives $\bar{B}_s \sim 2$ mT. For comparison, the radial component of the magnetic field at the CMB peaks at about 1 mT (Fig. 5.5).

Onset of Thermal Convection

We now turn to the question of the initiation of thermal convection in a rotating planetary core. Linear stability analysis shows that at low Ekman number the critical Rayleigh number $\mathrm{Ra}_{\mathrm{cr}}$ and the horizontal length scale ℓ_c of the first unstable mode scale as

$$\mathrm{Ra}_{\mathrm{cr}} \sim E^{-4/3}, \tag{5.99}$$

$$\ell_c \sim H E^{1/3}. \tag{5.100}$$

The linear stability analysis is challenging (Roberts 1968; Busse 1970; Jones et al. 2000; Dormy et al. 2004); to try to understand this result at a lower mathematical cost, we will take a heuristic approach and derive the Ekman number dependency of $\mathrm{Ra}_{\mathrm{cr}}$ from simple physical arguments.

We have seen above that geostrophic motions in a spherical container are restricted to zonal flows of the form $\mathbf{u} = u_\phi(s, t)\mathbf{e}_\phi$. This obviously cannot carry heat radially; thermal convection therefore necessarily involves deviations from geostrophy. This can result from buoyancy forces, inertia, viscous forces, or Lorentz force. We will not consider here the effect of the Lorentz force since we are interested in the initiation of convection in a planetary core, which here is assumed to predate the generation of the magnetic field. Furthermore, we can safely neglect inertia at the onset of convection since it is quadratic in $\mathbf{u}$ while the Coriolis and viscous forces are linear in $\mathbf{u}$. We therefore cannot rely on inertia to break the geostrophic balance, and we are left with viscous and buoyancy forces.

Let us first consider the vorticity equation (5.95), which, neglecting the inertia-derived terms, is

$$0 = 2\Omega\frac{\partial \mathbf{u}}{\partial z} + \alpha g \nabla T \times \mathbf{e}_r + \nu\nabla^2\boldsymbol{\zeta}. \tag{5.101}$$

Denoting by U the characteristic scale of convective velocities, ℓ the characteristic horizontal length scale of convective motions, and δT the characteristic scale of horizontal temperature variations, the first term is on the order of $\Omega U/H$, the baroclinic term on the order of $\alpha g\delta T/\ell$, and the viscous term on the order of $\nu U/\ell^3$ (the vorticity $\boldsymbol{\zeta} = \nabla \times \mathbf{u}$ is on the order of U/ℓ). Since buoyancy is the driver of the flow, we expect the baroclinic vorticity production to be always of importance, and balanced by either the viscous term or the Coriolis term, or both.

A balance between baroclinic production of vorticity and diffusion of vorticity gives

$$U \sim \frac{\alpha g}{\nu}\ell^2\delta T, \tag{5.102}$$

which is basically a Stokes velocity. Balancing baroclinic production of vorticity and stretching of the planetary vorticity gives

$$U \sim \frac{\alpha g}{\Omega}\frac{H}{\ell}\delta T, \tag{5.103}$$

(a *thermal wind* balance).

The two above velocity scales are equal when the three terms in Eq. (5.101) are of equal importance, which happens when the flow length scale is on the order of $\ell_c \sim (\nu H/\Omega)^{1/3} = HE^{1/3}$. At $\ell \gg \ell_c$, the viscosity term is small compared to the planetary vorticity stretching term and the relevant velocity scaling is Eq. (5.103); at $\ell \ll \ell_c$, the viscosity term is large compared to the planetary vorticity stretching term and the relevant velocity scaling is Eq. (5.102). Since the velocity scale increases with ℓ while $\ell < \ell_c$ [Eq. (5.102)], and then decreases with increasing ℓ [Eq. (5.103)], ℓ_c also happen to be the convection length scale at which the velocity would be maximal.

One key point here is that the $2\Omega\frac{\partial \mathbf{u}}{\partial z} \sim \alpha g \nabla T \times \mathbf{e}_r$ balance would yield a velocity field $\mathbf{u}$ with no radial component. This will not carry heat radially. In the absence of inertia, radial convection would therefore require viscous effects to be important. This may seem counter-intuitive, but viscous effects are actually necessary to initiate a Rayleigh–Bénard-type convection in a rotating sphere ! This would suggest that an horizontal scale smaller than ℓ_c is necessary for the initiation of convection. Since in addition the velocity scale at $\ell < \ell_c$ decreases with decreasing ℓ, we may expect that the optimal scale for the initiation of convection is on the order of ℓ_c: radial motions are damped by rotation effects at larger scale, while smaller scale motions are slower due to viscous friction.

Let us now consider a liquid parcel of size ℓ and temperature T larger than the surrounding temperature $\bar{T}(r)$ (temperature excess $\delta T = T - \bar{T}$). $\bar{T}(r)$ is the background conductive temperature profile. The radial velocity of the parcel is given by

$$u_r \sim \frac{\alpha g \ell^2}{\nu}(T - \bar{T}) \tag{5.104}$$

since, as argued above, viscous effects are necessary to the initiation of convection in a rotating sphere. To see how the velocity of the parcel evolves, take its Lagrangian derivative:

$$\frac{Du_r}{Dt} \sim \frac{\alpha g \ell^2}{\nu}\left(\frac{DT}{Dt} - \frac{D\bar{T}}{Dt}\right). \tag{5.105}$$

The first term in the parenthesis can be estimated from the heat equation,

$$\frac{DT}{Dt} = \kappa \nabla^2 T \sim -\kappa \frac{T - \bar{T}}{\ell^2}, \tag{5.106}$$

while the second term is

$$\frac{D\bar{T}}{Dt} = \frac{\partial \bar{T}}{\partial t} + \mathbf{u} \cdot \nabla T = u_r \frac{d\bar{T}}{dr} \sim \frac{\alpha g \ell^2}{\nu}\left(T - \bar{T}\right)\frac{d\bar{T}}{dr}. \tag{5.107}$$

Putting everything together, this gives

$$\frac{Du_r}{Dt} \sim \frac{\alpha g \ell^2}{\nu}\left(-\kappa \frac{T - \bar{T}}{\ell^2} - \frac{\alpha g \ell^2}{\nu}\left(T - \bar{T}\right)\frac{d\bar{T}}{dr}\right), \tag{5.108}$$

$$\sim \frac{\alpha g \ell^2 \left(T - \bar{T}\right)}{\nu \ell^2/\kappa}\left(-\frac{\alpha g \ell^4}{\kappa \nu}\frac{d\bar{T}}{dr} - 1\right). \tag{5.109}$$

This can be re-arranged to give an estimate of the growth rate of the vertical velocity:

$$\frac{1}{u_r}\frac{Du_r}{Dt} \sim \frac{\kappa}{\ell^2}\left(-\frac{\alpha g \ell^4}{\kappa \nu}\frac{d\bar{T}}{dr} - 1\right). \tag{5.110}$$

With $\frac{d\bar{T}}{dr} \sim -\Delta T/H$, this gives

$$\frac{1}{u_r}\frac{Du_r}{Dt} \sim \frac{\kappa}{\ell^2}\left(\frac{Ra}{(H/\ell)^4} - 1\right). \tag{5.111}$$

A parcel displaced upward (resp. downward) will keep rising (resp. sinking) if $\frac{1}{u_r}\frac{Du_r}{Dt} > 0$, i.e. if the term within the parenthesis in Eq. (5.111) is positive. This condition can be recast as a condition for the Rayleigh number, which must exceed a critical value $\mathrm{Ra}_{\mathrm{cr}}$ which is a function of the length scale ℓ of the perturbation:

$$Ra > \mathrm{Ra}_{\mathrm{cr}}(\ell) \sim \left(\frac{H}{\ell}\right)^4. \tag{5.112}$$

Perturbations with the largest length scale ℓ will have the lowest critical Rayleigh number and will then be favoured. In non-rotating convection, the only limit to ℓ is the size H of the convecting layer, so the lowest $\mathrm{Ra}_{\mathrm{cr}}$ will correspond to perturbations with $\ell \sim H$. The initiation of convection then simply requires Ra to exceed some critical value, which depends only on the geometry and boundary conditions (it is for example $27\pi^4/4 \simeq 657.5$ for Rayleigh-Bénard convection in a plane layer with free-slip boundaries). In a rotating sphere with radial gravity, driving radial motion requires initiating the convection with a horizontal length scale $\lesssim \ell_c = HE^{1/3}$. Since the critical Rayleigh number is a decreasing function of ℓ, we expect the length scale of the fastest growing mode to be $\sim\ell_c$, which gives

$$\mathrm{Ra}_{\mathrm{cr}} \sim E^{-4/3}, \tag{5.113}$$

as predicted by linear stability analysis.

Compressibility Effect

In the above analysis, we have left aside compressibility effects on the initiation of convection. Rather than the Boussinesq version of the heat transfer equation (5.85), we now consider its more general form

$$\frac{DT}{Dt} = \kappa\nabla^2 T + \frac{\alpha T}{\rho c_p}\frac{DP}{Dt} + \dot{\epsilon} : \tau, \tag{5.114}$$

where $\dot{\epsilon} : \tau$ is the viscous dissipation (τ is the stress tensor and $\dot{\epsilon}$ the deformation rate tensor). Equation (5.114) can be obtained from the entropy balance (for its derivation, see e.g. Ricard 2007).

Schwarzschild's Stability Criterion for Thermal Convection

To see the effect of the pressure term, let us consider a parcel of fluid displaced upward with negligible viscous dissipation and heat diffusion (thus following an isentropic path). It follows directly from Eq. (5.114) that the parcel's temperature will vary with its pressure as

$$\frac{DT}{DP} = \frac{\alpha T}{\rho c_p} \equiv \frac{\partial T_{\text{ad}}}{\partial P}, \tag{5.115}$$

which corresponds to adiabatic heating (cooling) due to compression (decompression).[7] This is equivalent to a radial gradient, the *adiabatic gradient*, given by

$$\frac{\partial T_{\text{ad}}}{\partial r} = -\frac{\alpha T g}{c_p}. \tag{5.116}$$

The adiabatic temperature difference ΔT_{ad} across the outer core is obtained by integrating one of Eqs. (5.115) or (5.116) (here assuming that α/c_p is constant and that $g(r) = g_{\text{cmb}} r / r_{\text{cmb}}$):

$$\Delta T_{\text{ad}} = T_{\text{icb}} \left[\exp\left(-\int_{r_{\text{icb}}}^{r_{\text{cmb}}} \frac{\alpha g}{c_p} dr \right) - 1 \right], \tag{5.117}$$

$$= T_{\text{icb}} \left[\exp\left(-\frac{1}{2} \frac{\alpha g_{\text{cmb}}}{c_p} \frac{r_{\text{cmb}}^2 - r_{\text{icb}}^2}{r_{\text{cmb}}} \right) - 1 \right], \tag{5.118}$$

where T_{icb} is the temperature at the Inner Core Boundary. In Earth's core, the adiabatic temperature difference is $\Delta T_{\text{ad}} \sim 1000$ K, so this is a significant effect.

If displaced fast enough for negligible heat diffusion, a parcel of fluid displaced vertically from its initial state will follow an adiabatic path. Its temperature T will then vary with pressure P according to Eq. (5.115). If the background temperature gradient $\bar{T}$ is steeper than the adiabatic gradient, i.e. if

$$\frac{\partial \bar{T}}{\partial P} > \frac{\partial T_{\text{ad}}}{\partial P}, \tag{5.119}$$

then the temperature of a parcel of fluid displaced upward will become larger than the background temperature, with the temperature difference increasing with the upward displacement of the parcel (Fig. 5.10**a**). The parcel will thus become less and less dense than the surrounding, and will keep rising. In contrast, if

$$\frac{\partial \bar{T}}{\partial P} < \frac{\partial T_{\text{ad}}}{\partial P}, \tag{5.120}$$

[7]The evolution of the parcel is actually even an isentropic process, since the parcel's evolution is both adiabatic (no transfer of heat and mass with its surrounding) and reversible (no friction).

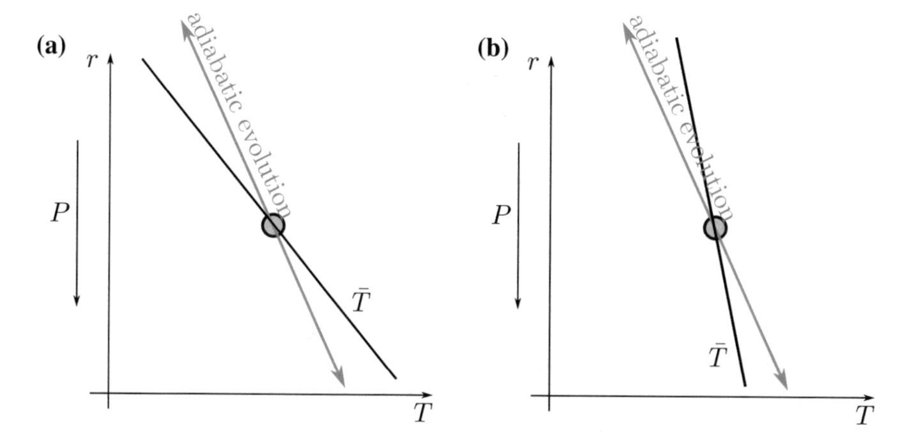

Fig. 5.10 Schwarzschild's criterion for thermal convection in a compressible fluid: a parcel of fluid displaced vertically and evolving adiabatically would find itself at a temperature higher than its surrounding if $\partial\bar{T}/\partial P > \partial T_{\text{ad}}/\partial P$ (figure **a**), and lower than its surrounding if $\partial\bar{T}/\partial P < \partial T_{\text{ad}}/\partial P$ (figure **b**)

then a parcel of fluid displaced upward will have a temperature lower than the background temperature (Fig. 5.10**b**). It will thus be denser than the surrounding fluid, and its buoyancy will eventually send it back toward its initial position. A temperature profile satisfying the condition (5.120) is thus stable against thermal convection.

Equation (5.119) is thus a *necessary* condition for thermal convection, known as *Schwarzschild's stability criterion.*[8] It is not a *sufficient* condition since it assumes that fluid parcels follow adiabatic and reversible paths when displaced upward or downward. In a real fluid, the parcels would exchange heat with the surrounding by diffusion, and viscous dissipation would be non-zero.

Critical Rayleigh Number with Compressibility Effects

Going back to our heuristic derivation of the critical Rayleigh number, we consider again the evolution of a fluid parcel displaced from its initial position, but now assume that the evolution of its temperature is governed by

$$\frac{DT}{Dt} = \kappa\nabla^2 T + \frac{\alpha T}{\rho c_p}\frac{DP}{Dt}, \tag{5.121}$$

ignoring the viscous dissipation term $\dot{\epsilon} : \tau$ included in Eq. (5.114). Viscous dissipation is quadratic in **u**, and can therefore be neglected when looking at the initiation of convection.

[8] Alternatively, Eq. (5.120) is a sufficient condition for stability.

In Eq. (5.121), the pressure derivative can be estimated as

$$\frac{DP}{Dt} \simeq u_r \frac{\partial P}{\partial r} \simeq -\rho g u_r \tag{5.122}$$

neglecting terms in pressure fluctuations. Equation (5.121) thus gives

$$\frac{DT}{Dt} \simeq -\kappa \frac{T - \bar{T}}{\ell^2} - u_r \frac{dT_{\mathrm{ad}}}{dr}. \tag{5.123}$$

Using this relation instead of Eq. (5.106) and going through the same steps [Eqs. (5.104)–(5.113)], the rate of change of the parcel's velocity is now given by

$$\frac{1}{u_r}\frac{Du_r}{Dt} \sim \frac{\kappa}{\ell^2}\left(-\frac{\alpha g \ell^4}{\kappa \nu}\frac{d\left(\bar{T} - T_{\mathrm{ad}}\right)}{dr} - 1\right) \tag{5.124}$$

instead of Eq. (5.110). Denoting by $\Delta T_{>\mathrm{ad}} = \Delta T - \Delta T_{\mathrm{ad}}$ the excess temperature difference above the adiabatic temperature across the outer core (or super-adiabatic temperature difference), we have $\frac{d(\bar{T}-T_{\mathrm{ad}})}{dr} \sim -\Delta T_{>\mathrm{ad}}/H$, and

$$\frac{1}{u_r}\frac{Du_r}{Dt} \sim \frac{\kappa}{\ell^2}\left(\frac{Ra_{>\mathrm{ad}}}{(H/\ell)^4} - 1\right), \tag{5.125}$$

where

$$Ra_{>\mathrm{ad}} = \frac{\alpha g H^3 \Delta T_{>\mathrm{ad}}}{\kappa \nu} \tag{5.126}$$

is the super-adiabatic Rayleigh number. The instability criterion is thus generalised by simply replacing the classical Rayleigh number by the super-adiabatic Rayleigh number. In a compressible fluid at low Ekman, convection will thus start if

$$Ra_{>\mathrm{ad}} > \mathrm{Ra}_{\mathrm{cr}} \sim E^{-4/3}. \tag{5.127}$$

Application to Earth's Core

Thermal Convection

How much super-adiabatic the core needs to be to meet this requirement? Using the definitions of $Ra_{>\mathrm{ad}}$ and E, the condition (5.127) can be recast as a condition for the super-adiabatic temperature difference:

$$\Delta T_{>\mathrm{ad}} \gtrsim \frac{\Omega^{4/3}\kappa}{\alpha g H^{1/3}\nu^{1/3}}, \tag{5.128}$$

Table 5.1 Representative values of some properties of the outer core

Radius of the core-mantle boundary	r_{cmb}	3480 km
Radius of the inner core boundary	r_{icb}	1221 km
Acceleration of gravity at CMB	g_{cmb}	10.68 $\mathrm{m.s^{-2}}$
Kinematic viscosity	ν	10^{-6} $\mathrm{m^2.s^{-1}}$
Thermal diffusivity	κ	5×10^{-6} $\mathrm{m^2.s^{-1}}$
Compositional diffusivity	κ_χ	$10^{-9\pm2}$ $\mathrm{m^2.s^{-1}}$
Magnetic diffusivity	η	1 $\mathrm{m^2.s^{-1}}$
Thermal expansion coefficient	α	10^{-5} $\mathrm{K^{-1}}$
Compositional expansion coefficient	β	1 $\mathrm{wt\%^{-1}}$

with parameter values from Table 5.1, the required $\Delta T_{>\mathrm{ad}}$ is only on the order of $10^{-7} - 10^{-6}$ K: this is tiny, compared for example to the adiabatic temperature variation across the core, which is $\sim$1000 K.

In practice, this means that Schwarzschild's criterion is a very good indicator of the likelihood of core convection. In a planetary core, the temperature difference across the core is not imposed. What is imposed is the heat flux Q_{cmb} at the Core-mantle Boundary, which is controlled by mantle convection. A requirement for thermal convection is therefore that the heat flux imposed by mantle convection is larger than the heat flux $Q^{\mathrm{ad}}_{\mathrm{cmb}}$ which can be carried by diffusion along an adiabatic temperature profile:

$$Q_{\mathrm{cmb}} > Q^{\mathrm{ad}}_{\mathrm{cmb}} = -4\pi r^2_{\mathrm{cmb}} k \frac{\partial T_{\mathrm{ad}}}{\partial r}, \tag{5.129}$$

$$= 4\pi r^2_{\mathrm{cmb}} \frac{\alpha\, g_{\mathrm{cmb}}\, T_{\mathrm{cmb}}}{c_p} k, \tag{5.130}$$

where k is the thermal conductivity of the core. The value of the thermal conductivity of the core is currently quite debated. Until quite recently, available estimates were in the range 30–60 $\mathrm{W.m^{-1}.K^{-1}}$ (Stacey and Anderson 2001; Stacey and Loper 2007), but much higher values (>150 $\mathrm{W.m^{-1}.K^{-1}}$) have since been proposed by several independent groups (de Koker et al. 2012; Pozzo et al. 2012; Gomi et al. 2013, 2016). Other groups (Seagle et al. 2013; Konôpková et al. 2016) still favour a relatively low value of k.

Finally, it is also instructive to estimate the horizontal scale ℓ_c at which convection is initiated. From Eq. (5.100), this is $\ell_c \sim H E^{1/3} \sim 2000\ \mathrm{km} \times (10^{-15})^{1/3} \sim 20$ m, a factor 10^5 smaller than the outer core shell thickness!

Compositional Convection

Though Eqs. (5.83)–(5.85) have been written for thermal convection, the same set of equations can be used to describe compositional convection if T is replaced by a concentration (in light elements) χ, α by a coefficient of compositional expansion β, and κ by a compositional diffusivity κ_χ. There is no direct compositional analog of the adiabatic gradient, so the condition for compositional convection is simply

$$Ra_\chi = \frac{\beta g H^3 \Delta\chi}{\kappa_\chi \nu} > \mathrm{Ra}_{\mathrm{cr}} \sim E^{-4/3}, \tag{5.131}$$

where $\Delta\chi$ is the light element concentration difference across the outer core. Written in terms of $\Delta\chi$, condition (5.131) is equivalent to the following condition:

$$\Delta\chi \gtrsim \frac{\Omega^{4/3}\kappa_\chi}{\beta g H^{1/3} \nu^{1/3}} \simeq 10^{-12}\ \mathrm{wt.\%}. \tag{5.132}$$

A tiny difference in composition can drive core convection.

Convective Dynamos

Governing Equations

In its simplest form, the set of governing equations for a convectively driven dynamo consists of the three equations governing rotating convection with the addition of the Lorentz force in the Navier–Stokes equation, plus the induction equation:

$$\rho\frac{D\mathbf{u}}{Dt} + 2\rho\Omega\times\mathbf{u} = -\nabla p + \alpha\rho T g \mathbf{e}_r + \frac{1}{\mu_0}(\nabla\times\mathbf{B})\times\mathbf{B} + \rho\nu\nabla^2\mathbf{u}, \tag{5.133}$$

$$\nabla\cdot\mathbf{u} = 0, \tag{5.134}$$

$$\frac{DT}{Dt} = \kappa\nabla^2 T, \tag{5.135}$$

$$\frac{D\mathbf{B}}{Dt} = (\mathbf{B}\cdot\nabla)\,\mathbf{u} + \eta\nabla^2\mathbf{B}. \tag{5.136}$$

This set of equations can be made dimensionless by using the characteristic scales already used in section "Rotating convection" in the case of thermal convection for lengths (H), time (H^2/ν), temperature (ΔT or qH/k) and pressure ($\rho\Omega\nu$). A magnetic field scale can, for example, be obtained by assuming a balance between the Coriolis acceleration and the Lorentz force. Writing the Lorentz force as $\mathbf{j}\times\mathbf{B}$ and estimating $\mathbf{j}$ from Ohm's law as $\mathbf{j}\sim\sigma\mathbf{u}\times\mathbf{B}$ gives a Lorentz force $\sim\sigma U B^2$. Balancing this with the Coriolis force which is $\sim\rho\Omega U$ gives a magnetic field scale

$B_0 = \sqrt{\rho\Omega/\sigma}$. Using this set of scales, we obtain the following dimensionless set of equations:

$$E\frac{D\mathbf{u}}{Dt} + 2\mathbf{e}_z \times \mathbf{u} = -\nabla P + \frac{E}{Pr} Ra\, T\mathbf{e}_r + (\nabla \times \mathbf{B}) \times \mathbf{B} + E\nabla^2\mathbf{u}, \tag{5.137}$$

$$\nabla \cdot \mathbf{u} = 0, \tag{5.138}$$

$$\frac{DT}{Dt} = \frac{1}{Pr}\nabla^2 T, \tag{5.139}$$

$$\frac{D\mathbf{B}}{Dt} = (\mathbf{B} \cdot \nabla)\,\mathbf{u} + \frac{1}{P_m}\nabla^2\mathbf{B}, \tag{5.140}$$

where the magnetic Prandtl number P_m is defined as the ratio of the kinematic viscosity ν to the magnetic diffusivity:

$$P_m = \frac{\nu}{\eta}. \tag{5.141}$$

P_m is about 10^{-6} in Earth's core: the magnetic field diffuses much faster than momentum, and we therefore expect the magnetic field to vary over larger length scales that the velocity field.

In addition to the *input* non-dimensional numbers (E, Ra, Pr, P_m), it is also often useful to consider *output* non-dimensional numbers based on measured dynamical quantities such as some averages of the velocity and magnetic field, $\langle u \rangle$ and $\langle B \rangle$. Examples of useful output non-dimensional numbers include the following:

$$\text{Reynolds number } Re = \frac{\text{inertia}}{\text{viscous forces}} = \frac{H\langle u \rangle}{\nu}, \tag{5.142}$$

$$\text{magnetic Reynolds number } R_m = \frac{\text{stretching of } \mathbf{B}}{\text{diffusion } \mathbf{B}} = \frac{H\langle u \rangle}{\eta} = P_m Re, \tag{5.143}$$

$$\text{Rossby number } Ro = \frac{\text{inertia}}{\text{Coriolis force}} = \frac{\langle u \rangle}{\Omega H} = E\,Re, \tag{5.144}$$

$$\text{Elsasser number } \Lambda = \frac{\text{Lorentz force}}{\text{Coriolis force}} = \frac{\sigma\langle B \rangle^2}{\rho\Omega}. \tag{5.145}$$

In this expression of the Elsasser number, the Lorentz force has been estimated from the expression $\mathbf{j} \times \mathbf{B}$ by taking $\mathbf{j} \sim \sigma\mathbf{u} \times \mathbf{B}$ from Ohm's law.

These numbers can be estimated for Earth's core as follows. If we accept that the geomagnetic field is produced by dynamo effect in the core, then the magnetic Reynolds number must be larger than $\mathcal{O}(10)$ (the critical magnetic Reynolds number for dynamo action in a spherical shell is typically $\sim$50). This implies that Re must be at least of $\mathcal{O}(10^7)$ and Rossby above $\mathcal{O}(10^{-8})$. The order of magnitude of the velocity corresponding to $R_m \sim 10$ is $10\eta/H \sim 5 \times 10^{-6}$ m.s^{-1}. If instead we take a velocity scale of 5×10^{-4} m.s^{-1} as obtained from core flow inversions, we obtain

$R_m \sim 10^3$, $Re \sim 10^9$, and $Ro \sim 10^{-6}$. With a magnetic field of ~4 mT (Gillet et al. 2010), the Elsasser number is $\Lambda \sim 10$.

The values of these numbers suggest that in the Navier–Stokes equation the dominant forces would be the Coriolis force, the Lorentz force, and presumably the buoyancy force (which is difficult to estimate from simple arguments, but which is likely non-negligible since it is the source of motion). This corresponds to the so-called MAC balance (Magnetic, Archimedes, Coriolis).

Solving numerically these equations in regimes which are relevant to the geodynamo is difficult; it has in fact not yet been possible to solve them with parameter values approaching that of the Earth. They are two main reasons for this:

1. The low viscosity of molten iron means that the velocity field likely develops small scale turbulent fluctuations, which would require a high spatial resolution and fine time-stepping to be fully resolved. We have seen in section "Rotating Convection" that thermal convection would initiate at a length scale on the order of $HE^{1/3}$, or about 20 m for $E \sim 10^{-15}$. This is 10^5 smaller than the outer core thickness. Resolving this scale in 3D numerical simulations would necessitate at least 10^5 grid points in each direction, or 10^{15} grid points in total. Today's most resolved numerical simulations of the geodynamo have a spatial resolution of about 2 km, which allows to reach $E = 10^{-7}$ (Schaeffer et al. 2017).
2. The magnetic Prandtl number is quite small in liquid metals (probably $\sim 10^{-6}$ in Earth's core), but much larger values are being used in dynamo simulations. Solving the induction equation with a low P_m present no intrinsic difficulty: a high magnetic diffusivity usually means smooth variations of the magnetic field and small magnetic fields gradients. What is difficult is to obtain dynamo action at a low P_m in numerical simulations. The reason for this is easily understood by noting that the magnetic Reynolds number can be written as the product of the Reynolds number and magnetic Prandtl number: $R_m = Re \times P_m$. Since dynamo operation requires reaching $R_m > \mathcal{O}(10)$, doing this with a low P_m requires a high value of Re. This implies the development of turbulent velocity fluctuations down to small length scales, which are difficult to resolve numerically.

Successes and Challenges

Though numerical dynamos are still quite far from Earth's conditions in terms of non-dimensional parameters, this does not mean that relevant numerical simulations cannot be done. In the past few decades, the geodynamo modelling community has been quite successful in strengthening the case for a convectively powered geodynamo.

As already discussed in section "The Geodynamo Hypothesis", a major achievement has been to demonstrate that dynamos can indeed be sustained by rotating convection (Zhang and Busse 1988; Glatzmaier and Roberts 1995; Kageyama et al. 1995). Since the first few numerical models of the dynamo (which were done at relatively high E and P_m), the non-dimensional parameter space has been explored toward lower E and P_m, and higher Ra. By performing geodynamo simulations at various values of the governing non-dimensional parameters, scaling laws for typical velocity and magnetic field strength have been obtained. When extrapolated to the core condition, the predictions of these scaling laws are in reasonable agreement with estimates of the core flow (from inversion of secular variation of the CMB magnetic field) and observed field strength (e.g. Starchenko and Jones 2002; Olson and Christensen 2006; Christensen and Aubert 2006; Christensen 2010). In addition, it has been shown that the magnetic field can be "Earth-like" in a well-defined region of the parameter space, which includes the estimated state of Earth's core (Christensen et al. 2010). "Earth-like" geodynamo simulations also often exhibit polarity reversals.

In spite of the important successes described above, the question of whether numerical simulations do reach a dynamical regime similar to planetary cores has been a constant preoccupation of the geodynamo community. In other words, are geodynamo simulations "Earth-like" for good reasons, or is it just coincidental? Numerical simulations are still far from the Earth in terms of non-dimensional numbers: the most massive numerical simulations of the geodynamo have reached $E = 10^{-7}$, $P_m = 0.1$, with a Rayleigh number 6×10^3 times supercritical (Schaeffer et al. 2017). Though quite impressive, this is still far from the Earth (seven or eight orders of magnitude in terms of E, five orders of magnitude in P_m). And most of available geodynamo simulations are significantly further away from Earth's core parameters values. The main question is whether viscous effects in state of the art simulations play a significant role or not. Christensen and Aubert (2006) have argued, from a large set of numerical dynamos ($E \in 10^{-6} - 3 \times 10^{-4}$, $P_m \in 0.06 - 10$), that the magnetic field strength and mean velocity follow scaling laws which are independent of diffusivities. This may suggest that an asymptotic regime has been reached, and that further decreasing the viscosity down to Earth's value would not change the dominant force balance, allowing to use the obtained scaling laws to extrapolate outputs of numerical simulations to planetary cores conditions. However, several authors have questioned whether viscosity effects in current geodynamo simulations are indeed small enough, and whether diffusivity-free scaling laws correctly describe the available set of numerical geodynamos (Stelzer and Jackson 2013; King and Buffett 2013; Cheng and Aurnou 2016). The ratio of magnetic to kinetic energies, which is estimated to be $\sim 10^3 - 10^4$ in Earth's core, is much smaller in numerical models at low ($\lesssim 1$) P_m: $\mathcal{O}(10)$ at best, and often below 1 (Schaeffer et al. 2017).

Energetics of the Geodynamo

Figure 5.11 shows schematically the energy flow of a dynamo powered by natural convection (thermal or compositional) or astronomical forcing (tides, precession, nutation, or libration). Natural convection converts the gravitational potential energy of unstable density gradients into kinetic energy; inertial instabilities excited by

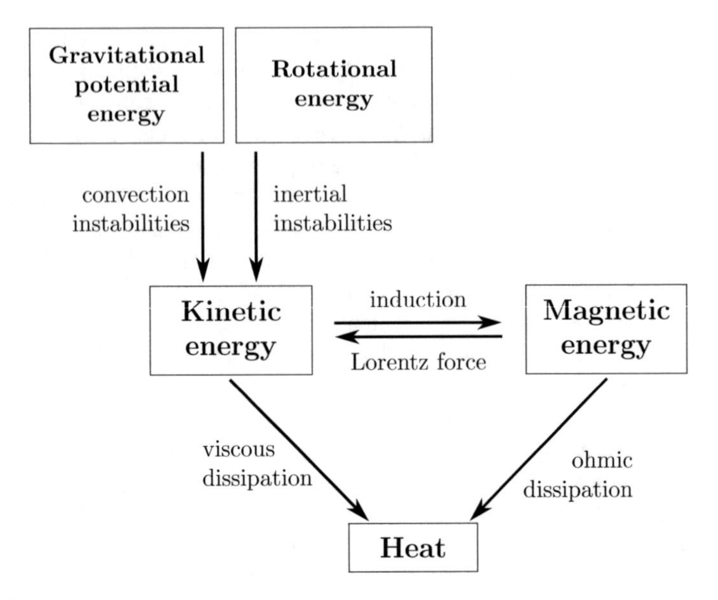

Fig. 5.11 Energy flow of the geodynamo

astronomical forcing converts rotational energy into kinetic energy. If the magnetic Reynolds number is large enough for dynamo action, a fraction of this kinetic energy is converted into magnetic energy through induction. Magnetic energy can be converted back into kinetic energy (through the work of the Lorentz force). Kinetic and magnetic energies are dissipated into heat through viscous and ohmic dissipation.

Dynamo action requires the rate of release of gravitational or rotational energies to be high enough to maintain the magnetic field against viscous and ohmic dissipation. The two energy reservoirs are huge: the gravitational potential energy of the Earth is $\simeq 2 \times 10^{32}$ J, and the rotational energy is $\simeq 2 \times 10^{29}$ J. Only a fraction of these would be enough to maintain Earth's magnetic field for the last 4 Gy if the amount of dissipation is in the range 0.1–3 TW as discussed later in this section. However, the fact that enough energy is available in Earth's system is by no means a sufficient condition for sustaining the geomagnetic field. Only a small fraction of this energy can actually be made available to sustain the geomagnetic field; the goal of the energetics approach of the geodynamo is to estimate this fraction.

An equation for the evolution of the magnetic energy $E_m = \int_{\mathcal{V}_\infty} \frac{B^2}{2\mu_0} d\mathcal{V}$ can be obtained from Ohm's law and Maxwell's equations, or equivalently from the dot product of the induction equation with $\mathbf{B}$. Integration over the whole space $\mathcal{V}_\infty$ gives

$$\underbrace{\frac{d}{dt}\int_{\mathcal{V}_\infty} \frac{B^2}{2\mu_0} d\mathcal{V}}_{\text{Rate of change of magnetic energy}} = -\underbrace{\int_{\mathcal{V}} \mathbf{u} \cdot \mathbf{f}_L d\mathcal{V}}_{\text{power of Lorentz force}} - \underbrace{\int_{\mathcal{V}} \frac{j^2}{\sigma} d\mathcal{V}}_{\text{Ohmic dissipation (=Joule heating)}}, \tag{5.146}$$

where $\mathbf{f}_L$ is the Lorentz force, and $\mathcal{V}$ the volume of the core. In this equation, the magnetic energy includes the contribution of the magnetic field outside the core. On the other hand, the integrals of the power of the Lorentz force and of ohmic dissipation can be restricted to the core volume $\mathcal{V}$ under the assumption that the electric conductivity is equal to 0 outside the core. This equation cannot be used alone to predict whether the geodynamo can be dynamically sustained, because the (unknown) velocity field $\mathbf{u}$ appears in the equation. We therefore need a second conservation equation involving $\mathbf{u}$, which is obtained by taking the dot product of Navier–Stokes equation with $\mathbf{u}$, and again integrating over the volume of the core. This gives an equation for the time derivative of the total kinetic energy of the core:

$$\frac{dE_k}{dt} = \int_{\mathcal{V}} (\mathbf{f}_L + \rho \mathbf{g} - \nabla P) \cdot \mathbf{u}\, d\mathcal{V} + \oint_{\mathcal{S}} (\underline{\mathcal{T}} \cdot \mathbf{u}) \cdot \mathbf{n}\, d\mathcal{S} - \Phi_\nu, \qquad (5.147)$$

where $\mathcal{S}$ is the surface area of the core-mantle boundary, $\underline{\mathcal{T}}$ is the deviatoric stress tensor and $\Phi_\nu = \int_{\mathcal{V}} \underline{\mathcal{T}} : \nabla \mathbf{u}\, d\mathcal{V}$ is the viscous dissipation. There is no power associated with the Coriolis force since $(\rho \Omega \times \mathbf{u}) \cdot \mathbf{u} = 0$.

Combining the kinetic and magnetic energy equations, we can get rid of the power of the Lorentz force and obtain:

$$\frac{d}{dt}(E_m + E_k) = \underbrace{\int_{\mathcal{V}} (\rho \mathbf{g} - \nabla P) \cdot \mathbf{u}\, d\mathcal{V} + \oint_{\mathcal{S}} (\underline{\mathcal{T}} \cdot \mathbf{u}) \cdot \mathbf{n}\, d\mathcal{S}}_{\text{available power} P_a} - \underbrace{(\Phi_\nu + \Phi_J)}_{\Phi} \qquad (5.148)$$

In this equation:

1. The term $\int_{\mathcal{V}} (\rho \mathbf{g} - \nabla P) \cdot \mathbf{u}\, d\mathcal{V}$ is the rate of work of non-hydrostatic processes. It is equal to 0 if the core is in a hydrostatic state. In planetary cores, deviations from a hydrostatic state can be due to density, pressure or gravity field fluctuations, which can arise from either natural convection or astronomical forcing. The integral is positive in the case of natural convection; it can be either positive or negative in the case of astronomical forcing.
2. The term $\oint_{\mathcal{S}} (\underline{\mathcal{T}} \cdot \mathbf{u}) \cdot \mathbf{n}\, d\mathcal{S}$ is the rate of work of deviatoric stresses (viscous stresses) at the CMB. It is different from zero if the mantle rotates at a different rate than the core (in particular due to precession/nutation/libration) or if the CMB is deformed (tides).
3. Φ is the total dissipation due to Joule heating (Ohmic dissipation) and viscous heating (viscous dissipation). It is always positive.

Starting from a state at rest with $E_k = E_m = 0$, the sum of the kinetic and magnetic energy would increase if the sum of the two first terms on the RHS of Eq. (5.148), which we denote by P_a, is positive. Increasing $\mathbf{u}$ and $\mathbf{B}$ will increase the dissipation Φ up to a point of saturation at which the rate of energy production is on average balanced by dissipation, i.e. $P_a \simeq \Phi$. In a statistically steady state, time averaging

Eq. (5.148) gives

$$\langle \Phi \rangle = P_a = \left\langle \int_{\mathcal{V}} (\rho \mathbf{g} - \nabla P) \cdot \mathbf{u} \, d\mathcal{V} + \oint_{\mathcal{S}} (\underline{\tau} \cdot \mathbf{u}) \cdot \mathbf{n} \, d\mathcal{S} \right\rangle . \tag{5.149}$$

From an energetic point of view, the geodynamo problem amounts to estimate whether the rate of energy release P_a is large enough to sustain Earth's magnetic field against dissipation. To answer this question, we need to estimate: (i) how dissipative is the geodynamo, and (ii) what is the available power P_a in the core.

How Dissipative is the Geodynamo?

The amount of ohmic dissipation in the core is given by

$$\Phi_J = \int_{\mathcal{V}} \frac{j^2}{\sigma} d\mathcal{V} = \int_{\mathcal{V}} \frac{|\nabla \times \mathbf{B}|^2}{\sigma \mu_0^2} d\mathcal{V}.$$

To estimate how dissipative is the geodynamo, we thus need to know how $\mathbf{B}$ varies spatially in the core. This is difficult. As discussed in section "The Geometry of Earth's Magnetic Field", direct observations of the core magnetic field are restricted to the poloidal field at the core-mantle boundary, up to spherical harmonic degree $l_{\max} = 13$, which corresponds to a spatial wavelength of about 1600 km. We have no direct observation of the harmonic components of the poloidal field at $l > 13$, and no constraint on the spatial variations of the toroidal part of the magnetic field. Numerical simulations of the geodynamo suggest that ohmic dissipation is dominated by the contributions of smaller scale components of the magnetic field (Roberts et al. 2003), so estimating the geodynamo dissipation from direct observations seems hopeless. Published estimates of Φ_J have been obtained from extrapolations of numerical or experimental dynamos results (Buffett 2002; Roberts et al. 2003; Christensen and Tilgner 2004; Christensen et al. 2010; Stelzer and Jackson 2013), and range from ~0.1 TW to several TW.

Estimating the Available Power

The energetics approach of the geodynamo problem has so far been restricted to the case of a convectively driven dynamo; the case of mechanical forcing has not been treated in a rigorous and usable way, in part because of the difficulty of estimating the $\int_{\mathcal{V}} (\rho \mathbf{g} - \nabla P) \cdot \mathbf{u} \, d\mathcal{V}$ term. We will thus restrict ourselves to the case of a convectively driven dynamo. In this situation the surface integral in equation (5.148) is equal to 0.

Two different methods have been classically used: either work on the entropy budget rather than on Eq. (5.148) (e.g. Gubbins et al. 2004; Labrosse 2015), or estimate directly the contribution of convective motions on P_a. These two approaches are equivalent (Lister 2003) and in both cases require assumptions on the state of the core, which is usually assumed to be continuously well mixed by convective motions. Here we use the later approach, which is perhaps a bit more physically intuitive, and follows the approach of Buffett et al. (1996). We will only sketch the derivation of the model; the details can be found in Buffett et al. (1996).

Denoting by ψ the gravitational potential, with $\mathbf{g} = -\nabla\psi$, one can show by using mass conservation that

$$P_a = \int_{\mathcal{V}} (-\rho\nabla\psi - \nabla P) \cdot \mathbf{u}\, d\mathcal{V} \tag{5.150}$$

$$= -\int_{\mathcal{V}} \psi(r) \frac{\partial \rho}{\partial t} d\mathcal{V} \tag{5.151}$$

where $\partial\rho/\partial t$ is the rate of change of the density due solely to convective re-arrangement. The next step is to calculate the change of density in the core associated to the convective re-distribution of heat and chemical elements, under the assumption that convection keeps the core well mixed and isentropic. By doing this, one can write P_a as

$$P_a = \underbrace{\left(\beta\rho F_{\mathrm{icb}} + \frac{\alpha}{c_p} Q_{\mathrm{icb}}^{>\mathrm{ad}}\right)\left(\bar{\psi} - \psi_{\mathrm{icb}}\right)}_{\text{Contribution of IC solidification}} + \underbrace{\frac{\alpha}{c_p} Q_{\mathrm{cmb}}^{>\mathrm{ad}}\left(\psi_{\mathrm{cmb}} - \bar{\psi}\right)}_{\text{Contribution of CMB flux}} \tag{5.152}$$

where $Q_{\mathrm{icb}}^{>\mathrm{ad}} = Q_{\mathrm{icb}} - Q_{\mathrm{icb}}^{\mathrm{ad}}$ and $Q_{\mathrm{cmb}}^{>\mathrm{ad}} = Q_{\mathrm{cmb}} - Q_{\mathrm{cmb}}^{\mathrm{ad}}$ are the super-adiabatic heat flux at the inner core boundary (ICB) and core-mantle boundary (CMB), $F_{\mathrm{icb}} = 4\pi r_{\mathrm{icb}}^2 \dot{r}_{\mathrm{icb}} c$ is the flux of light elements concentration (in wt.%.m^3.s^{-1}) at the ICB (c being the concentration in light elements of the outer core), $\bar{\psi}$ is the mass-averaged value of the gravitational potential in the outer core, and ψ_{cmb} and ψ_{icb} are the values of the gravitational potential at the CMB and ICB. The factors $\bar{\psi} - \psi_{\mathrm{cmb}}$ and $\psi_{\mathrm{icb}} - \bar{\psi}$ come from the fact that density perturbations originating from either the CMB or ICB are re-distributed over the whole core. The super-adiabatic heat flux at the ICB happens to be well approximated by the release of latent heat due to inner core solidification, $Q_{\mathrm{icb}}^{>\mathrm{ad}} \simeq 4\pi r_{\mathrm{icb}}^2 \rho L \dot{r}_{\mathrm{icb}}$, where $\dot{r}_{\mathrm{icb}}$ is the time derivative of the inner core radius r_{icb}, and L is the latent heat of iron at core conditions. Since F_{icb} is also proportional to $\dot{r}_{\mathrm{icb}}$, the whole ICB contribution is proportional to the rate of growth of the inner core, which is itself controlled by the rate at which heat is extracted from the core.

The terms in Eq. (5.152) are all tied to the core thermal evolution and controlled by the heat flux at the CMB. To use Eq. (5.152) in a predictive way, one needs to relate the rate of inner core growth to the heat flux at the CMB, which can be done by considering the energy balance of the core (coming from the first law of

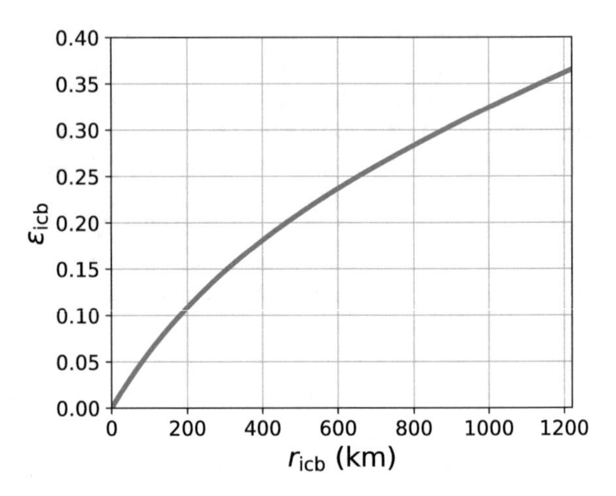

Fig. 5.12 *Dynamo efficiency* ϵ_{icb} associated with compositional and thermal buoyancy fluxes from the inner core boundary, as a function of inner core size (calculated from Buffett et al. (1996))

thermodynamics), which simply states that the heat flux extracted from the core at the CMB is equal to the sum of sensible heat released by the cooling of the core, latent heat of inner core crystallisation, and compositional energy (associated with the mixing in the core of light elements released by inner core solidification). Doing this allows to write Eq. (5.152) as

$$P_a = \epsilon_{\text{icb}} Q_{\text{cmb}} + \epsilon_{\text{cmb}} Q_{\text{cmb}}^{>\text{ad}}, \tag{5.153}$$

where ϵ_{icb} and ϵ_{cmb} are the so-called *dynamo efficiencies*.[9] Expressions for these efficiencies can be found in Buffett et al. (1996) or Lister (2003). One important point is that ϵ_{icb} is an increasing function of inner core size (Fig. 5.12), which means that the contribution of compositional convection increases with inner core size. An approximate expression of ϵ_{cmb} is

$$\epsilon_{\text{cmb}} \simeq \frac{1}{5} \frac{\alpha g_{\text{cmb}} r_{\text{cmb}}}{c_p} \simeq 0.1. \tag{5.154}$$

Having obtained estimates of ϵ_{icb} and ϵ_{cmb}, it is now possible to estimate the convective power P_a. One key parameter is the heat flux conducted along the adiabat, $Q_{\text{cmb}}^{\text{ad}}$, which depends mostly on the thermal conductivity of the outer core (see Eq. (5.130)). Figure 5.13a and b show P_a as a function of the CMB heat flux and inner core size, for $Q_{\text{cmb}}^{\text{ad}}$ equal to either 5 TW or 15 TW. The 5 TW and 15 TW values are representative of the low ($\sim$30 W.m^{-1}.K^{-1}) and high ($\sim$100 W.m^{-1}.K^{-1}) thermal conductivity estimates. Figure 5.13c and d show the evolution with time of inner core size r_{icb} and convective power P_a, for $Q_{\text{cmb}}^{\text{ad}}$ equal to either 5 TW or 15

[9]The CMB efficiency is usually defined in another way, such that $P_a = (\epsilon_{\text{icb}} + \epsilon_{\text{cmb}})\, Q_{\text{cmb}}$. This has the advantage of linking P_a directly to Q_{cmb}, but then ϵ_{cmb} is itself a function of Q_{cmb} (it includes a factor $Q_{\text{cmb}}^{>\text{ad}}/Q_{\text{cmb}}$).

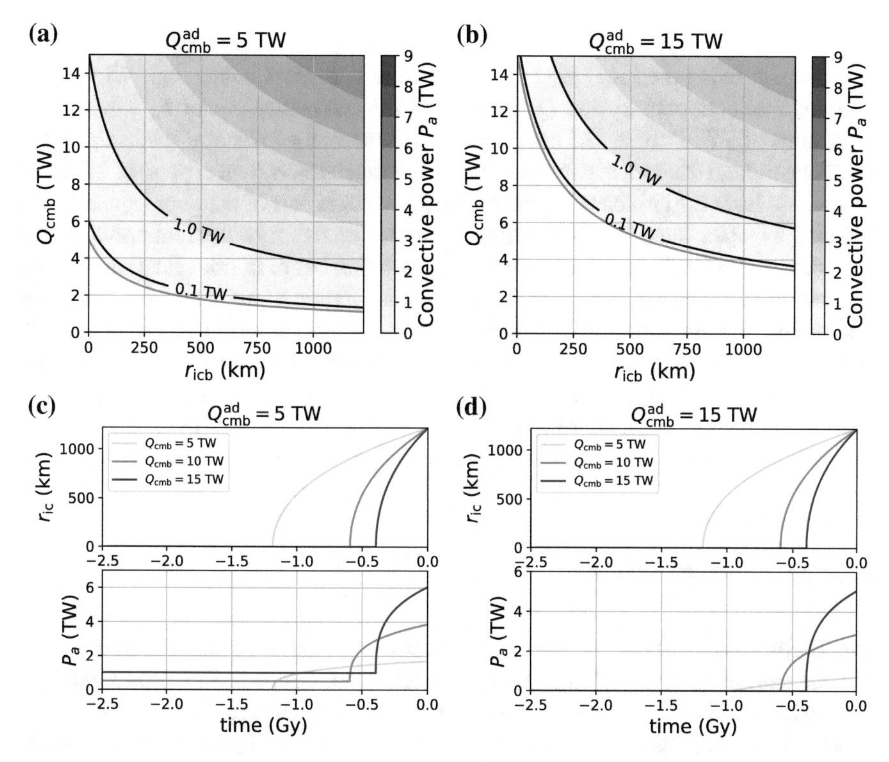

Fig. 5.13 **a** Convective power P_a as a function of r_{icb} and Q_{cmb}, with $Q^{ad}_{cmb} = 5$ TW; **b** same as **a** with $Q^{ad}_{cmb} = 15$ TW; **c** evolution with time of inner core size r_{icb} and convective power P_a, for $Q^{ad}_{cmb} = 5$ TW and Q_{cmb} equal to 5, 10, or 15 TW; **d** same as **c** with $Q^{ad}_{cmb} = 15$ TW

TW, and imposed CMB heat flux Q_{cmb} equal to 5, 10, or 15 TW, the current CMB heat flux being thought to be in the range 5–15 TW (Lay et al. 2006). On Fig. 5.13a, b, the grey line corresponds to the limit above which $P_a > 0$.

One can see that a positive P_a requires the CMB heat flux to be above a threshold value, which is equal to Q^{ad}_{cmb} at $r_{icb} = 0$, and decreases with increasing r_{icb} due to the increasing importance of the ICB buoyancy flux. From Eq. (5.153), a dynamo dissipating at a given rate Φ_0 requires a CMB heat flux

$$Q_{cmb}(P_a = \Phi_0) = \frac{\epsilon_{cmb} Q^{ad}_{cmb} + \Phi_0}{\epsilon_{cmb} + \epsilon_{icb}}. \tag{5.155}$$

The CMB heat flux above which $P_a > 0$ is given by setting $\Phi_0 = 0$ in Eq. (5.155). On Fig. 5.13a, b are also shown the Q_{cmb} values corresponding to $\Phi_0 = 0.1$ TW and 1 TW.

One can see that in the current state of the core, the convective power can easily be of a few terawatts if the CMB heat flux is in the range 5–15 TW as currently believed (Lay et al. 2006). There is therefore no difficulty in powering the geodynamo with the inner core at its current size. However, powering the geodynamo happens to be more problematic when the inner core is smaller, and before its nucleation. With a low core thermal conductivity and $Q_{\mathrm{cmb}}^{\mathrm{ad}} = 5$ TW, the convective power P_a would be of a few tens of TW. Driving the geodynamo with thermal convection seems therefore possible, though the convective power would be much lower than at present. Whether this would have a significant impact on the large-scale part of the geomagnetic field remains an open question. If the high estimates of the core thermal conductivity are correct, the CMB heat flux may well be sub-adiabatic. In this situation thermal convection is not possible, and driving the dynamo requires another source of motion and energy. Possible additional sources of energy include exsolution of light elements from the core (O'Rourke and Stevenson 2016; Badro et al. 2016; O'Rourke et al. 2017) and astronomical forcing (Andrault et al. 2016). In addition, adding some radioactive heating in the core (possible due to ^{40}K) would help by decreasing the rate of cooling of the core and increasing the age of the inner core (Labrosse 2015).

Inner Core Dynamics

The Earth's inner core is the deepest layer of our planet: a 1221 km-radius sphere of solid iron-alloy surrounded by molten metal. Its existence was unknown until the first observations of seismic reflexions at the inner core boundary by Inge Lehmann, in one of the shortest-title paper ever: P' (Lehmann 1936). The arrivals of P-waves in the core shadow zone, where P-waves are refracted away by the presence of the 3600-km-radius core, have been explained by the existence of a new discontinuity inside the core, the inner core boundary.

Since these first observations, the study of seismic waves travelling through the inner core and normal modes sampling the deepest layers have provided a blurry image of the inner core structure (Fig. 5.14). Birch (1940) and Jacobs (1953) have proposed that the inner core is a solid sphere of the same metal constituting the outer core, while the actual solidity has been demonstrated only several years later by Dziewonski and Gilbert (1971). Poupinet et al. (1983) were the first to note the different propagation velocities of waves travelling parallel to the rotation axis and perpendicular to it. This anisotropy of seismic properties has since been extensively studied, demonstrating the existence of a complex structure of the inner core. Among the most robust and surprising features of the inner core structure, we can cite two: a strong anisotropy for the bulk of the inner core, and an uppermost layer of the inner core with a strong hemispherical dichotomy in P-waves velocity, but no detectable anisotropy.

The existence of anisotropy in the inner core is an evidence for crystal orientation within the bulk of the inner core (iron crystals being elastically anisotropic). Such crystal orientation is the main motivation for studying flows in the inner core, as lattice-preferred orientation (LPO) may be deformation-induced (Karato 2012). Thus, it is thought that the observed structure may be an evidence for flows within the inner core, likely to be combined with initial crystallisation-induced LPO.

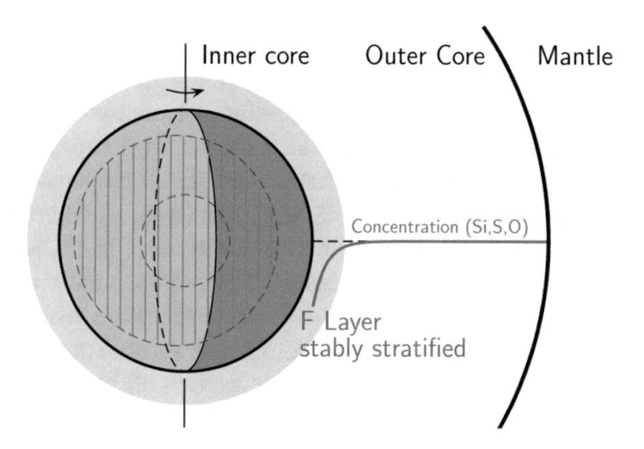

Fig. 5.14 A schematic view of the structure of the inner core. In addition to a global elastic anisotropy oriented parallel to the axis of rotation of the Earth (grey lines), the inner core has radial and horizontal variations of its seismic properties. There is a strong asymmetry between its western and eastern hemispheres (approximately defined by the Greenwich meridian), which have different seismic waves propagation velocity, attenuation, and degree of anisotropy. Anisotropy is weak or non-existent near the surface of the inner core, and increases in depth. Finally, the inner core is surrounded by a layer that appears to be stably stratified, the F-layer

We will discuss here the constitutive equations and some of the aspects of inner core dynamics, focusing on the flow forced in the inner core by the magnetic field diffused from the outer core.

Constitutive Equations

We consider an incompressible fluid in a spherical domain of radius r_{icb}, with a Newtonian rheology and uniform viscosity η. Neglecting inertia and rotation, the equation of continuity and the conservation of momentum are written as

$$\nabla \cdot \mathbf{u} = 0, \tag{5.156}$$

$$\mathbf{0} = -\nabla p' + \frac{\eta}{\rho_s}\nabla^2 \mathbf{u} + \mathbf{F}, \tag{5.157}$$

where $\mathbf{F}$ denotes volume forces, p' is the dynamic pressure (including gravity potential) and $\mathbf{u}$ the velocity field.

Among the volume forces, the buoyancy force is written as $\mathbf{F}_{\text{buoyancy}} = \Delta\rho\mathbf{g}$, where $\Delta\rho$ is the density difference compared to a neutral density profile and $\mathbf{g} = g_{\text{icb}} r / r_{\text{icb}} \mathbf{e}_r$ the acceleration of gravity. g_{icb} is the acceleration of gravity at the surface of the inner core.

In this model, the density variations only play a role in the buoyancy term, and they can be related to variations in temperature or concentration in light elements (Si, S, O, ...) compared to a reference state. We take as the reference temperature profile the adiabatic temperature profile $T_{\text{ad}}(r)$ anchored at the melting temperature at the radius of the inner core r_{icb}. The deviation of the temperature field compared to this reference profile is the potential temperature $\Theta = T - T_{\text{ad}}$. The potential composition field is defined as $C = c^s - c^s_{\text{icb}}$, where c^s is the concentration of light elements in the inner core and c^s_{icb} its value at the inner core boundary. The density variations are thus $\rho\alpha_T\Theta$ or $\rho\alpha_C C$ for respectively thermal or compositional stratification, where α_T and α_C are the thermal and compositional expansion coefficients. As both potential temperature and potential composition are solutions of an advection–diffusion equation and are both set to zero by construction at the inner core boundary, we will consider a general equation for a quantity χ, representing either of these quantities. χ is solution of

$$\frac{\partial\chi}{\partial t} + \mathbf{u}\cdot\nabla\chi = \kappa_\chi\nabla^2\chi + S_\chi(t), \tag{5.158}$$

where κ_χ is the diffusivity and $S\chi(t)$ a source term built from the evolution of the reference profile as

$$S_T = \kappa_T\nabla^2 T_{\text{ad}} - \frac{\partial T_{\text{ad}}}{\partial t}, \tag{5.159}$$

$$S_C = -\frac{dc^s_{\text{icb}}}{dt}. \tag{5.160}$$

The continuity and momentum equations can be solved using a poloidal–toroidal decomposition of the velocity field $\mathbf{u} = \nabla\times(T\mathbf{r}) + \nabla\times\nabla\times(P\mathbf{r})$, where $\mathbf{r} = r\mathbf{e}_r$ is the position vector and T and P respectively the toroidal and poloidal components. In the following, we will only consider boundary conditions with a zero vertical vorticity and volume forces without toroidal components. The flow is thus expected to have only non-zero poloidal component, and applying $\mathbf{r}\cdot(\nabla\times\nabla\times)$ to Eq. (5.157), we obtain

$$0 = -(\nabla^2)L^2P + \mathbf{r}\cdot(\nabla\times\nabla\times\mathbf{F}), \tag{5.161}$$

where L^2 is the Laplace horizontal operator defined as

$$L^2 = -\frac{1}{\sin\theta}\frac{\partial}{\partial\theta}\left(\sin\theta\frac{\partial}{\partial\theta}\right) - \frac{1}{\sin^2\theta}\frac{\partial^2}{\partial\phi^2}. \tag{5.162}$$

One could note that for a volume force of the form $\mathbf{F} = F_r r\mathbf{e}_r$ such as the buoyancy forces, the second term of the right-hand side of the equation simplifies as $\mathbf{r}\cdot(\nabla\times\nabla\times\mathbf{F}) = L^2F_r$. Splitting the volume force term as one term for the buoyancy forces and one term for the other forces, we have $\mathbf{F} = \rho\alpha_\chi\chi r g_{\text{icb}}\mathbf{e}_r/r_{\text{icb}} + \mathbf{F}_{\text{volume}}$. Expanding the scalar fields P and χ with horizontal spherical harmonics Y_l^m satisfying $L^2Y_l^m = -l(l+1)Y_l^m$ of degree l and order m, as $P = \sum_{l,m} P_l^m Y_l^m$ and

$\chi = \sum_{l,m} \chi_l^m Y_l^m$, the equation of interest is eventually written for each (l, m) as

$$\mathrm{D}_l^2 P_l^m - \rho \alpha_\chi g_{\mathrm{icb}} \frac{r}{r_{\mathrm{icb}}} \chi_l^m - \frac{f_l^m}{l(l+1)} = 0, \tag{5.163}$$

where f_l^m is the spherical harmonics decomposition of the poloidal component of the volume force $\mathbf{F}$, and D_l a second-order differential operator defined as

$$\mathrm{D}_l = \frac{\mathrm{d}}{\mathrm{d}r^2} + \frac{2}{r}\frac{\mathrm{d}}{\mathrm{d}r} - \frac{l(l+1)}{r^2}. \tag{5.164}$$

Boundary Conditions

The inner core boundary is a crystallisation front, where the iron-alloy of the outer core freezes due to the slow secular cooling of our planet. Its exact position is determined by the intersection of the melting temperature profile of the iron-alloy and the temperature profile in the core. Any solid material pushed dynamically further from this intersection would melt, while any liquid pushed inward would freeze. The timescale τ_ϕ involved in the freezing or melting of a small topography at the inner core boundary can be estimated from the timescale needed by outer core convection to extract the latent heat released by crystallisation. We note h the topography at the inner core boundary.

From continuity of stress at the ICB, the mechanical boundary conditions are written at $r = r_{\mathrm{icb}}(t)$. We consider that the dynamical topography h is small (compared to the horizontal wavelength) and that the vector normal to the boundary is close to the radial unit vector. The tangential and normal components of the stress tensor are written as

$$\tau_{r\theta} = \eta \left[r \frac{\partial}{\partial r}\left(\frac{u_\theta}{r}\right) + \frac{1}{r}\frac{\partial u_r}{\partial \theta} \right], \tag{5.165}$$

$$\tau_{r\phi} = \eta \left[r \frac{\partial}{\partial r}\left(\frac{u_\phi}{r}\right) + \frac{1}{r \sin\theta}\frac{\partial u_r}{\partial \phi} \right], \tag{5.166}$$

$$\tau_{rr} = 2\eta \frac{\partial u_r}{\partial r} - p, \tag{5.167}$$

where p here is the total pressure. The viscosity of the outer core being much smaller than the viscosity of the inner core, we can assume tangential stress-free conditions. The boundary conditions are then written as $\tau_{r\theta}(r = r_{\mathrm{icb}}) = \tau_{r\phi}(r = r_{\mathrm{icb}}) = 0$ and continuity of τ_{rr} across the ICB, which for a small topography amounts to state that the normal stress on the inner core side at $r = r_{\mathrm{icb}}$ is equilibrated by the weight of the topography:

$$\underbrace{2\eta\frac{\partial u_r}{\partial r} - p'}_{\text{normal stress}} = \underbrace{\Delta\rho g h}_{\substack{\text{topography}\\ \text{weight}}}, \tag{5.168}$$

p' being the dynamical pressure on the inner core side of the ICB.

To close the system of equations, we consider the time evolution of the topography. The topography can be formed by deformation of the inner core boundary by the underlying flow $u_r - \dot{r}_{\text{icb}}$ and is eroded by phase change, such that we can write that $\mathrm{D}h/\mathrm{D}t = u_r - \dot{r}_{\text{icb}} + V_r$, where V_r is the velocity of phase change in the radial direction. V_r at first order is $-h/\tau_\phi$, where τ_ϕ is a typical timescale for the phase change. Considering a dynamical equilibrium for the topography, we obtain $u_r - \dot{r}_{\text{icb}} = h/\tau_\phi$ and the continuity of normal stress is written as

$$-\Delta\rho g_{\text{icb}}\tau_\phi(u_r - \dot{r}_{\text{icb}}) - 2\eta\frac{\partial u_r}{\partial r} + p' = 0. \tag{5.169}$$

Non-dimensionalisation and Final Set of Equations

The governing equations are made dimensionless using characteristic scales for time, length, velocity, pressure and χ (potential temperature or composition) as, respectively, the diffusion time $r_{\text{icb}}^2/(6\kappa_\chi)$, its radius r_{icb}, $\kappa_\chi/r_{\text{icb}}$, $\eta\kappa_\chi/r_{\text{icb}}^2$ and $S_\chi r_{\text{icb}}^2/(6\kappa)$. Using the same symbols for dimensionless quantities, the non-dimensional set of equations is

$$\nabla\cdot\mathbf{u} = 0, \tag{5.170}$$

$$-\nabla p' + Ra\,\chi\,\mathbf{r} + \nabla^2\mathbf{u} + \mathbf{F}_{\text{volume}} = 0, \tag{5.171}$$

$$\frac{\partial\chi}{\partial t} = \nabla^2\chi - \mathbf{u}\cdot\nabla\chi + 6. \tag{5.172}$$

The dimensionless number Ra is a Rayleigh number expressed as

$$Ra = \frac{\alpha_\chi \rho g_{\text{icb}} S_\chi r_{\text{icb}}^5}{6\kappa_\chi^2\eta}. \tag{5.173}$$

The momentum equation (5.171) can also be written for the poloidal decomposition in spherical harmonics as

$$\mathrm{D}_l^2 P_l^m - Ra\,\chi_l^m - \frac{f_l^m}{l(l+1)} = 0. \tag{5.174}$$

The last step is to express the boundary conditions in term of the poloidal decomposition and in non-dimensional form. Noting that $u_r = L^2 P/r$ and that the horizontal integration of the momentum equation taken at $r = 1$ gives $-p' + \partial(r\nabla^2 P)/\partial r =$ cste, the stress free condition takes the form

$$\frac{\mathrm{d}^2 P_l^m}{\mathrm{d}r^2} + [l(l+1) - 2]\frac{P_l^m}{r^2} = 0, \tag{5.175}$$

and the normal stress balance is

$$r^2 \frac{\mathrm{d}^3 P_l^m}{\mathrm{d}r^3} - 3l(l+1)\frac{\mathrm{d}P_l^m}{\mathrm{d}r} = \left[l(l+1)\mathcal{P} - \frac{6}{r}\right] P_l^m, \tag{5.176}$$

where $\mathcal{P}$ is a dimensionless number comparing the timescale of viscous relaxation of the boundary $\eta/\Delta\rho g_{\mathrm{icb}} r_{\mathrm{icb}}$ and the time scale of phase change τ_ϕ, *de facto* quantifying the permeability of the inner core boundary. It is defined as

$$\mathcal{P} = \frac{\Delta\rho g_{\mathrm{icb}} r_{\mathrm{icb}} \tau_\phi}{\eta}. \tag{5.177}$$

Unstable or Stable Stratification in the Inner Core?

The core crystallises from the center outward because the solidification temperature of the core mixture increases with depth faster than the (adiabatic) core geotherm (Jacobs 1953). One consequence of this solidification mode is that the inner core is cooled from above, a configuration which is potentially prone to thermal convection. Thermal convection further requires the inner core temperature profile to be super-adiabatic, which depends on a competition between extraction of the inner core internal heat by diffusion and advection, and cooling at the ICB. Equation (5.158) shows that super-adiabaticity in the inner core (i.e. a potential temperature increasing with depth) requires S_T to be positive. Fast cooling and a low inner core thermal diffusivity ($S_T > 0$) promotes super-adiabaticity; slow cooling and high thermal diffusivity ($S_T < 0$) results in a stable thermal stratification.

The expression of S_T (Eq. 5.159) can be rewritten by writing the time derivative of the temperature at the ICB as a function of the rate of inner core growth (Deguen and Cardin 2011). This gives

$$S_T = -\frac{1}{r_{\mathrm{icb}}} \left.\frac{\mathrm{d}T_{\mathrm{ad}}}{\mathrm{d}r}\right|_{\mathrm{icb}} \left(\left[\frac{\mathrm{d}T_s}{\mathrm{d}T_{\mathrm{ad}}} - 1\right] r_{\mathrm{icb}} \dot{r}_{\mathrm{icb}} - 3\kappa_T \right), \tag{5.178}$$

where $\mathrm{d}T_s/\mathrm{d}T_{\mathrm{ad}}$ is the ratio of the Clapeyron slope to the adiabat. It is then straight-forward that the term S_T is positive only if

$$\frac{dr_{\mathrm{icb}}^2}{dt} > \frac{6\kappa_T}{\frac{\mathrm{d}T_s}{\mathrm{d}T_{\mathrm{ad}}} - 1}. \tag{5.179}$$

Unfortunately, the uncertainties on the thermal conductivity and inner core growth rate are such that the sign of S_T is not known with much certainty. The high value of the thermal conductivity currently favoured (de Koker et al. 2012; Pozzo et al. 2012; Gomi et al. 2013, 2016) results in a stable thermal stratification.

The inner core may also have developed a compositional stratification. The concentration in light elements of newly crystallised solid, c^s_{icb}, is linked through the partition coefficient D to the concentration in the liquid from which it crystallises, c^l_{icb}, as

$$c^s_{\text{icb}} = Dc^l_{\text{icb}}, \tag{5.180}$$

while its derivative with respect to inner core size is

$$\frac{dc^s_{\text{icb}}}{dr_{\text{icb}}} = D\frac{dc^l_{\text{icb}}}{dr_{\text{icb}}} + c^l_{\text{icb}}\frac{dD}{dr_{\text{icb}}}, \tag{5.181}$$

$$= Dc^l_{\text{icb}}\left[\frac{d\ln c^l_{\text{icb}}}{dr_{\text{icb}}} + \frac{d\ln D}{dr_{\text{icb}}}\right]. \tag{5.182}$$

A stable compositional stratification would develop if c^s_{icb} increases with increasing inner core size (more light elements in the upper part of the inner core); conversely, an unstable stratification would develop if c^s_{icb} decreases with increasing inner core size. The first term on the right-hand-side term of Eq. (5.182) is very likely positive, due to the gradual enrichment of the outer core in light elements expelled during crystallisation. The second term depends on how D varies with pressure and temperature along the (P, T) path defined by the evolution of the position of the inner core boundary. Ab initio calculations (Gubbins et al. 2013) suggest that it is negative, and of the same order of magnitude as the first term on the r.h.s. of Eq. (5.182). The relative importance of the two terms depends on the exact composition of the inner core (Gubbins et al. 2013; Labrosse 2014), which is not very well constrained. Again, it is difficult to be definitive: given our current knowledge of the composition of the core and of the partitioning behaviour of its light elements, stable and unstable compositional stratifications seem equally plausible.

Natural (thermal or compositional) convection in the inner core has been studied in details (e.g. Weber and Machetel 1992; Wenk et al. 2000; Alboussière et al. 2010; Monnereau et al. 2010; Deguen and Cardin 2011; Cottaar and Buffett 2012; Deguen et al. 2013; Mizzon and Monnereau 2013; Deguen et al. 2018). In the limit $\mathcal{P} \to 0$, the convection instability takes the form of a translation of the inner core, with melting on one hemisphere and solidification on the other (Alboussière et al. 2010; Monnereau et al. 2010; Deguen et al. 2013; Mizzon and Monnereau 2013; Deguen et al. 2018). We will focus here on the case of neutral or stable stratification, and consider the flow forced by the Lorentz force associated with the magnetic field diffused in the inner core from the outer core.

Deformation Induced by the Lorentz Force

As discussed in section "The Geodynamo Hypothesis", the flow in the outer core sustains a magnetic field extending upward to the surface of the Earth but also inward inside the inner core. The magnetic Reynolds number of the inner core is likely small: assuming, for example, a velocity 10^{-10} m.s^{-1} gives a magnetic Reynolds number on the order of 10^{-5}. This shows that the magnetic field is only diffused inside the inner core, with no net advection or generation of the field. A diffused magnetic field in the inner core will add two terms in the set of equations: the Lorentz force in the momentum equation, and Joule heating in the energy equation. We are interested here in the flow driven by the Lorentz force in the inner core (Lasbleis et al. 2015).

Geodynamo simulations often exhibits a strong toroidal magnetic field close to the inner core boundary. As we are interested in the largest effect on the inner core dynamics, we consider here only low-order toroidal components of the magnetic field at the ICB, which have the largest penetration length scale.

We thus add in the momentum equation the Lorentz force due to a purely toroidal and axisymmetric magnetic field of degree two at the ICB $\mathbf{B}|_{\text{icb}} = B_0 \sin\theta\cos\theta\mathbf{e}_\phi$ (Buffett and Bloxham 2000). Imposing this field at the ICB and solving for its diffusion inside the inner core assuming it does not vary with time ($\boldsymbol{\nabla}^2\mathbf{B} = 0$), we obtain $\mathbf{B} = B_0 r^2/r_{\text{icb}}^2 \cos\theta\sin\theta\mathbf{e}_\phi$. This field is associated to an electric current density $\mathbf{j} = \frac{1}{\mu_0}\boldsymbol{\nabla}\times\mathbf{B}$, with μ_0 the magnetic permeability. The associated Lorentz force is $\mathbf{f}_L = \mathbf{j}\times\mathbf{B}$ which non-potential part (magnetic tension) can be written as

$$\tilde{\mathbf{f}}_L = \frac{B_0^2}{\mu_0 r_{\text{icb}}}\frac{r^3}{r_{\text{icb}}^3}\left[f_r\mathbf{e}_r + f_\theta\mathbf{e}_\theta\right], \tag{5.183}$$

with f_r and f_θ two functions of θ expressed as

$$f_r(\theta) = 3\cos^4\theta - \frac{15}{7}\cos^2\theta + \frac{4}{35}, \tag{5.184}$$

$$f_\theta(\theta) = \cos\theta\sin\theta\left(\frac{4}{7} - 3\cos^2\theta\right). \tag{5.185}$$

Injecting this force into the dimensionless Stokes equation, we obtain

$$\mathbf{0} = -\boldsymbol{\nabla}p + Ra\,\chi\,\mathbf{r} + \boldsymbol{\nabla}^2\mathbf{u} + M\tilde{\mathbf{f}}_L, \tag{5.186}$$

where

$$M = \frac{B_0^2 r_{\text{icb}}^2}{\mu_0\eta\kappa} \tag{5.187}$$

is similar to a Hartmann number, quantifying the ratio of the Lorentz force to the viscous force, and $\tilde{\mathbf{f}}_L$ is defined as in (5.183) without the prefactor $\frac{B_0^2}{\mu_0 r_{\text{icb}}}$.

Equation (5.186) is solved using a poloidal decomposition and horizontal spherical harmonics decomposition. The term corresponding to the Lorentz force gives

$$\mathbf{r} \cdot (\nabla \times \nabla \times \tilde{\mathbf{f}}_L) = 8r^2(1 - 3\cos^2\theta) = -\frac{16}{\sqrt{5}} r^2 Y_2^0, \tag{5.188}$$

where $Y_2^0 = \sqrt{5}(3\cos^2\theta - 1)/2$. The momentum equation can thus be written as an equation for the spherical harmonics components P_l^m and t_l^m of respectively the poloidal component of the velocity and the temperature as

$$D_l^2 P_l^m - Ra\, t_l^m + \frac{16}{\sqrt{5}l(l+1)} M r^2 \delta_{2l}\delta_{0m} = 0, \quad l \geq 1, \tag{5.189}$$

where δ is the Kronecker symbol.

Neutral Stratification

We first consider the neutral stratification end-member where $Ra = 0$. In that case, the only force driving flows in the system is the Lorentz force, and we do not need to solve for the temperature or composition fields. The flow results from a balance between the Lorentz and viscous forces. Since the characteristic length scale of velocity variations must be the size of the inner core (1 in dimensionless form), we have

$$\underbrace{\nabla^2 \mathbf{u}}_{\sim u} \sim \underbrace{M\tilde{\mathbf{f}}_L}_{\sim M}, \tag{5.190}$$

which implies that the magnitude of the velocity field should be proportional to M.

We can now solve analytically the flow field for a neutral stratification. With $Ra = 0$, Eq. (5.189) reduces for $l = 2$ and $m = 0$ to

$$D_2^2 P_l^m + \frac{8}{3\sqrt{5}} M r^2 = 0, \tag{5.191}$$

which we solve with the boundary conditions at $r = 1$ described in section "Boundary Conditions":

$$\frac{\mathrm{d}^2 P_2^0}{\mathrm{d}r} + 4\frac{P_2^0}{r^2} = 0, \tag{5.192}$$

$$r\frac{\mathrm{d}r^3 P_2^0}{\mathrm{d}r^3} - 18\frac{1}{r}\frac{\mathrm{d}P_2^0}{\mathrm{d}r} = \left(\mathcal{P} - \frac{1}{r^2}\right) 6P_2^0. \tag{5.193}$$

Equation (5.191) is a fourth order non-homogeneous differential equation, which solution can be obtained by solving the homogeneous equation and noticing that

$$P_2^0 = -\frac{1}{3^3 7\sqrt{5}} M r^6 \tag{5.194}$$

is one solution of the complete equation. Searching for a polynomial solution, we find that r^α is solution of the homogeneous equation $D_2^2 P_2^0 = 0$ if α is a zero of the polynomial expression $[\alpha(\alpha+1)-6][(\alpha-2)(\alpha-1)-6]$. We then obtain the general solution of equation (5.191) as

$$P_2^0(r) = -\frac{M}{3^3 7\sqrt{5}} r^6 + Ar^{-3} + Br^{-1} + Cr^2 + Dr^4. \tag{5.195}$$

A and B must be equal to 0 for the velocity to remain finite at $r = 0$. C and D are obtained from the boundary conditions at $r = 1$, and we finally obtain

$$P_2^0(r) = \frac{1}{3^3 7\sqrt{5}} M \left(-r^6 + \frac{14}{5} r^4 - \frac{9}{5} r^2 + \frac{1}{19+5\mathcal{P}} \left[\frac{204}{5} r^4 - \frac{544}{5} r^2 \right] \right). \tag{5.196}$$

From the expression of P_2^0, we can now obtain the expressions for the velocity field from

$$u_r = 3\frac{P_2^0}{r} Y_2^0, \tag{5.197}$$

$$u_\theta = \frac{1}{r}\frac{d}{dr}(r\, P_2^0)\frac{\partial Y_2^0}{\partial \theta}. \tag{5.198}$$

Defining the r.m.s. velocity as

$$V_{\text{rms}}^2 = \frac{3}{4\pi} \int_0^{2\pi} \int_0^{\pi} \int_0^1 (u_r^2 + u_\theta^2) \sin\theta r^2 \, dr \, d\theta \, d\phi, \tag{5.199}$$

we obtain

$$V_{\text{rms}} = M \frac{4}{189} \sqrt{\frac{34}{715}} \frac{\sqrt{74029 - 1576\, P + 76\, P^2}}{19 + 5\, P}. \tag{5.200}$$

This expression for the RMS velocity gives insight on the effect of $\mathcal{P}$ on the global dynamics. As predicted at the beginning of the subsection, the velocity is indeed a linear function of M, with the boundary conditions modifying the prefactor. For $\mathcal{P} \to 0$, $V_{\text{RMS}} \sim 0.066M$, and for $\mathcal{P} \to \infty$, $V_{\text{RMS}} \sim 0.008M$. Permeable boundary conditions ($\mathcal{P} \to 0$) give velocities about one order of magnitude higher than impermeable boundary conditions ($\mathcal{P} \to \infty$) (Fig. 5.15).

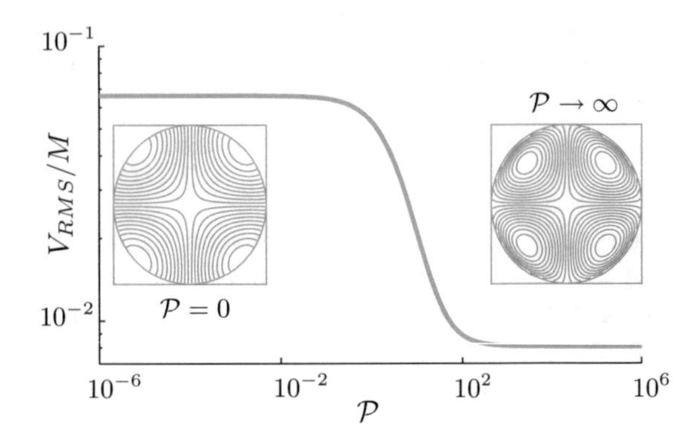

Fig. 5.15 Neutral stratification: r.m.s. velocity as a function of the parameter $\mathcal{P}$, and meridional cross-section of the streamlines of the two end-members $\mathcal{P} \to 0$ and $\mathcal{P} \to \infty$. (Modified from Lasbleis et al. (2015).)

Stable Stratification

If the inner core is stably stratified ($Ra < 0$), the buoyancy forces resulting from the deformation of constant density surfaces would tend to oppose further deformation, and inhibit vertical motions. We can thus anticipate that the flow obtained in the limit of $Ra = 0$ can be significantly altered by a strong stratification.

To estimate to what extent the neutral stratification solution can be altered by the presence of a stable density stratification, we consider the vorticity equation (obtained by taking the curl of Eq. (5.186)), which is

$$\mathbf{0} = -Ra\frac{\partial \chi}{\partial \theta}\mathbf{e}_\phi + M\nabla \times \tilde{\mathbf{f}}_L + \nabla^2\boldsymbol{\zeta}, \tag{5.201}$$

where $\boldsymbol{\zeta} = \nabla \times \mathbf{u}$ is the vorticity. Because the form of the magnetic field considered here forces a degree 2 flow, and because in non-dimensional form χ varies between 0 and 1, the θ derivative of χ is $\lesssim 1$. The magnitude of the baroclinic vorticity production $-Ra\frac{\partial\chi}{\partial\theta}$ is therefore $\lesssim -Ra$. Since the vorticity production associated with the Lorentz force is on the order of M, a stable stratification can affect significantly the flow only if $-Ra \gtrsim M$. As one of the main unknown for inner core dynamics is the viscosity, it is interesting to note that the ratio M/Ra,

$$\frac{M}{Ra} = \frac{B_0^2}{\mu_0 \Delta\rho\, g_{\text{icb}}\, r_{\text{icb}}}, \tag{5.202}$$

does not depend on the viscosity, so that the boundary between a strongly stratified regime and a neutral stratification does not depends on the viscosity.

Figure 5.16 shows the temperature and vorticity fields obtained by solving numerically Eqs. (5.172) and (5.189) at $M = 10^4$ and $Ra = -10^4$, -6×10^4, -10^5, and -10^6. The stratification has a negligible effect on the flow at $Ra = -10^4$, but already has a significant effect at $Ra = -6 \times 10^4$, which is consistent with the criterion we just derived ($-Ra \gtrsim M$ for a strong effect of the stratification). At $Ra = -10^5$ and -10^6, the flow induced by the Lorentz force is essentially con-

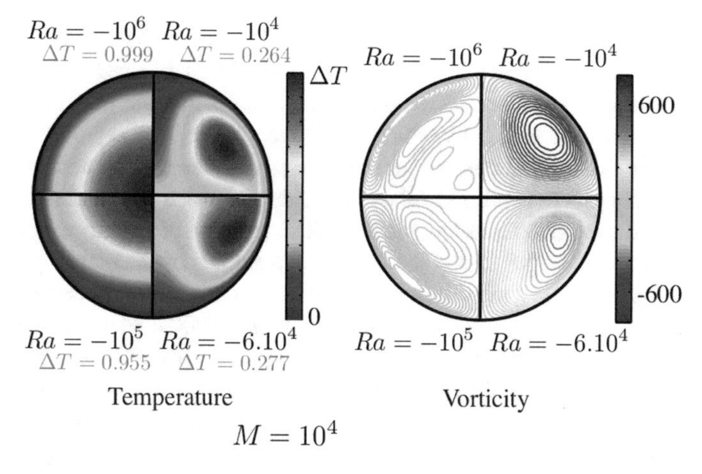

Fig. 5.16 Meridional cross-sections of the temperature and the vorticity fields for $M = 10^4$ and four different values of the Rayleigh number. At low Ra, the flow is similar to the neutral stratification case. If $-Ra \gtrsim M$ (strong stratification) the flow is confined in a layer at the top of the inner core. (Modified from Lasbleis et al. (2015).)

fined to a thin layer below the inner core boundary, in which the flow direction is essentially horizontal. Lasbleis et al. (2015) found that the thickness of this layer is $\propto (-Ra)^{-1/6}$ and the strain rate in the layer is $\propto M(-Ra)^{-1/3}$. For application to Earth's inner core, we take $\Delta\rho \sim -1$ kg.m^{-3} and $B_0 \sim 4$ mT (Gillet et al. 2010), and find $M/(-Ra) \sim 10^6$, which would imply that stratification strongly affects the flow forced by the Lorentz force.

Core Formation

The final section of this chapter concerns the formation of Earth's core. The Earth formed about 4.56 Gy ago through the accretion of solar nebula material, a process which is estimated to have taken a few tens of millions of years. Accretion in the solar system went through different dynamical phases which involved increasingly energetic and catastrophic impacts and collisions (Lunine et al. 2011; Walsh et al. 2011). The last phase of accretion, in which most of the Earth mass was accreted, involved extremely energetic collisions between already differentiated planetary embryos ($\sim$100–1000 km size), which resulted in widespread melting and the formation of magma oceans.

The basic ingredients of a terrestrial planet—an iron-rich metal and silicates—are immiscible, and can separate under the action of gravity to form an iron-rich core surrounded by a rocky mantle. The timescale of phase separation is much smaller than the accretion timescale if at least one of the two phases is liquid, and it is thus believed that core formation has been concomitant with Earth's accretion. Importantly, the metal added to the Earth by each impact had no reason to be in thermodynamic equilibrium with the silicate mantle of the growing Earth, which implies that heat and chemical elements would have been transferred from one phase to the other when the two phases were in contact. Chemical elements and heat released during accretion have thus been partitioned between the core and mantle, in a way which depends on the exact physical mechanism by which metal and silicate have separated.

This has important implications for the state of the planet at the end of its accretion, and its subsequent evolution. The partitioning between core and mantle of the heat released during accretion has set the initial temperature contrast between the mantle and core, a key parameter for the early dynamics of the planet, with implications for the possibility of forming a basal magma ocean (Labrosse et al. 2007), the existence of an early dynamo (Williams and Nimmo 2004; Monteux et al. 2011), and the subsequent thermal and magnetic evolution of the planet. The partitioning of chemical elements between the core and mantle has been used to constrain the timing of differentiation and the physical conditions under which it occurred (e.g. Yin et al. 2002; Halliday 2004; Wood et al. 2006; Rudge et al. 2010). It has also important implications for a number of geodynamical issues. One example is the identity and abundance of light elements and radioactive elements in the core (Corgne et al. 2007; Badro 2015), which depend on the conditions (pressure P, temperature T, oxygen fugacity fO_2) at which metal and silicate have interacted for the last time.

The large impacts which dominated the last stages of Earth formation injected enormous amounts of kinetic energy into the magma oceans, creating highly turbulent environments in which it has been conjectured that the cores of the bodies impacting the Earth would fragment down to centimetre scale, at which metal and silicates can efficiently exchange chemical elements and heat (Stevenson 1990; Karato and Rama Murthy 1997; Rubie et al. 2003). The metal is envisioned to disperse in the magma ocean and equilibrate with the silicates, before raining out. It would then collect at the base of the magma ocean, and finally migrate toward the core as large diapirs (Stevenson 1990; Karato and Rama Murthy 1997; Monteux et al. 2009; Samuel et al. 2010) or by the propagation of iron dykes (Stevenson 2003), at which point further chemical equilibration is unlikely to be significant.

Most geochemical models of core formation are based on this so-called *iron rain* scenario, but the fluid dynamics involved is actually poorly understood, even at a qualitative level. Efficient chemical exchange requires a high metal–silicates interfacial area-to-volume ratio, which requires fragmentation or stretching of the metal down to ~cm size. Differentiation of terrestrial bodies started early, and it is now recognised that most of the mass of the Earth was accreted from already differentiated planetary bodies, with cores of their own. Whether or not these large volumes of iron (~100–1000 km) would indeed fragment down to cm scale at which chemical equilibration can occurs therefore remains an open question, and a matter of much speculation (Dahl and Stevenson 2010; Deguen et al. 2011; Samuel 2012; Deguen et al. 2014; Wacheul et al. 2014).

We have yet no well-tested and self-consistent theory of fragmentation and thermal or compositional equilibration in the context of metal–silicate separation in a magma ocean. For this reason, we have here made the choice of focusing on the basic concepts and mechanisms which we believe are important to know and understand to tackle this problem, rather than trying to give a definitive answer to this question.

Problem Set-Up and Non-dimensional Numbers

To keep things reasonably simple, we will ignore here the dynamics of the impact itself and consider the fate of an initially spherical mass of molten iron falling into silicates, which can be either solid or molten. This molten iron mass may be either the core of a planetary body impacting the Earth or a fragment of the core if it has been significantly dispersed by the impact. As an additional simplifying hypothesis, we will even assume that the metal mass has no initial velocity.

The volume of molten iron is assumed to be close to spherical, and has a radius R_0. We denote by ρ_m and $\rho_s = \rho_m - \Delta\rho$ the densities of metal and silicates, and by η_m and η_s their viscosities. The metal and silicates phases are immiscible and we denote by γ the interfacial tension of the metal–silicates interface. We denote by g the acceleration of gravity.

The evolution of the metal mass depends on seven dimensional parameters ($R_0, \rho_m, \rho_s, \eta_m, \eta_s, \gamma$) involving 3 fundamental units (length, weight, and time). According to Vashy-Buckingham's theorem, the number of independent non-dimensional numbers to be used to describe the problem is equal to $7 - 3 = 4$. One possible set is the following:

$$\text{Bond number: } Bo = \frac{\Delta\rho\, g\, R_0^2}{\gamma}, \tag{5.203}$$

$$\text{Grashof number: } Gr = \frac{\Delta\rho}{\rho_s}\frac{g R_0^3}{\nu_s^2}, \tag{5.204}$$

$$\text{density ratio: } \frac{\rho_m}{\rho_s}, \tag{5.205}$$

$$\text{viscosity ratio: } \frac{\eta_m}{\eta_s}. \tag{5.206}$$

The Bond number is a measure of the relative importance of buoyancy and interfacial tension. The Grashof number is basically a Reynolds number obtained by taking Stokes' velocity as a velocity scale. Additional useful numbers include Reynolds and Weber numbers, which can be defined as

$$\text{Reynolds number: } Re = \frac{\rho_s U R_0}{\eta_s}, \tag{5.207}$$

$$\text{Weber number: } We = \frac{\rho_s U^2 R_0}{\gamma}, \tag{5.208}$$

where U is a velocity scale to be defined. The Reynolds and Weber numbers compare inertia to viscous forces and interfacial tension, respectively. The density ratio is close to 2.

If, in addition, we consider heat or mass transfer between metal and silicates, then two additional dimensional parameters enter the problem: the diffusivities (thermal or compositional) κ_m and κ_s in the metal and silicates. Since there is no additional fundamental units, Vashy-Buckingham's theorem implies that two additional non-dimensional numbers must be used. One possible choice is to use the ratio of the diffusivities and a Péclet number:

$$\text{diffusivity ratio: } \frac{\kappa_m}{\kappa_s}, \tag{5.209}$$

$$\text{Péclet number: } Pe = \frac{U R_0}{\kappa_s}. \tag{5.210}$$

In what follows, we will assume that $\kappa_s = \kappa_m$ for the sake of simplicity.

Preliminary Considerations

Terminal Velocity

Since we have chosen to focus on the case of a metal mass falling with no initial velocity, a relevant velocity scale is its terminal velocity, reached when the buoyancy force ($\sim \Delta \rho g R_0^3$) is balanced by the drag on the surface of the metal mass. Two different scalings can be obtained depending on whether the drag is dominated by viscous stresses ($\sim \eta_s U / R_0$) or dynamic pressure ($\sim \rho_s U^2$). The ratio of the dynamic pressure and viscous stresses contributions to the total drag is

$$\frac{\text{dynamic pressure}}{\text{viscous stress}} \sim \frac{\rho_s U^2}{\eta_s U / R_0} = Re. \tag{5.211}$$

The drag on the metal mass is obtained by multiplying the dominant stress by the surface area of the metal mass: it is on the order of $\eta_s U R_0$ if the drag is dominated by the contribution of viscous stresses (low Re), and on the order of $\sim \rho_s U^2 R_0^2$ if it is dominated by the contribution of dynamic pressure (high Re). The force balance on the metal mass can thus be written as

$$\underbrace{\Delta \rho g R_0^3}_{\text{buoyancy}} \sim \underbrace{\max\left(\rho_s U^2 R_0^2, \eta_s U R_0\right)}_{\text{drag}}. \tag{5.212}$$

The terminal velocity obtained from this balance is

$$U \sim \min\left(\frac{\Delta \rho g R_0^2}{\eta_s}, \left(\frac{\Delta \rho}{\rho_s} g R_0\right)^{1/2}\right). \tag{5.213}$$

The first scaling gives Stokes' settling velocity, and corresponds to the low Re limit; the second scaling is the so-called newtonian scaling, and corresponds to the high Re limit. The two velocities are on the same order of magnitude when

$$Gr \sim 1, \tag{5.214}$$

which defines the boundary between the two scalings (here $\nu_s = \eta_s/\rho_s$). The terminal velocity is thus

$$U \sim \frac{\Delta\rho g R_0^2}{\eta_s} \quad \text{if } Gr \ll 1, \tag{5.215}$$

$$U \sim \left(\frac{\Delta\rho}{\rho_s} g R_0\right)^{1/2} \quad \text{if } Gr \gg 1. \tag{5.216}$$

With these scalings, the Reynolds and Weber numbers based on the terminal velocity are given by

$$\begin{cases} Re & \sim Gr \\ We & \sim Gr\,Bo \end{cases} \quad \text{if } Gr \ll 1, \tag{5.217}$$

$$\begin{cases} Re & \sim Gr^{1/2} \\ We & \sim Bo \end{cases} \quad \text{if } Gr \gg 1. \tag{5.218}$$

Maximal Stable Size of a Falling Drop

Interfacial tension (unit $J.m^{-2}$) can be interpreted as an energy per unit area. Deforming an interface in a way which results in an increase of the interfacial area costs energy, and interfacial tension effects will tend to minimise the surface area of the interface. If no other force acts on a drop, interfacial tension would keep it spherical, hence minimising its surface area.

Interfacial tension can also be seen as a force per unit of length (it can be verified that $J.m^{-2} = N.m^{-1}$): if a piece of an interface is divided into two parts, the force imparted by one part of the surface on the other is parallel to the interface and has a magnitude given by the product of the interfacial tension with the length of the curve separating the two parts of the surface. If integrated over a curved surface, one can also show that interfacial tension induces a pressure jump across the interface equal to

$$\Delta P = \gamma\left(\frac{1}{R_1} + \frac{1}{R_2}\right), \tag{5.219}$$

where R_1 and R_2 are the principal radii of curvature. This pressure jump is called Laplace's pressure. Across a spherical interface (the surface of a drop or bubble of radius R_0), Laplace's pressure is equal to $2\gamma/R_0$.

A falling drop can be deformed by the stresses imparted by the surrounding fluid onto the drop, or in other words by the fluid drag. If the total drag on the drop is F_{drag}, the mean stress on the surface of the drop is $\sim F_{\text{drag}}/R_0^2$. One can expect significant deformation of the drop if the hydrodynamic stress variations due to the drag exceed Laplace's pressure:

$$\frac{F_{\text{drag}}}{R_0^2} \gtrsim \frac{\gamma}{R_0}. \tag{5.220}$$

If the drop reached its terminal velocity, then the drag must be equal to the total buoyancy of the drop, $F_{\text{drag}} \sim \Delta\rho g R_0^3$. Combining this with Eqs. (5.220), we find that strong deformation of the drop will happen if its radius is larger than a critical radius R_c given by

$$R_c \sim \left(\frac{\gamma}{\Delta\rho g}\right)^{1/2}. \tag{5.221}$$

This length is also known as a *capillary length*: it is the length scale over which buoyancy effects dominate over surface tension effects. Interfacial tension will keep the drop close to spherical if its radius is smaller than R_c. Equation (5.221) is equivalent to writing that deformation is significant if the Bond number of the drop is large compared to 1. The interfacial tension between metal and silicates is on the order of 1 J.m^{-2} and $\Delta\rho \sim 4000$ kg.m^{-3}. With $g \sim 10$ m.s^{-2}, we thus have $R_c \sim 5$ mm.

Strong drop deformation may happen before reaching terminal velocity, and in this case the above scaling will not be the most relevant. If drag is dominated by viscous effects ($F_{\text{drag}} \sim \eta_s U R_0$), then we find that deformation of the drop may happen if its velocity is larger than a critical velocity

$$U_c \sim \frac{\gamma}{\eta}. \tag{5.222}$$

This criterion is of limited use since if the drag is dominated by viscous effect (which means that $Re \ll 1$), then the drop velocity will very quickly reach a terminal velocity equal to Stokes' settling velocity. Using Stokes' velocity for U in Eq. (5.222) gives Eq. (5.220).

If instead drag is dominated by the contribution of dynamic pressure ($F_{\text{drag}} \sim \rho_s U^2 R_0^2$), then we find that strong deformation requires the drop Weber number is large:

$$We = \frac{\rho_s U^2 R_0}{\gamma} \gtrsim 1. \tag{5.223}$$

This criterion reduces to Eq. (5.221) when the drop reaches its terminal velocity ($U \sim \sqrt{(\Delta\rho/\rho_s) g R_0}$ since in this limit $Re \gg 1$).

The Low Reynolds Limit: Diapirism

A first relevant limit of the problem described in section "Problem Set-Up and Non-dimensional Numbers" corresponds to molten metal diapirs travelling through a solid, or partially molten, silicate layer. The radius of these diapirs may be similar to the size of the core of the impactors, say somewhere between 1 km and 1000 km. The acceleration of gravity is smaller or equal to its current value in Earth's mantle, $g \sim 10$ m.s^{-2}. The viscosity η_s of the silicates is a strong function of temperature, and can also be significantly decreased if the silicate layer is partially molten. A reasonable range is 10^{15}–10^{21} Pa.s. The viscosity of molten iron is $\sim 10^{-2}$ Pa.s.

With this parameter values, $Bo \gtrsim 10^{10}$, $Gr \lesssim 10^{-4}$, $\rho_m/\rho_s \sim 2$, $\eta_m/\eta_s \lesssim 10^{-18}$. The Grashof number being small, we are well into the low Reynolds regime: viscous forces dominate over inertia in the silicates. The limit of low Re and high Bo has been studied numerically in the context of core formation (Samuel and Tackley 2008; Monteux et al. 2009; Samuel et al. 2010), and experimentally in other contexts (e.g. Ribe 1983; Bercovici and Kelly 1997). In this limit surface tension is unimportant, but the volume of metal is kept roughly spherical because $Re \ll 1$. Since viscous forces are so important in the silicates, the flow around the metal mass is limited to spatial scales on the order of R_0, which limits the deformation of the metal mass (in other words, small scale perturbations of the metal–silicate interface shape are damped viscously).

The falling velocity is thus simply given by Stokes' velocity [Eq. (5.215)]. The law of heat or mass transfer between the diapir and its surrounding is also well known (e.g. Clift et al. 1978; Ribe 2007; Ulvrová et al. 2011: the heat flux is given by

$$\text{heat flux} = a 4\pi R_0^2 k_s \frac{\bar{T} - T_s(z)}{R_0} Pe^{1/2} \tag{5.224}$$

where k_s is the thermal conductivity in the silicates, $\bar{T}$ the mean temperature in the diapir, $T_s(z)$ the temperature of the surrounding silicate layer at depth z, and a is a constant on the order of 1. A similar expression can be written for chemical elements transfer.

The heat balance of the diapir writes

$$\rho_m c_{p,m} \frac{4\pi}{3} R_0^3 \frac{d\bar{T}}{dt} = -4\pi R_0^2 a k_s \frac{\bar{T} - T_s(z)}{R_0} Pe^{1/2}, \tag{5.225}$$

where $c_{p,m}$ is the heat capacity of the metal phase. Transforming the time derivative into a derivative with respect to the distance z travelled by the diapir (using $d(...)/dt = U d(...)/dz$) and re-arranging gives

$$\frac{d\bar{T}}{dz} + \frac{\bar{T}}{\ell} = \frac{T_s(z)}{\ell}, \tag{5.226}$$

where the characteristic length ℓ is given by

$$\ell = \frac{1}{3a} \frac{\rho_m c_{p,m}}{\rho_s c_{p,s}} R_0 Pe^{1/2}. \tag{5.227}$$

ℓ is the characteristic distance over which the temperature of the diapir responds to changes of the surrounding temperature, the *thermal equilibration distance*. The general solution of equation (5.226) is

$$\bar{T} = T_0 \mathrm{e}^{-z/\ell} + \int_0^z \frac{\mathrm{e}^{(z'-z)/\ell}}{\ell} T_s(z') dz', \tag{5.228}$$

where T_0 is the initial temperature of the diapir. In practice, the amount of heat transfer is small because ℓ is typically larger than the mantle thickness. The compositional diffusivity in the solid silicates being perhaps four orders of magnitude smaller that the thermal diffusivity, exchange of chemical elements would be even smaller (the equilibration distance being $\propto Pe^{1/2} \propto \kappa^{-1/2}$). Diapirs migrating through a solid part of the mantle would therefore exchange a negligible amount of heat and chemical elements with the surrounding mantle.

The High Reynolds Limit: Metal–Silicates Separation in a Magma Ocean

Let us now consider the case of a volume of molten metal falling into molten rocks—a *magma ocean*. The parameter values are similar to what we have considered when discussing the case of diapirism, except that the viscosity of the silicates is much smaller, on the order of 10^{-1} Pa.s. With these parameter values, $Bo \gtrsim 10^{10}$, $Gr \gtrsim 10^{22}$, $\rho_m/\rho_s \sim 2$, $\eta_m/\eta_s \lesssim 10^{-18}$. Since $Gr \gg 1$, the relevant velocity scale is the newtonian scaling given by Eq. (5.216). This gives $Re \sim Gr^{1/2} \gtrsim 10^{11}$ and $We \sim Bo \gtrsim 10^{10}$. The very large values of Re, Bo, and We imply that neither viscous forces nor interfacial tension can keep the metal volume spherical: a molten mass of metal falling into a magma ocean should suffer significant deformation, possibly resulting in its fragmentation into drops.

Observations from Laboratory Experiments

The very large values of the Bond and Weber numbers imply that interfacial tension should be unimportant for the large-scale dynamics. This suggests to first look at the case of infinite Bond and Weber numbers, which corresponds to the limit of miscible fluids.

Figure 5.17a shows snapshots from an experiment in which a negatively buoyant volume of an aqueous solution of sodium iodide is released into fresh water. The volume of the (dyed) negatively buoyant fluid is seen to increase as it falls, which indicates that it entrains and incorporates ambient fluid, resulting in its gradual dilution. Measurements show that the mean radius of the dyed mixture increases linearly with the distance from the point of release. This is what is known as a *turbulent thermal* in the fluid mechanics and atmospheric science communities (e.g. Batchelor

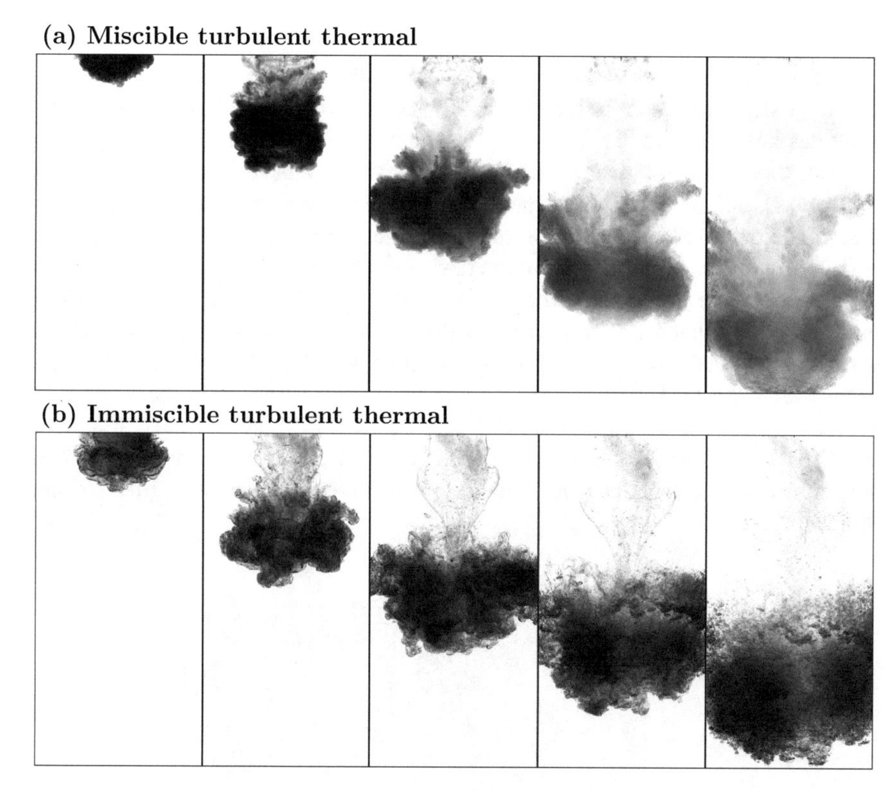

Fig. 5.17 **a** A 169 mL volume of an aqueous solution of NaI ($\rho = 1502$ kg.m^{-3}) falling into fresh water, at Re $= 4 \times 10^4$, $\rho_m/\rho_s = 1.5$, $\eta_m/\eta_s = 1$. The time interval between each image is 0.3 s. **b** A 169 mL volume of an aqueous solution of NaI ($\rho = 1601$ kg.m^{-3}) falling into a low-viscosity silicon oil ($\rho = 821$ kg.m^{-3}, $\eta = 8.2 \times 10^{-4}$ Pa.s), at $Bo = 3.4 \times 10^4$, Re $= 5.5 \times 10^4$, $\rho_m/\rho_s = 1.95$, $\eta_m/\eta_s = 1.2$ (Deguen et al. 2019). The time interval between each image is 0.2 s

1954; Scorer 1957; Woodward 1959). The name *thermal* is inherited from the usage of glider pilots, for whom a thermal is an isolated mass of warm air rising through the lower atmosphere. Though the buoyancy in atmospheric thermals is due to temperature differences, the nature of the source of buoyancy (thermal or compositional) happens to be of secondary importance and introduces no qualitative difference. The term thermal has since been used to denote an isolated buoyant mass of a fluid rising or falling (depending on the sign of the buoyancy), irrespectively of the nature of the source of buoyancy. Here we will also often use the term *buoyant cloud* instead of thermal.

Figure 5.17b shows snapshots from a similar experiment in which a negatively buoyant volume of sodium iodide is now released into a low-viscosity silicon oil. The NaI solution and the silicon oil are *immiscible*, so we are one important step closer to the core-mantle differentiation configuration. The experimental fluids and configuration have been chosen so as to maximise the values of the Bond and

Reynolds numbers, which are $Bo = 3.4 \times 10^4$ and $\mathrm{Re} = 5.5 \times 10^4$. The density ratio is $\rho_m/\rho_s = 1.95$ (close to metal–silicate), and the viscosity ratio is $\eta_m/\eta_s = 1.2$. The large-scale evolution of the negatively buoyant volume is strikingly similar to what has been observed in the *miscible* experiment: the volume of the negatively buoyant fluid increases linearly with distance, which indicates that it entrains and incorporates silicon oil. PIV measurements on a similar experiment (Fig. 5.18) show that the velocity field has a vortex ring structure, with most of the entrainment of silicon oil probably occurring from the rear of the cloud.

In contrast, the small scale structure in the immiscible experiment is qualitatively different from what we can observe in the miscible experiment. In the miscible experiment, the negatively buoyant solutions *mixes* with the entrained water, diffusion of the NaI salt allowing homogenisation at the molecular scale. In the immiscible experiment, the NaI solution of course does not mix with the entrained silicon oil since the two liquids are immiscible. A close inspection of the last snapshot of the immiscible experiment reveals that the dense phase has been fragmented into droplets.

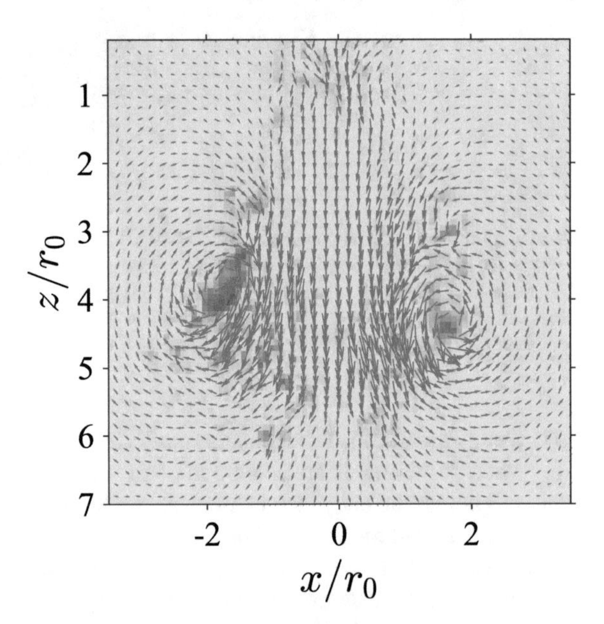

Fig. 5.18 Velocity field obtained from PIV measurements in an experiment in which a 169 mL of aqueous solution of NaI ($\rho = 1280$ kg.m^{-3}) is released in a 1cst viscosity silicon oil. The colorscale gives the vorticity field (red is clockwise, blue counterclockwise). The concentration of the NaI solution has been chosen so that its optical index matches that of the silicon oil, in order to avoid optical distorsion. The non-dimensional parameter values are $\mathrm{Bo} = 2 \times 10^4$, $\mathrm{Re} = 4.2 \times 10^4$, $\rho_m/\rho_s = 1.56$, $\eta_m/\eta_s = 1.2$

Large-Scale Dynamics: Turbulent Entrainment Model

We consider the evolution of a mass of negatively buoyant fluid falling into another one (Fig. 5.19). The mass of negatively buoyant fluid has an initially spherical shape and an initial radius R_0, and has a density $\rho_a + \Delta\rho$, where ρ_a is the density of the surrounding fluid. We denote by u_z the vertical velocity of the center of mass of the negatively buoyant cloud, and by $R(z)$ its spatial extension. Following Batchelor (1954), we assume that far from the source u_z and R only depend on either distance z or time t, and on its total amount of buoyancy defined as

$$B = g\frac{\Delta\rho}{\rho_a}V_0,$$

where V_0 is the initial volume of the released buoyant mass. We thus assume that surface tension has no effect on the evolution of u_z and R. Dimensional analysis then shows that

$$\begin{cases} u_z \sim B^{1/2}z^{-1} \\ R \sim z \end{cases} \quad \text{or, equivalently} \quad \begin{cases} u_z \sim B^{1/4}t^{-1/2} \\ R \sim B^{1/4}t^{1/2} \end{cases},$$

which predicts that the spatial extension of the cloud increases linearly with z: the cloud must therefore entrain ambient fluid. The prediction that the mean velocity decreases as z^{-1}, or, equivalently, as R^{-1}, is consistent with the fact that the total buoyancy of the cloud is conserved, but not its volume. The buoyancy is "diluted" by the incorporation of neutrally buoyant ambient fluid to the cloud.

A more physical (and more general) way to obtain the evolution of a turbulent thermal has been given by Morton et al. (1956), who based their analysis on the assumption that the rate of entrainment of ambient fluid within the buoyant cloud is simply proportional to the mean vertical velocity u_z and to the surface area of the cloud (Fig. 5.19). This is the basic assumption of the *turbulent entrainment* models used to describe the dynamics of turbulent clouds, plumes, and jets.

This assumption implies that the time derivative of the cloud volume is given by

$$\frac{d}{dt}\left(\frac{4\pi}{3}R^3\right) = 4\pi R^2 \alpha\, u_z, \tag{5.229}$$

where α is the *entrainment coefficient*. Noting that $d(...)/dt = u_z d(...)/dz$, where z is the vertical position of the center of mass of the cloud, integration of equation (5.229) gives

$$R = R_0 + \alpha z, \tag{5.230}$$

which is consistent with the prediction of dimensional analysis.

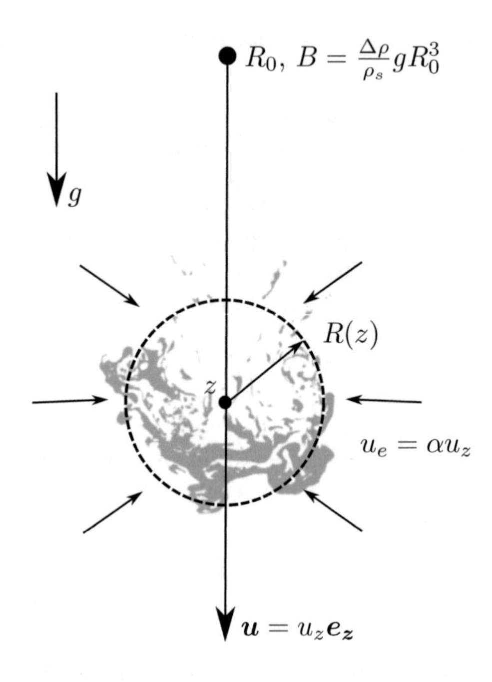

Fig. 5.19 The *turbulent thermal* model: a volume of fluid with initial radius R_0 and density $\rho_a + \Delta\rho$ is released at $z = 0$ in a fluid of density ρ_a. The thermal has a mean vertical velocity u_z. Its mean radius R increases with the distance z from the source due to entrainment of ambient fluid at a rate $u_e = \alpha u_z$

Conservation of momentum allows to obtain a predictive law for the vertical velocity u_z of the center of mass of the cloud. Ignoring fluid drag on the cloud and a possible loss of buoyancy in the wake of the cloud, conservation of momentum simply states that

$$\frac{d}{dt}\left(\frac{4\pi}{3}R^3\bar{\rho}u_z\right) = \frac{4\pi}{3}R_0^3\Delta\rho g \tag{5.231}$$

where $\bar{\rho}$ is the mean density of the cloud, given by

$$\bar{\rho} = \rho_a + \Delta\rho\frac{R_0^3}{R^3} = \rho_a + (1+\alpha z)^{-3}\Delta\rho. \tag{5.232}$$

Conservation of mass implies that

$$\frac{d}{dt}\left(\frac{4\pi}{3}R^3\bar{\rho}\right) = 4\pi\rho_a R^2\alpha u_z. \tag{5.233}$$

Using this relation in Eq. (5.231), using the transformation $d(...)/dt = u_z d(...)/dz = \alpha u_z d(...)/dR$, and re-arranging gives

$$\frac{du_z^2}{dR} + \frac{6}{R}\frac{\rho_a}{\bar{\rho}}u_z^2 = \frac{2g}{\alpha}\frac{\Delta\rho}{\bar{\rho}}\frac{R_0^3}{R^3}, \tag{5.234}$$

a linear first order ordinary differential equation with varying coefficients. The solution, written for u_z as a function of z, is

$$u_z = \left(\frac{g}{2\alpha^3} \frac{\Delta\rho}{\rho_a} R_0^3 \right)^{1/2} \mathcal{F}\left(\frac{R_0}{\alpha z}, \frac{\Delta\rho}{\rho_a} \right) \frac{1}{z}, \tag{5.235}$$

where

$$\mathcal{F}\left(\frac{R_0}{\alpha z}, \frac{\Delta\rho}{\rho_a} \right) = \frac{\left[1 + 4\frac{R_0}{\alpha z} + 6\left(\frac{R_0}{\alpha z} \right)^2 + 4\left(1 + \frac{\Delta\rho}{\rho_a} \right)\left(\frac{R_0}{\alpha z} \right)^3 \right]^{1/2}}{1 + 3\frac{R_0}{\alpha z} + 3\left(\frac{R_0}{\alpha z} \right)^2 + \left(1 + \frac{\Delta\rho}{\rho_a} \right)\left(\frac{R_0}{\alpha z} \right)^3}. \tag{5.236}$$

The function $\mathcal{F}$ tends toward 1 at large $\alpha z/R_0$. The full solution is thus consistent with the dimensional analysis prediction when far from the source. Close to the source, the velocity is given (at first order in $\alpha z/R_0$), by

$$u_z = \left(2\frac{\Delta\rho}{\rho_a} g z \right)^{1/2}. \tag{5.237}$$

Though the turbulent thermal model just described has been developed to model *miscible* flows, experiments suggest that it can also be applied to immiscible fluids in situations where the Bond and Weber numbers are large (Deguen et al. 2014; Landeau et al. 2014; Wacheul et al. 2014; Wacheul and Le Bars 2017). A qualitative comparison of Fig. 5.17a, b suggests that it is indeed the case. The similarity between the miscible and immiscible experiments also holds on a quantitative level: from a series of experiments similar to that presented on Fig. 5.17, we have been able to show that the evolution of both R and u_z are very well described by the turbulent entrainment model with $\alpha = 0.25 \pm 0.05$, similar to miscible flows (Deguen et al. 2014; Landeau et al. 2014). This demonstrates that there is indeed no effect of surface tension on the large-scale part of the flow.

Fragmentation

Qualitative Observations form Experiments

In the experiment shown on Fig. 5.17b and in similar experiments, most of the fragmentation of the dense liquid occurs during a relatively short time span. In Fig. 5.17b, the dense phase is essentially continuous until the third snapshot, and almost entirely fragmented into drops at the fourth snapshot. The analysis of images obtained with a high-speed camera (1kHz) shows that drops formation results from two mechanisms:

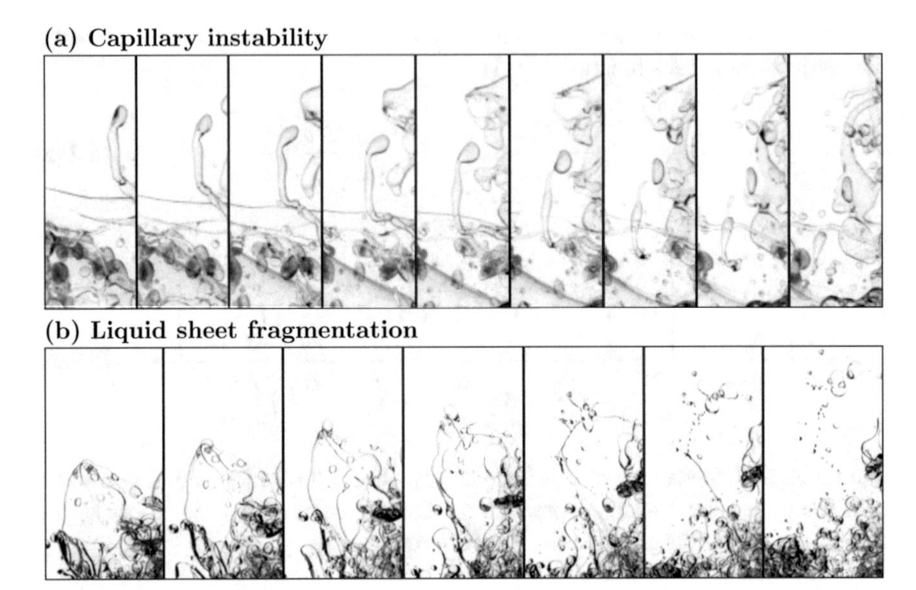

Fig. 5.20 Two fragmentations mechanisms observed in the experiment shown in Fig. 5.17b. **a** Fragmentation of a liquid (NaI solution) ligament. The time interval between two images is $\Delta t = 10$ ms, and the width of each image is 1 cm. **b** Fragmentation of liquid film. The time interval between two images is $\Delta t = 20$ ms, and the width of each image is 1.8 cm

1. the fragmentation of stretched cylindrical ligaments of aqueous solution through the Rayleigh–Plateau capillary instability, as shown in Fig. 5.20a (the mechanism of the Rayleigh–Plateau instability is explained below),
2. the fragmentation of thin liquid films, as shown in Fig. 5.20b. In this regime, thin films of aqueous solutions are stretched by the flow before eventually being punctured. The film then quickly retracts, the liquid forming the film gathering into ligaments which then fragment into drops due to the Rayleigh–Plateau instability.

These two modes of fragmentation are classically observed in fragmentation problems in a variety of contexts. In fact, liquid fragmentation necessitates a capillary instability, irrespectively of the nature of the flow (Villermaux 2007). What varies from one problem to another is the sequence of mechanisms resulting in the formation of ligaments which can fragment as a result of the Rayleigh–Plateau capillary instability. In experiments such as shown in Fig. 5.17b, the observed sequence is the following: (i) the interface is destabilised and deformed by the combined effect of shear and Rayleigh–Taylor instabilities; (ii) three-dimensional structures generated by the destabilisation of the interface are stretched and stirred by the mean flow and velocity field fluctuations; (iii) stirring produces ligaments and films, which will then break up and produce a population of drops.

Rayleigh–Plateau Capillary Instability

In many situations, the deformation of an interface results in an increase of its surface area, and hence of its interfacial energy. In this case deformation is not energetically favoured, and mechanical work therefore has to be provided to deform the interface. This is for example the case of initially planar interface: any perturbation of the interface results in an increase of its surface area and energy.

In contrast, the deformation of a cylindrical interface can, under certain conditions, result in a decrease of its surface area, and hence of its interfacial energy. Take a cylinder of one liquid into another, of length L and radius R_0. Its surface area is $2\pi R_0 L$ and its interfacial energy is $2\pi R_0 L\gamma$. It is easy to see that the cylindrical shape is not very favourable from an energetic point of view: if the liquid of the cylinder is re-arranged to form a sphere of the same volume ($\pi R_0^2 L$), the sphere will have a radius equal to $\left[(3/4)R_0^2 L\right]^{1/3}$, and a surface area equal to $4(3/4)^{2/3}\pi R_0^{4/3} L^{2/3}$, which is smaller than the cylinder surface area if the length of the cylinder is larger than $(9/2)R_0$. This shows that the fragmentation into drops of liquid cylinder is energetically favoured if the ratio of its length and radius is larger than 9/2.

Is fragmentation dynamically possible? To see if it is, let us consider again a liquid cylinder of radius R_0, and assume now that its surface is perturbed from its initial shape as

$$R(x) = \bar{R} + \epsilon \sin\left(2\pi \frac{x}{\lambda}\right), \tag{5.238}$$

where x is the coordinate along the axis of symmetry of the cylinder, and ϵ and λ the amplitude and wavelength of the perturbation (Fig. 5.21). Note that conservation

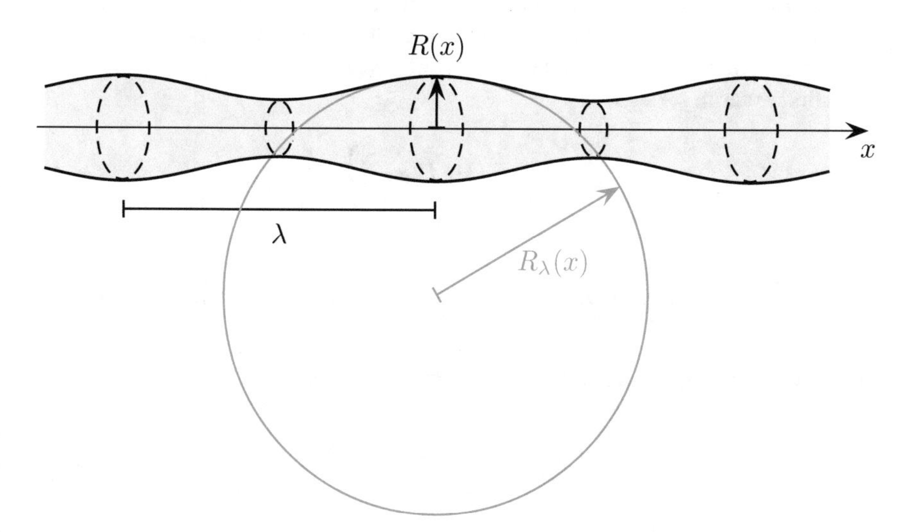

Fig. 5.21 A liquid cylinder (blue) with its surface perturbed by an axisymmetric sinusoidal perturbation

of mass implies that $\bar{R} < R_0$: comparing the volume of a section of length λ of the unperturbed and perturbed states indeed shows that

$$\bar{R} = R_0 \left(1 - \frac{\epsilon^2}{4R_0^2}\right)^{1/2}. \tag{5.239}$$

Calculating the surface area of the perturbed cylinder shows that the perturbation induces a decrease of the surface area (and hence of energy) if $\lambda > 2\pi R_0$, which suggests that the cylindrical shape may be unstable to perturbations with wavelengths larger than $2\pi R_0$.

To see how the instability works, let us consider the two principal curvatures of the interface (Fig. 5.21). One is the curvature associated with the radius of the cylinder, $1/R(x)$, and the other is the curvature associated with the longitudinal perturbation, $1/R_\lambda(x)$, which is equal to the divergence in the x-direction of the normal $\mathbf{n}$ of the interface. The Laplace pressure jump across the interface is equal to

$$\Delta P = \gamma \left(\frac{1}{R(x)} + \frac{1}{R_\lambda(x)}\right). \tag{5.240}$$

The contribution of R_λ is positive where $R > R_0$, and negative where $R < R_0$. It thus produces a pressure gradient from regions of large R to regions of small R, which can drive a flow from large to small R that would decrease the amplitude of the radius perturbation. It thus has a stabilising effect. In contrast, the pressure jump associated with $1/R$ is larger in regions of small R. It thus produces a pressure gradient from small to large R, which may drive a flow that would increase the amplitude of the perturbation. The amplitude of the radius perturbation can therefore grow if the pressure gradient associated with the curvature $R(x)$ is larger in magnitude that the pressure gradient associated with the curvature $R_\lambda(x)$.

At first order in ϵ, one find

$$\frac{1}{R} = \frac{1}{R_0}\left[1 - \frac{\epsilon}{R_0}\sin\left(2\pi\frac{x}{\lambda}\right)\right] + \mathcal{O}(\epsilon^2), \tag{5.241}$$

and

$$\frac{1}{R_\lambda} = \frac{-1}{\sqrt{1+(dR/dx)^2}}\frac{d^2R}{dx^2} = \frac{4\pi^2}{\lambda^2}\epsilon\sin\left(2\pi\frac{x}{\lambda}\right) + \mathcal{O}(\epsilon^2), \tag{5.242}$$

which gives a pressure jump across the interface given by

$$\Delta P = \gamma\left[\frac{1}{R_0} + \epsilon\left(\frac{4\pi^2}{\lambda^2} - \frac{1}{R_0^2}\right)\sin\left(2\pi\frac{x}{\lambda}\right)\right] + \mathcal{O}(\epsilon^2). \tag{5.243}$$

Its gradient along x is given by

$$\frac{\partial \Delta P}{\partial x} = \epsilon \frac{2\pi}{\lambda} \gamma \left(\frac{4\pi^2}{\lambda^2} - \frac{1}{R_0^2} \right) \cos \left(2\pi \frac{x}{\lambda} \right) + \mathcal{O}(\epsilon^2). \tag{5.244}$$

This shows that the pressure gradient inside the liquid cylinder is from small to large R if $\frac{4\pi^2}{\lambda^2} < \frac{1}{R_0^2}$, and from large to small R instead. In other words, the initial perturbation will grow if

$$\lambda > \lambda_c = 2\pi R_0, \tag{5.245}$$

which will eventually lead to the fragmentation of the cylinder into drops. This dynamical criterion is slightly more restrictive that the energy criterion, since $2\pi \simeq 6.26 > 9/2 = 4.5$.

Things get more complicated in situations where liquid ligaments are deformed and stretched by the ambient flow. Fragmentation can be significantly protracted by stretching effects (Taylor 1934; Tomotika 1936; Mikami et al. 1975; Eggers and Villermaux 2008), which can be understood as follows. Let us consider a stretched ligament, the surface of which is modulated by a longitudinal perturbation of wavelength λ. The stretching will affect the disturbance, which will see its wavelength increase in proportion to the amount of stretching. If the perturbation wavelength initially corresponds to the optimal wavelength for the growth of the capillary instability, increasing the wavelength will decrease the rate of growth of the disturbance.

Chemical and Heat Transfer at the Drop Scale

If the metal phase ends up being fragmented into drops of size equal or smaller than the maximal stable size $R_c \sim \sqrt{\gamma/(\Delta\rho g)}$ [Eq. (5.221)], then thermal and chemical equilibration of the metal phase with the surrounding silicates is not an issue (e.g. Stevenson 1990; Karato and Rama Murthy 1997; Rubie et al. 2003; Ulvrová et al. 2011, as shown below.

One can show (Lherm and Deguen 2018) that the timescale of chemical equilibration of a falling metal drop with its surrounding is given by

$$\tau_{\text{eq}} \sim \frac{R^2}{6\kappa_s} D_{\text{m/s}} \text{Pe}_{\text{s}}^{-1/2} \left[1 + \frac{1}{D_{\text{m/s}}} \left(\frac{\kappa_s}{\kappa_m} \right)^{1/2} \right], \tag{5.246}$$

where κ_s and κ_m are the compositional diffusivities in the silicates and metal, $D_{\text{m/s}}$ the metal/silicates partitioning coefficient of a given chemical element, and $Pe_s = UR/\kappa_s$. The distance ℓ_{eq} fallen by the drop during a time τ_{eq} is given by

$$\ell_{\text{eq}} = \tau_{\text{eq}} U \sim \frac{R}{6} D_{\text{m/s}} \text{Pe}_{\text{s}}^{1/2} \left[1 + \frac{1}{D_{\text{m/s}}} \left(\frac{\kappa_s}{\kappa_m} \right)^{1/2} \right] \tag{5.247}$$

In the case of siderophile elements ($D_{\mathrm{m/s}} \gg 1$), this is approximated by

$$\ell_{\mathrm{eq}} \sim \frac{R}{6} D_{\mathrm{m/s}} {\mathrm{Pe}_{\mathrm{s}}}^{1/2}. \tag{5.248}$$

For a metal drop falling into a magma ocean, R_c is about 5 mm. The corresponding Grashof number is

$$Gr = \frac{\Delta\rho}{\rho_s} \frac{g R_c^3}{\nu_s^2} \sim \left(\frac{g}{10\,\mathrm{m.s^{-2}}}\right) \times \left(\frac{10^{-2}\,\mathrm{Pa.s}}{\eta_s}\right)^2 \times 10^5. \tag{5.249}$$

Magma ocean viscosity is estimated to be in the range 10^{-3}–10^{-2} Pa.s, which implies that the drop is in the newtonian regime and has a terminal velocity given by $\sim\sqrt{(\Delta\rho/\rho_s g R_c}$. With a compositional diffusivity $\kappa_s \sim 10^{-9}$ m.s^{-2}, this gives a Péclet number around 10^6 and we thus have $(R/6){\mathrm{Pe}_{\mathrm{s}}}^{1/2} \simeq 1$ m. Siderophile elements like Nickel or Cobalt have partitioning coefficients $D_{\mathrm{m/s}}$ around 10^3 at low pressure, and as low as $\sim$10 when approaching the pressure of the core-mantle boundary. This would give $\ell_{\mathrm{eq}} \sim 1$ km at low pressure, and $\ell_{\mathrm{eq}} \sim 10$ m at high pressure. This is in both cases much smaller than typical magma ocean depth, which implies that drops of metal a few mm in size will readily equilibrate with the surrounding molten silicates.

However, the above conclusion rests on the assumption that the metal phase fragments into drops of a few mm in radius. We have no well-tested fragmentation model that can be used in the context of core formation, so whether the metal phase would fragment or not is still an open question. It is also possible that efficient equilibration does not require fragmentation of the metal phase. The metal phase would necessarily be intensely stirred and stretched before fragmentation, and this may allow for efficient chemical transfer between the metal and silicates phase (Lherm and Deguen 2018).

Acknowledgements This project has received funding from the European Research Council (ERC) under the European Unions Horizon 2020 research and innovation programme (grant agreement No 716429).

References

Alboussière, T., Deguen, R., & Melzani, M. (2010). Melting induced stratification above the Earth's inner core due to convective translation. *Nature*, *466*, 744–747.

Alfè, D., Gillan, M. J., & Price, G. D. (2000). Constraints on the composition of the Earth's core from ab initio calculations. *Nature*, *405*, 172–175.

Andrault, D., Monteux, J., Le Bars, M., & Samuel, H. (2016). The deep Earth may not be cooling down. *Earth and Planetary Science Letters*, *443*, 195–203.

Badro, J., Fiquet, G., Guyot, F., Gregoryanz, E., Occelli, F., Antonangeli, D., et al. (2007). Effect of light elements on the sound velocities in solid iron: Implications for the composition of Earth's core. *Earth and Planetary Science Letters*, *254*, 233–238.

Badro, J., Brodholt, J. P., Piet, H., Siebert, J., & Ryerson, F. J. (2015). Core formation and core composition from coupled geochemical and geophysical constraints. *Proceedings of the National Academy of Sciences*, *112*(40), 12310–12314.
Badro, J., Siebert, J., & Nimmo, F. (2016). An early geodynamo driven by exsolution of mantle components from Earth's core. *Nature*, *536*(7616), 326–328.
Batchelor, G. K. (1954). Heat convection and buoyancy effects in fluids. *Quarterly Journal of the Royal Meteorological Society*, *80*, 339–358.
Bercovici, D., & Kelly, A. (1997). The non-linear initiation of diapirs and plume heads. *Physics of the Earth and Planetary Interiors*, *101*, 119–130.
Berhanu, M., Monchaux, R., Fauve, S., Mordant, N., Pétrélis, F., Chiffaudel, A., et al. (2007). Magnetic field reversals in an experimental turbulent dynamo. *EPL (Europhysics Letters)*, *77*(5), 59001.
Birch, F. (1940). The alpha-gamma transformation of iron at high pressures, and the problem of the Earth's magnetism. *American Journal of Science*.
Birch, F. (1952). Elasticity and constitution of the Earth's interior. *Journal of Geophysical Research*, *57*, 227.
Birch, F. (1964). Density and composition of mantle and core. *Journal of Geophysical Research*, *69*, 4377.
Bondi, H., & Lyttleton, R. A. (1948). On the dynamical theory of the rotation of the earth: I. the secular retardation of the core. In *Mathematical Proceedings of the Cambridge Philosophical Society* (Vol. 44, pp. 345–359). Cambridge University Press.
Braginsky, S. I. (1963). Structure of the F layer and reasons for convection in the Earth's core. *Doklady Akademii Nauk SSSR English Translation*, *149*, 1311–1314.
Braginsky, S. I. (1970). Torsional magnetohydrodynamics vibrations in the Earth's core and variations in day length. *Geomagnetism and Aeronomy*, *10*, 3–12.
Buffett, B. A. (2002). Estimates of heat flow in the deep mantle based on the power requirements for the geodynamo. *Geophysical Research Letters*, *29*(12), 120000–1.
Buffett, B. A., & Bloxham, J. (2000). Deformation of Earth's inner core by electromagnetic forces. *Geophysical Research Letters*, *27*, 4001–4004.
Buffett, B. A., Huppert, H. E., Lister, J. R., & Woods, A. W. (1996). On the thermal evolution of the Earth's core. *Journal of Geophysical Research*, *101*, 7989–8006.
Bullard, E. (1949). The magnetic field within the Earth. *Proceedings of the Royal Society of London A*, *197*(1051), 433–453.
Bullard, E. (1950). The transfer of heat from the core of the Earth. *Geophysical Journal International*, *6*, 36–41.
Bullard, E., & Gellman, H. (1954). Homogeneous dynamos and terrestrial magnetism. *Philosophical Transactions of the Royal Society of London A*, *247*(928), 213–278.
Busse, F. H. (1970). Thermal instabilities in rapidly rotating systems. *Journal of Fluid Mechanics*, *44*, 441–460.
Cabanes, S., Schaeffer, N., & Nataf, H.-C. (2014). Turbulence reduces magnetic diffusivity in a liquid sodium experiment. *Physical Review Letters*, *113*(18), 184501.
Cheng, J. S., & Aurnou, J. M. (2016). Tests of diffusion-free scaling behaviors in numerical dynamo datasets. *Earth and Planetary Science Letters*, *436*, 121–129.
Christensen, U. R., & Wicht, J. (2015). Numerical dynamo simulations. *Treatise on Geophysics (Second Edition)*, *8*, 245–277.
Christensen, U. R., Aubert, J., & Hulot, G. (2010). Conditions for Earth-like geodynamo models. *Earth and Planetary Science Letters*, *296*(3–4), 487–496.
Christensen, U. R. (2010). Dynamo scaling laws and applications to the planets. *Space Science Reviews*, *152*(1–4), 565–590.
Christensen, U. R., & Aubert, J. (2006). Scaling properties of convection-driven dynamos in rotating spherical shells and application to planetary magnetic fields. *Geophysical Journal International*, *166*(1), 97–114. ISSN 1365-246X.

Christensen, U. R., & Tilgner, A. (2004). Power requirement of the geodynamo from ohmic losses in numerical and laboratory dynamos. *Nature*, *429*, 169–171.

Clift, R., Grace, J. R., & Weber, M. E. (1978). *Bubbles, drops and particles*. New York: Academic Press.

Corgne, A., Keshav, S., Fei, Y., & McDonough, W. F. (2007). How much potassium is in the Earth's core? New insights from partitioning experiments. *Earth and Planetary Science Letters*, *256*, 567–576.

Cottaar, S., & Buffett, B. (2012). Convection in the Earth's inner core. *Physics of the Earth and Planetary Interiors*, *198–199*, 67–78.

Dahl, T. W., & Stevenson, D. J. (2010). Turbulent mixing of metal and silicate during planet accretion—and interpretation of the Hf–W chronometer. *Earth and Planetary Science Letters*, *295*, 177–186.

de Koker, N., Steinle-Neumann, G., & Vlcek, V. (2012). Electrical resistivity and thermal conductivity of liquid Fe alloys at high P and T, and heat flux in Earth's core. *Proceedings of the National Academy of Sciences*, *109*(11), 4070–4073.

Deguen, R. (2012). Structure and dynamics of Earth's inner core. *Earth and Planetary Science Letters*, *333–334*, 211–225.

Deguen, R., & Cardin, P. (2011). Thermo-chemical convection in Earth's inner core. *Geophysical Journal International*, *187*, 1101–1118.

Deguen, R., Olson, P., & Cardin, P. (2011). Experiments on turbulent metal-silicate mixing in a magma ocean. *Earth and Planetary Science Letters*, *310*, 303–313.

Deguen, R., Alboussière, T., & Cardin, P. (2013). Thermal convection in Earth's inner core with phase change at its boundary. *Geophysical Journal International*, *194*(3), 1310–1334.

Deguen, R., Landeau, M., & Olson, P. (2014). Turbulent metal-silicate mixing, fragmentation, and equilibration in magma oceans. *Earth and Planetary Science Letters*, *391*, 274–287.

Deguen, R., Alboussière, T., & Labrosse, S. (2018). Double-diffusive translation of Earth's inner core. *Geophysical Journal International*, *214*(1), 88–107.

Deguen, R., Risso, F., Keita, M. Double-diffusive translation of Earth's inner core, in prep..

Dormy, E., Soward, A. M., Jones, C. A., Jault, D., & Cardin, P. (2004). The onset of thermal convection in rotating spherical shells. *Journal of Fluid Mechanics*, *501*, 43–70.

Dziewonski, A. M., & Gilbert, F. (1971). Solidity of the inner core of the Earth inferred from normal mode observations. *Nature*, *234*, 465–466.

Eggers, J., & Villermaux, E. (2008). Physics of liquid jets. *Reports on Progress in Physics*, *71*(3), 036601.

Gailitis, A., Lielausis, O., Platacis, E., Dement'ev, S., Cifersons, A., Gerbeth, G., et al. (2001). Magnetic field saturation in the riga dynamo experiment. *Physical Review Letters*, *86*(14), 3024.

Gillet, N., Jault, D., Canet, E., & Fournier, A. (2010). Fast torsional waves and strong magnetic field within the Earths core. *Nature*, *465*(7294), 74–77.

Gilman, P. A., & Miller, J. (1981). Dynamically consistent nonlinear dynamos driven by convection in a rotating spherical shell. *The Astrophysical Journal Supplement Series*, *46*, 211–238.

Glatzmaier, G. A. (1984). Numerical simulations of stellar convective dynamos. I. The model and method. *Journal of Computational Physics*, *55*(3), 461–484.

Glatzmaier, G. A. (1985a). Numerical simulations of stellar convective dynamos. II-field propagation in the convection zone. *The Astrophysical Journal*, *291* 300–307.

Glatzmaier, G. A. (1985b). Numerical simulations of stellar convective dynamos III. At the base of the convection zone. *Geophysical & Astrophysical Fluid Dynamics*, **31**(1–2), 137–150.

Glatzmaier, G. A., & Roberts, P. H. (1995). A three-dimensional self-consistent computer simulation of a geomagnetic field reversal. *Nature*, *377*, 203–209. https://doi.org/10.1038/377203a0.

Gomi, H., Ohta, K., Hirose, K., Labrosse, S., Caracas, R., Verstraete, M. J., et al. (2013). The high conductivity of iron and thermal evolution of the Earth's core. *Physics of the Earth and Planetary Interiors*, *224*, 88–103.

Gomi, H., Hirose, K., Akai, H., & Fei, Y. (2016). Electrical resistivity of substitutionally disordered hcp Fe-Si and Fe-Ni alloys: Chemically-induced resistivity saturation in the Earth's core. *Earth and Planetary Science Letters*, *451*, 51–61.
Gubbins, D. (1977). Energetics of the Earth's core. *Journal of Geophysics Zeitschrift Geophysik*, *43*, 453–464.
Gubbins, D., Alfè, D., Masters, G., Price, G. D., & Gillan, M. (2004). Gross thermodynamics of two-component core convection. *Geophysical Journal International*, *157*, 1407–1414.
Gubbins, D., Alfè, D., & Davies, C. J. (2013). Compositional instability of Earth's solid inner core. *Geophysical Research Letters*, *40*, 1–5.
Halliday, A. (2004). Mixing, volatile loss and compositional change during impact-driven accretion of the Earth. *Nature*, *427*(6974), 505–509. ISSN 0028-0836. https://doi.org/10.1038/nature02275.
Herzenberg, A. (1958). Geomagnetic dynamos. *Philosophical Transactions of the Royal Society of London*, *250*(986), 543–583.
Hirose, K., Morard, G., Sinmyo, R., Umemoto, K., Hernlund, J., Helffrich, G., et al. (2017). Crystallization of silicon dioxide and compositional evolution of the Earth's core. *Nature*, *543*(7643), 99.
Holme, R. (2015). Large-scale flow in the core. *Treatise on Geophysics*, *8*, 91–113.
Jacobs, J. A. (1953). The Earth's inner core. *Nature*, *172*, 297.
Jaupart, C., & Mareschal, J.-C. (2010). *Heat generation and transport in the Earth*. Cambridge: Cambridge University Press.
Jephcoat, A., & Olson, P. (1987). Is the inner core of the Earth pure iron? *Nature*, *325*, 332–335.
Jones, C. A. (2011). Planetary magnetic fields and fluid dynamos. *Annual Review of Fluid Mechanics*, *43*, 583–614.
Jones, C. A., Soward, A. M., & Mussa, A. I. (2000). The onset of thermal convection in a rapidly rotating sphere. *Journal of Fluid Mechanics*, *405*, 157–179.
Kageyama, A., Sato, T., & (Complexity Simulation Group) (1995). Computer simulation of a magnetohydrodynamic dynamo. II. *Physics of Plasmas*, *2*(5), 1421–1431.
Karato, S., & Rama Murthy, V. (1997). Core formation and chemical equilibrium in the Earth—I. Physical considerations. *Physics of the Earth and Planetary Interiors*, *100*(1–4), 61–79. ISSN 00319201. https://doi.org/10.1016/S0031-9201(96)03232-3.
Karato, S. (2012). *Deformation of Earth materials: an introduction to the rheology of solid Earth*. Cambridge: Cambridge University Press.
King, E. M., & Buffett, B. A. (2013). Flow speeds and length scales in geodynamo models: The role of viscosity. *Earth and Planetary Science Letters*, *371*, 156–162.
Konôpková, Z., McWilliams, R. S., Gómez-Pérez, N., & Goncharov, A. F. (2016). Direct measurement of thermal conductivity in solid iron at planetary core conditions. *Nature*, *534*(7605), 99–101.
Labrosse, S. (2014). Thermal and compositional stratification of the inner core. *Comptes Rendus Geoscience*, *346*, 119–129.
Labrosse, S. (2015). Thermal evolution of the core with a high thermal conductivity. *Physics of the Earth and Planetary Interiors*, *247*, 36–55.
Labrosse, S., Hernlund, J. W., & Coltice, N. (2007). A crystallizing dense magma ocean at the base of the Earth's mantle. *Nature*, *450*(7171), 866–869. ISSN 0028-0836. https://doi.org/10.1038/nature06355.
Landeau, M., Deguen, R., & Olson, P. (2014). Experiments on the fragmentation of a buoyant liquid volume in another liquid. *Journal of Fluid Mechanics*, *749*, 478–518.
Larmor, J. (1919). How could a rotating body such as the sun become a magnet. *Report of the British Association for the Advancement of Science*, 159–160.
Lasbleis, M., & Deguen, R. (2015). Building a regime diagram for the Earth's inner core. *Physics of the Earth and Planetary Interiors*, *247*, 80–93.
Lasbleis, M., Deguen, R., Cardin, P., & Labrosse, S. (2015). Earth's inner core dynamics induced by the Lorentz force. *Geophysical Journal International*, *202*, 548–563.

Lay, T., Hernlund, J., Garnero, E. J., & Thorne, M. S. (2006). A post-perovskite lens and D"heat flux beneath the central Pacific. *Science*, *314*, 1272–1276.

Lehmann, I. (1936). P'. *Bureau Central Sismologique International*, *14*, 87–115.

Lherm, V., & Deguen, R. (2018). Small-scale metal/silicate equilibration during core formation: The influence of stretching enhanced diffusion on mixing. *Journal of Geophysical Research: Solid Earth*, *123*(12), 10–496.

Lister, J. R. (2003). Expressions for the dissipation driven by convection in the Earth's core. *Physics of the Earth and Planetary Interiors*, *140*, 145–158. https://doi.org/10.1016/S0031-9201(03)00169-9.

Loper, D. E. (1978). The gravitationally powered dynamo. *Geophysical Journal International*, *54*, 389–404.

Lowes, F. J., & Wilkinson, I. (1963). Geomagnetic dynamo: a laboratory model. *Nature (London)*, *198*, 4886.

Lunine, J. I., O'brien, D. P., Raymond, S. N., Morbidelli, A., Quinn, T., & Graps, A. L. (2011). Dynamical models of terrestrial planet formation. *Advanced Science Letters*, *4*(2), 325–338.

Malkus, W. V. R. (1963). Precessional torques as the cause of geomagnetism. *Journal of Geophysical Research*, *68*(10), 2871–2886.

Malkus, W. V. R. (1968). Precession of the Earth as the cause of geomagnetism: Experiments lend support to the proposal that precessional torques drive the Earth's dynamo. *Science*, *160*(3825), 259–264.

Maus, S., Rother, M., Stolle, C., Mai, W., Choi, S., Lühr, H., Cooke, D., & Roth, C. (2006). Third generation of the Potsdam Magnetic Model of the Earth (POMME). *Geochemistry, Geophysics, Geosystems*, *7*(7).

Mikami, T., Cox, R. G., & Mason, S. G. (1975). Breakup of extending liquid threads. *International Journal of Multiphase Flow*, *2*(2), 113–138.

Mizzon, H., & Monnereau, M. (2013). Implication of the lopsided growth for the viscosity of Earth's inner core. *Earth and Planetary Science Letters*, *361*, 391–401.

Monchaux, R., Berhanu, M., Bourgoin, M., Moulin, M., Odier, P., Pinton, J.-F., et al. (2007). Generation of a magnetic field by dynamo action in a turbulent flow of liquid sodium. *Physical Review Letters*, *98*(4), 044502.

Monnereau, M., Calvet, M., Margerin, L., & Souriau, A. (2010). Lopsided growth of Earth's inner core. *Science*, *328*, 1014–1017.

Monteux, J., Ricard, Y., Coltice, N., Dubuffet, F., & Ulvrova, M. (2009). A model of metal–silicate separation on growing planets. *Earth and Planetary Science Letters*, *287*(3–4), 353–362. ISSN 0012821X. https://doi.org/10.1016/j.epsl.2009.08.020.

Monteux, J., Jellinek, A. M., & Johnson, C. L. (2011). Why might planets and moons have early dynamos? *Earth and Planetary Science Letters*, *310*(3), 349–359.

Morton, B. R., Taylor, G., & Turner, J. S. (1956). Turbulent gravitational convection from maintained and instantaneous sources. *Proceedings of the Royal Society of London. Series A, Mathematical and Physical Sciences*, *234*, 1–23.

Nataf, H.-C., & Gagnière, N. (2008). On the peculiar nature of turbulence in planetary dynamos. *Comptes Rendus Physique*, *9*(7), 702–710.

Olson, P. (2013). Experimental dynamos and the dynamics of planetary cores. *Annual Review of Earth and Planetary Sciences*, *41*, 153–181.

Olson, P., & Christensen, U. R. (2006). Dipole moment scaling for convection-driven planetary dynamos. *Earth and Planetary Science Letters*, *250*(3–4), 561–571.

O'Rourke, J. G., & Stevenson, D. J. (2016). Powering Earth's dynamo with magnesium precipitation from the core. *Nature*, *529*(7586), 387–389.

O'Rourke, J. G., Korenaga, J., & Stevenson, D. J. (2017). Thermal evolution of Earth with magnesium precipitation in the core. *Earth and Planetary Science Letters*, *458*, 263–272.

Ponomarenko, Y. B. (1973). Theory of the hydromagnetic generator. *Journal of Applied Mechanics and Technical Physics*, *14*(6), 775–778.

Poupinet, G., Pillet, R., & Souriau, A. (1983). Possible heterogeneity of the Earth's core deduced from PKIKP travel times. *Nature*, *305*, 204–206.

Pozzo, M., Davies, C., Gubbins, D., & Alfè, D. (2012). Thermal and electrical conductivity of iron at Earth's core conditions. *Nature*, *485*, 355–358.

Ribe. N. M. (2007). Analytical Approaches to mantle dynamics. In Schubert, G. (Ed.), *Treatise on geophysics* (Vol. 7).

Ribe, N. M. (1983). Diapirism in the Earth's mantle: Experiments on the motion of a hot sphere in a fluid with temperature dependent viscosity. *Journal of Volcanology and Geothermal Research*, *16*, 221–245.

Ricard, Y. (2007). Physics of mantle convection. In Schubert, G. (Ed.), *Treatise on geophysics* (Vol. 7). Elsevier.

Roberts, P. H. (1968). On the thermal instability of a rotating-fluid sphere containing heat sources. *Philosophical Transactions of the Royal Society of London A*, *263*(1136), 93–117.

Roberts, G. O. (1972). Dynamo action of fluid motions with two-dimensional periodicity. *Philosophical Transactions of the Royal Society of London A*, *271*(1216), 411–454.

Roberts, P. H., Jones, C. A., & Calderwood, A. (2003). Energy fluxes and ohmic dissipation in the Earth's core. In Jones C. A., Soward A. M., & Zhang K. (Eds.), *Earth's core and lower mantle*. Taylor & Francis.

Rubie, D. C., Melosh, H. J., Reid, J. E., Liebske, C., & Righter, K. (2003). Mechanisms of metal–silicate equilibration in the terrestrial magma ocean. *Earth and Planetary Science Letters*, *205*(3–4), 239–255. ISSN 0012821X. https://doi.org/10.1016/S0012-821X(02)01044-0.

Rubie, D. C., Jacobson, S. A., O'Brien, D. P., Young, Ed. D., de Vries, J., Nimmo, F., et al. (2015). Accretion and differentiation of the terrestrial planets with implications for the compositions of early-formed solar system bodies and accretion of water. *Icarus*, *248*, 89–108.

Rudge, J. F., Kleine, T., & Bourdon, B. (2010). Broad bounds on Earth's accretion and core formation constrained by geochemical models. *Nature Geoscience*, *3*, 439–443.

Samuel, H. (2012). A re-evaluation of metal diapir breakup and equilibration in terrestrial magma oceans. *Earth and Planetary Science Letters*, *313*, 105–114.

Samuel, H., & Tackley, P. J. (2008, June). Dynamics of core formation and equilibration by negative diapirism. *Geochemistry, Geophysics, Geosystems*, *9*, 6011–+. https://doi.org/10.1029/2007GC001896.

Samuel, H., Tackley, P. J., & Evonuk, M. (2010). Heat partitioning in terrestrial planets during core formation by negative diapirism. *Earth and Planetary Science Letters*, *290*, 13–19. https://doi.org/10.1016/j.epsl.2009.11.050.

Schaeffer, N., Jault, D., Nataf, H.-C., & Fournier, A. (2017). Turbulent geodynamo simulations: A leap towards Earth's core. *Geophysical Journal International*, *211*(1), 1–29.

Scorer, R. S. (1957). Experiments on convection of isolated masses of buoyant fluid. *Journal of Fluid Mechanics Digital Archive*, *2*(06), 583–594. https://doi.org/10.1017/S0022112057000397.

Seagle, C. T., Cottrell, E., Fei, Y., Hummer, D. R., & Prakapenka, V. B. (2013). Electrical and thermal transport properties of iron and iron-silicon alloy at high pressure. *Geophysical Research Letters*, *40*(20), 5377–5381.

Stacey, F. D., & Anderson, O. L. (2001). Electrical and thermal conductivities of Fe-Ni-Si alloy under core conditions. *Physics of the Earth and Planetary Interiors*, *124*, 153–162.

Stacey, F. D., & Loper, D. E. (2007). A revised estimate of the conductivity of iron alloy at high pressure and implications for the core energy balance. *Physics of the Earth and Planetary Interiors*, *161*, 13–18.

Starchenko, S. V., & Jones, C. A. (2002). Typical velocities and magnetic field strengths in planetary interiors. *Icarus*, *157*(2), 426–435.

Stelzer, Z., & Jackson, A. (2013). Extracting scaling laws from numerical dynamo models. *Geophysical Journal International*, *193*(3), 1265–1276.

Stevenson, D. J. (1990). Fluid dynamics of core formation. *Origin of the earth* (pp. 231–249). Oxford: Oxford University Press.

Stevenson, D. J. (2003). Planetary science: Mission to Earth's core—a modest proposal. *Nature*, *423*(6937), 239–240.
Stieglitz, R., & Müller, U. (2001). Experimental demonstration of a homogeneous two-scale dynamo. *Physics of Fluids*, *13*(3), 561–564.
Sumita, I., & Bergman, M. I. (2015). Inner-core dynamics. In Schubert G. (Ed.), *Treatise on geophysics* (Vol. 8, pp. 297–316). Elsevier.
Taylor, G. I. (1934). The formation of emulsions in definable fields of flow. *Proceedings of the Royal Society of London A*, *146*(858), 501–523.
Tomotika, S. (1936). Breaking up of a drop of viscous liquid immersed in another viscous fluid which is extending at a uniform rate. *Proceedings of the Royal Society of London A*, *153*(879), 302–318.
Ulvrová, M., Coltice, N., Ricard, Y., Labrosse, S., Dubuffet, F., Velímskỳ, J., et al. (2011). Compositional and thermal equilibration of particles, drops and diapirs in geophysical flows. *Geochemistry Geophysics Geosystems*, *12*(10), 1–11.
Verhoogen, J. (1961). Heat balance of the Earth's core. *Geophysical Journal of the Royal Astronomical Society*, *4*, 276–291.
Villermaux, E. (2007). Fragmentation. *Annual Review of Fluid Mechanics*, *39*, 419–446.
Wacheul, J.-B., & Le Bars, M. (2017). Experiments on fragmentation and thermo-chemical exchanges during planetary core formation. *Physics of the Earth and Planetary Interiors*.
Wacheul, J.-B., Le Bars, M., Monteux, J., & Aurnou, J. M. (2014). Laboratory experiments on the breakup of liquid metal diapirs. *Earth and Planetary Science Letters*, *403*, 236–245.
Walsh, K. J., Morbidelli, A., Raymond, S. N., O'Brien, D. P., & Mandell, A. M. (2011). A low mass for Mars from Jupiter/'s early gas-driven migration. *Nature*, *475*(7355), 206–209.
Weber, P., & Machetel, P. (1992). Convection within the inner-core and thermal implications. *Geophysical Research Letters*, *19*, 2107–2110.
Wenk, H.-R., Baumgardner, J. R., Lebensohn, R. A., & Tomé, C. N. (2000). A convection model to explain anisotropy of the inner core. *Journal of Geophysical Research*, *105*, 5663–5678.
Williams, J.-P., & Nimmo, F. (2004, February). Thermal evolution of the Martian core: Implications for an early dynamo. *Geology*, *32* 97–+.
Wood, B., Walter, M., & Wade, J. (2006). Accretion of the Earth and segregation of its core. *Nature*, *441*(7095), 825–833. ISSN 0028-0836. https://doi.org/10.1038/nature04763.
Woodward, B. (1959). The motion in and around isolated thermals. *Quarterly Journal of the Royal Meteorological Society*, *85*(364), 144–151.
Yin, Q., Jacobsen, S. B., Yamashita, K., Blichert-Toft, J., Telouk, P., & Albarede, F. (2002). A short timescale for terrestrial planet formation from Hf/W chronometry of meteorites. *Nature*, *418*(6901), 949–952. ISSN 0028-0836. https://doi.org/10.1038/nature00995.
Zhang, K., & Busse, F. H. (1988). Finite amplitude convection and magnetic field generation in a rotating spherical shell. *Geophysical & Astrophysical Fluid Dynamics*, *44*(1–4), 33–53.

Chapter 6
A Brief Introduction to Turbulence in Rotating and Stratified Fluids

Benjamin Favier

Abstract This chapter discusses basic aspects of turbulent flows relevant for the small-scale fluid dynamics of planets and stars. We particularly focus on how geometrical confinement, rotation, and stratification affect the nature of turbulent motions at different spatial scales. We introduce a hierarchy of models from the celebrated theory of Kolmogorov valid for homogeneous and isotropic turbulence to gradually more realistic models including rotation and stratification effects. Emphasis is put on simple physical processes and qualitative observations and not on rigorous mathematical derivations.

Introduction

Most of the fluid layers of planets and stars are in a turbulent state. This is a direct consequence of the very large spatial extent of these fluid domains so that molecular viscosity is virtually negligible at these scales. A cascade mechanism is therefore required to bring kinetic energy from the scales at which it is injected (usually by direct forcing or by some instability mechanism) to those where dissipative mechanisms are efficient. In the classical homogeneous isotropic case, this leads to the self-similar theory of turbulence postulated by Kolmogorov (1941). Fluid layers inside planets and stars are however constrained by at least four fundamental aspects, two of which will be discussed in this chapter: background rotation, stable density stratification, compressibility, and magnetic field.

This chapter will only consider non-electrically conducting fluids so that the subtle interaction between fluid motions and magnetic fields, a topic called magnetohydrodynamics (see the chapter by Deguen and Lasbleis of this book and Davidson (2013) for more details), will not be discussed. We will also focus on incompressible dynamics.

B. Favier (✉)
Aix Marseille University, CNRS, Centrale Marseille, IRPHE UMR, 7342 Marseille, France
e-mail: favier@irphe.univ-mrs.fr

M. Le Bars and D. Lecoanet (eds.), *Fluid Mechanics of Planets and Stars*,
CISM International Centre for Mechanical Sciences 595,
https://doi.org/10.1007/978-3-030-22074-7_6

Adding rotation and stratification to the turbulence problem is two edged: on the one hand, linear waves are now supported which reintroduces linear dynamics to the fundamentally nonlinear problem of turbulence. On the other hand, additional dimensionless parameters mean that different dynamical regimes are expected depending on the relative importance between the turbulence and the linear effects.

This chapter is by no mean a complete overview of the vast literature about turbulence in rotating and stratified fluids. The goal is primary to link fundamental models of turbulence such as homogeneous and isotropic turbulence to slightly more realistic models aiming at modeling small-scale flows in geophysical and astrophysical systems. The applications of these general concepts to more specific and complex flows inside planets and stars can be found in the other chapters and in specialized books and articles (e.g., Pedlosky (1992); Vallis (2006); Clarke et al. (2007); Davidson (2013); Alexakis and Biferale (2018)).

3D Homogeneous Isotropic Turbulence

This is the canonical model and probably the most natural starting point when considering turbulence. Most of our current understanding of turbulence was developed using the following assumptions: homogeneity and isotropy (and to a lesser extent, stationarity) (Batchelor 1953). While these assumptions are highly unrealistic at first glance, they allow for a systematic mathematical description of the statistics of a turbulent flow. Additionally, even if the large scales of realistic turbulent flows are neither homogeneous nor isotropic, these idealized properties might be recovered at the smallest spatial scales of the flow, far from any anisotropic and inhomogeneous energy injection mechanism.

Let us consider a velocity field $\boldsymbol{u}$ function of the position $\boldsymbol{x}$ and time t. A standard quantity to statistically describe this flow field is the velocity correlation tensor defined by

$$R_{ij}(\boldsymbol{x}, \boldsymbol{r}, t) = \left\langle u_i(\boldsymbol{x}, t) u_j(\boldsymbol{x} + \boldsymbol{r}, t) \right\rangle, \tag{6.1}$$

where brackets indicate an ensemble average. Assuming the flow to be homogeneous implies that the statistics of the flow do not depend on position $\boldsymbol{x}$ but only on the separation vector $\boldsymbol{r}$:

$$R_{ij}(\boldsymbol{x}, \boldsymbol{r}, t) = R_{ij}(\boldsymbol{r}, t). \tag{6.2}$$

In other words, there are no spatial gradients in any statistical quantity describing the flow. This discards physical boundaries or interfaces since the statistics of the flow would then depend on the distance to that boundary or interface, making the system effectively inhomogeneous. Assuming the flow to be additionally isotropic implies that only the norm of the separation vector matters

$$R_{ij}(\boldsymbol{r}, t) = R_{ij}(r, t) \quad \text{with} \quad r = |\boldsymbol{r}|. \tag{6.3}$$

There are no preferential direction required to statistically describe the flow, which is not true as soon as rotation or stratification is introduced. Finally, it is often assumed that the flow is stationary so that

$$R_{ij}(r,t) = R_{ij}(r)\ . \tag{6.4}$$

All these assumptions greatly simplify the statistical description of a turbulent flow, as will become apparent in the next sections. However, one should never forget that these assumptions are almost never satisfied in realistic flows and should systematically be questioned.

Let us now define the governing equations. The velocity field satisfies the Navier–Stokes equations for an incompressible fluid

$$\frac{\partial \boldsymbol{u}}{\partial t} + \boldsymbol{u}\cdot\nabla\boldsymbol{u} = -\nabla P + \nu\nabla^2\boldsymbol{u} + \boldsymbol{F} \tag{6.5}$$

$$\nabla\cdot\boldsymbol{u} = 0\ , \tag{6.6}$$

where P is the pressure divided by the constant fluid density, ν is the constant kinematic viscosity, and $\boldsymbol{F}$ is some external forcing. The only dimensionless number characteristic of these equations is the Reynolds number Re, measuring the relative importance of the nonlinear inertial term to viscous forces

$$\frac{|\boldsymbol{u}\cdot\nabla\boldsymbol{u}|}{|\nu\nabla^2\boldsymbol{u}|} \approx \frac{UL}{\nu} \equiv Re, \tag{6.7}$$

where we have introduced U a typical velocity and L a typical length. In many contexts, and this is the case for most of the fluid layers of planets and stars, $Re \gg 1$ and the flow is inevitably turbulent. At a spatial scale L characteristic of the flow, viscous forces are negligible compared to the inertia of the fluid.

The scalar product between Eq. (6.5) and $\boldsymbol{u}$, integrated over a given volume V (assuming homogeneity or appropriate boundary conditions), leads to the following kinetic energy equation:

$$\frac{dK}{dt} = \frac{d}{dt}\int_V \frac{1}{2}\boldsymbol{u}^2 \mathrm{d}V = -\int_V \underbrace{2\nu S_{ij}S_{ij}\,\mathrm{d}V}_{\epsilon} + \int_V \boldsymbol{u}\cdot\boldsymbol{F}\mathrm{d}V, \tag{6.8}$$

where S_{ij} is the rate of strain tensor and we have introduced ϵ the rate of dissipation of turbulent kinetic energy. As a consequence, steady states are obtained when the work done by the forcing is balanced by viscous dissipation. Neglecting viscosity altogether, which might look like a good idea for high Reynolds number flows, leads to an unbounded growth of the kinetic energy.

The Zeroth Law of Turbulence

This is the single most important empirical observation about turbulence. The zeroth law of turbulence is the following:

> The rate of dissipation of turbulent kinetic energy is finite
> and independent of viscosity for $Re \to \infty$.

No matter how small the viscosity is, providing that it is not zero, ϵ will remain finite. There are several empirical observations which are directly related to the zeroth law of turbulence. The drag coefficient of an object moving in a viscous fluid is eventually independent of the Reynolds number when the latter is large enough. The energy loss in a pipe flow with a sudden change in the section is independent of viscosity at large Reynolds numbers (the so-called Borda–Carnot energy loss equation).

At first look, the fact that the flow can dissipate energy even when the viscosity goes to zero can be surprising. The rate of dissipation of kinetic energy is $\epsilon = 2\nu S_{ij} S_{ij}$, where $S_{ij} = 1/2(\partial u_i/\partial x_j + \partial u_j/\partial x_i)$. In order for ϵ to remain constant as $\nu \to 0$, the velocity gradients must diverge $\partial u_i/\partial x_j \to \infty$. We therefore expect very large velocity gradients in the flow in order to dissipate energy. Assuming that ϵ depends on u and l only (and not on viscosity), where u is a typical velocity and l a typical length scale, we obtain the classical dimensional scaling

$$\epsilon \sim u^3/l \, . \tag{6.9}$$

Richardson's Cascade and Kolmogorov's 2/3rd Law

The cascade mechanism, which brings energy from large to small scales, was postulated by Richardson (1922) and summarized in the famous poem

> Big whirls have little whirls that feed on their velocity,
> And little whirls have lesser whirls and so on to viscosity.

The main hypothesis is that the energy flux from large to small scales takes the form of a long chain of inertial transfers. This is observed in many turbulent flows where large-scale circulations collapse into small-scale disorder (see a numerical example in Fig. 6.1). The flux of energy Π across any spatial scale r must be equal to ϵ (far from forcing and dissipation) and is equal to the kinetic energy u_r^2 at scale r divided by the turnover time r/u_r at the same scale:

$$\Pi(r) = \frac{u_r^2}{r/u_r} = \frac{u_r^3}{r} = \epsilon \quad \Rightarrow \quad u_r^2 \sim (\epsilon r)^{2/3} \, . \tag{6.10}$$

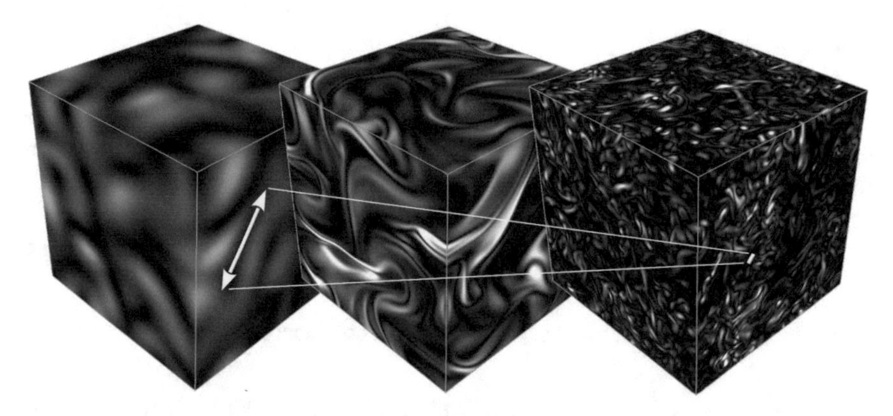

Fig. 6.1 Time evolution of the enstrophy in a standard tri-periodic direct numerical simulation of the decaying (i.e. without external forcing) Navier–Stokes equations. The initial condition contains only large-scale structures which collapse into small-scale vorticity filaments as time increases. Dark colors correspond to low enstrophy values while bright colors correspond to large enstrophy values

This is the so-called Kolmogorov's 2/3rd law, which is usually written for the longitudinal velocity structure function as

$$\left\langle (\delta u_r)^2 \right\rangle = C_K\, (\epsilon r)^{2/3} \tag{6.11}$$

where $\delta u_r(\boldsymbol{r}) = u(\boldsymbol{x}+\boldsymbol{r}) - u(\boldsymbol{x})$ and C_K is a universal constant.

Let us assume that the large scales have a velocity u and a length scale l, while the dissipative scales have a velocity u_η and a length scale η. Viscous dissipation balances inertia only at the dissipation scale so that the Reynolds number at that scale is of order unity:

$$\frac{u_\eta \eta}{\nu} \sim 1\,. \tag{6.12}$$

The dissipation rate is $\epsilon \equiv 2\nu S_{ij} S_{ij} \sim \nu (u_\eta/\eta)^2$ leading to

$$\eta \sim \left(\frac{\nu^3}{\epsilon}\right)^{1/4} \quad \text{and} \quad u_\eta \sim (\nu\epsilon)^{1/4} \tag{6.13}$$

where η is often called the Kolmogorov scale. This is the scale at which energy is dissipated and beyond which the flow becomes smooth due to dominant viscous effects. Using $\epsilon \sim u^3/l$, one can make the link between large and small scales as a function of the Reynolds number:

$$\frac{\eta}{l} \sim \left(\frac{ul}{\nu}\right)^{-3/4} = Re^{-3/4} \quad \text{and} \quad \frac{u_\eta}{u} \sim \left(\frac{ul}{\nu}\right)^{-1/4} = Re^{-1/4}\,. \tag{6.14}$$

Spectral Statistics

Turbulence dynamics is often described in Fourier space, which is a natural way of studying the energy transfers across different scales in a homogeneous context. The forward and backward Fourier transforms are

$$\hat{\boldsymbol{u}}(\boldsymbol{k}) = \frac{1}{(2\pi)^3}\int \boldsymbol{u}(\boldsymbol{x})e^{-i\boldsymbol{k}\cdot\boldsymbol{x}}\mathrm{d}\boldsymbol{x} \quad \text{and} \quad \boldsymbol{u}(\boldsymbol{x}) = \int \hat{\boldsymbol{u}}(\boldsymbol{k})e^{i\boldsymbol{k}\cdot\boldsymbol{x}}\mathrm{d}\boldsymbol{k}\,. \tag{6.15}$$

The velocity correlation tensor can be defined in both spaces as

$$\hat{R}_{ij}(\boldsymbol{k}) = \frac{1}{(2\pi)^3}\int R_{ij}(\boldsymbol{r})e^{-i\boldsymbol{k}\cdot\boldsymbol{x}}\mathrm{d}\boldsymbol{x} \quad \text{and} \quad R_{ij}(\boldsymbol{r}) = \langle u_i(\boldsymbol{x}+\boldsymbol{r})u_j(\boldsymbol{x})\rangle\,. \tag{6.16}$$

Starting from the total kinetic energy of the system, one can then define the standard isotropic energy spectrum $E(k)$ as

$$K = \frac{1}{2}\int \hat{R}_{ii}(\boldsymbol{k})\mathrm{d}\boldsymbol{k} = \int e(\boldsymbol{k})\mathrm{d}\boldsymbol{k} = \int_0^\infty \underbrace{\left[\int_0^\pi \int_0^{2\pi} e(\boldsymbol{k})k^2 \sin\theta \mathrm{d}\theta\mathrm{d}\phi\right]}_{E(k)} \mathrm{d}k, \tag{6.17}$$

where the wave vector $\boldsymbol{k}$ has been written in spherical coordinates $(k.\theta, \phi)$. For isotropic velocity fields,

$$E(k) = 2\pi k^2 \hat{R}_{ii}(k)\,. \tag{6.18}$$

Note that while this definition is valid for any homogeneous turbulent flow, it is important to remember that an angular average has been performed, so that anistropic flows, such as rotating or stratified flows, are not well described by the inherently isotropic quantity $E(k)$.

Assuming that $E(k)$ depends on ϵ and k only (and not on viscosity) leads to the celebrated Kolmogorov energy spectrum

$$E(k) \sim \epsilon^{2/3}k^{-5/3}\,. \tag{6.19}$$

While this slope is indeed observed in many experiments and simulations (Frisch 1995), it is important to remember that this quantity only carries a fraction of the information contained in the full velocity field. In particular, the phase information, fundamental to reconstruct the velocity field in physical space, is lost. As an example, we show in Fig. 6.2 two velocity fields with the same Kolmogorov-like energy spectrum. One is a solution of the Navier–Stokes equations while the second is not.

The evolution equation for the energy spectrum is given by the so-called Lin equation

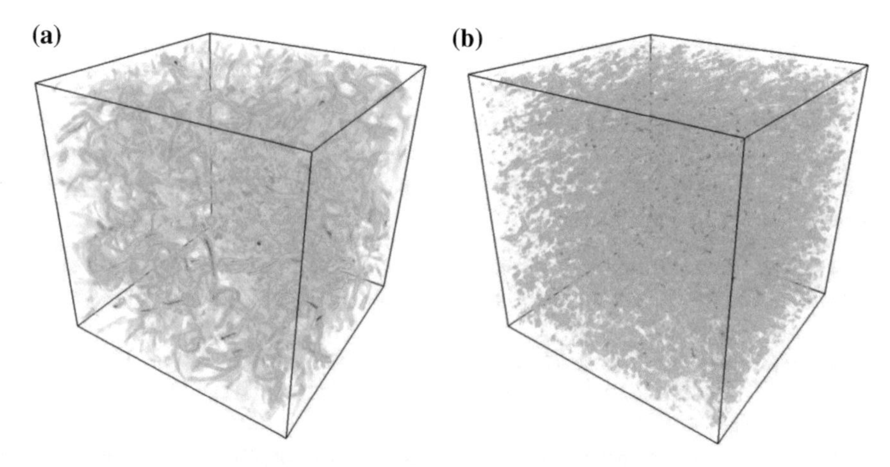

Fig. 6.2 **a** Volume rendering of the enstrophy in a standard tri-periodic direct numerical simulation of the forced Navier–Stokes equations. **b** Same as (**a**) but the flow field is not a solution of the Navier–Stokes equations. It is incompressible and has the same power spectrum $E(k)$ as the flow showed in (**a**), but the phase information has been randomized. Figures taken from Favier (2009)

$$\left(\frac{\partial}{\partial t}+\underbrace{2\nu k^2}_{\text{Dissipation}}\right)E(k,t)=\underbrace{T(k,t)}_{\text{Transfer}}+\underbrace{P(k,t)}_{\text{Production}}, \tag{6.20}$$

where $T(k)$ is the energy flux through wave number k which can be expressed in terms of third-order velocity moments (Frisch 1995; Davidson 2013). By definition, the transfer term does not contribute to the overall energy budget, so that

$$\int_0^\infty T(k)\mathrm{d}k=0\,. \tag{6.21}$$

In a stationary state, we have the balance

$$\int_0^\infty 2\nu k^2 E(k)\mathrm{d}k=\int_0^\infty P(k)\mathrm{d}k=\epsilon \tag{6.22}$$

which is the same as Eq. (6.8).

Even if Kolmogorov theory is widely used in many contexts, one should remember that it holds when the following assumptions are satisfied:

- Homogeneity (which excludes physical boundaries, local bursts of turbulence, mixing layers, interfaces, etc.).
- Isotropy (which excludes rotating and stratified flows).
- Locality (which excludes nonlocal energy transfers across separated length scales).
- $Re \to \infty$ (which is fortunately verified by geophysical flows).

2D Homogeneous Isotropic Turbulence

The second canonical model to study turbulence, and which can also be seen as a first step toward geophysical and astrophysical turbulent flows, is two-dimensional (2D) turbulence. There are many situations where a 3D velocity field becomes effectively quasi-2D, for example:

- Rotation: rapid rotation dynamically leads to flows invariant along the rotation axis (see Taylor–Proudman theorem and section 3D Homogeneous Turbulence in Rotating Fluids).
- Magnetic field: an intense externally imposed magnetic field can lead to an anisotropic dissipation and partial bi-dimensionalization (Alexakis 2011; Favier et al. 2011).
- Geometric confinement: a fluid confined to a layer of depth h can be considered 2D over horizontal length scales $L \gg h$ (Smith et al. 1996).

Governing Equations

Let us assume that the velocity field is purely 2D in the (x, y)-plane

$$\boldsymbol{u}(x, y, t) = \big(u(x, y, t), v(x, y, t), 0\big) , \tag{6.23}$$

the vorticity is then

$$\boldsymbol{\omega} = \nabla \times \boldsymbol{u} = (0, 0, \zeta) \quad \text{with} \quad \zeta = \frac{\partial v}{\partial x} - \frac{\partial u}{\partial y} . \tag{6.24}$$

Note that the vortex stretching term $\boldsymbol{\omega} \cdot \nabla \boldsymbol{u}$ in the vorticity equation, so important in 3D and at the origin of the generation of small-scale intense velocity gradients, is identically zero. This means that the vorticity is conserved along fluid trajectories (without viscosity of course), which will have dramatic consequences on the energy fluxes.

The general equation for the vorticity is

$$\frac{\partial \zeta}{\partial t} + \boldsymbol{u} \cdot \nabla \zeta = \nu \nabla^2 \zeta . \tag{6.25}$$

Introducing the stream function ψ such that

$$\boldsymbol{u} = -\nabla \times (\psi \boldsymbol{e}_z) = -\nabla \psi \times \boldsymbol{e}_z , \tag{6.26}$$

we have

$$\frac{\partial \zeta}{\partial t} + J(\psi, \zeta) = \nu \nabla^2 \zeta, \tag{6.27}$$

where

$$J(\psi,\zeta)=\frac{\partial\psi}{\partial x}\frac{\partial\zeta}{\partial y}-\frac{\partial\zeta}{\partial x}\frac{\partial\psi}{\partial y}\quad\text{and}\quad\zeta=\nabla^2\psi\,. \tag{6.28}$$

The total kinetic energy over a domain S is

$$K=\frac{1}{2}\int_S \boldsymbol{u}^2\mathrm{d}S=\frac{1}{2}\int_S(\nabla\psi)^2\,\mathrm{d}S=-\frac{1}{2}\int_S\psi\zeta\mathrm{d}S \tag{6.29}$$

and one can define a new quantity called enstrophy defined as

$$Z=\frac{1}{2}\int_S\zeta^2\mathrm{d}S=\frac{1}{2}\int_S\left(\nabla^2\psi\right)^2\mathrm{d}S\,. \tag{6.30}$$

In the absence of forcing, these two quantities evolve according to the following conservation equations:

$$\frac{dK}{dt}=-2\nu Z=-\epsilon\quad\text{and}\quad\frac{dZ}{dt}=-\nu\int_S|\nabla\zeta|^2\mathrm{d}S\,. \tag{6.31}$$

Both kinetic energy and enstrophy are therefore inviscid invariants of the 2D Navier–Stokes equations. A key observation is that the enstrophy Z is bounded by its initial value. Since $\epsilon=2\nu Z$ and Z is bounded, it is not possible for ϵ to remain constant in the limit of vanishing viscosity. The zeroth law of turbulence is not applicable, there is no dissipation anomaly in 2D turbulence! It is therefore expected that $\epsilon\to 0$ when $\nu\to 0$.

Forward Cascade of Enstrophy

The vorticity of fluid parcels is conserved in the absence of viscosity. As a uniform vorticity blob is being stretched by the underlying turbulent flow, its transverse scale is decreasing, thus increasing vorticity gradients. This leads to the conclusion that the enstrophy is cascading toward small scales at a constant flux η. The typical turnover time at a wave number k with energy $E(k)$ is

$$\tau(k)\sim\left(k^3E(k)\right)^{-1/2}, \tag{6.32}$$

so that the dissipation rate of enstrophy $Z=\int k^2E(k)\mathrm{d}k$ per unit mass is given by

$$\eta\sim\frac{k^3E(k)}{\tau(k)}\sim\left(k^3E(k)\right)^{3/2}\quad\Rightarrow\quad E(k)\sim\eta^{2/3}k^{-3}\,. \tag{6.33}$$

This prediction is valid only for scales smaller than the energy injection scale and corresponds to a direct cascade of enstrophy.

Inverse Cascade of Energy

If the enstrophy is cascading toward small scales, what is a fate of the kinetic energy, which is another invariant of the inviscid equations? Let us define the energy spectrum centroid

$$k_c = \frac{\int kE(k)\mathrm{d}k}{\int E(k)\mathrm{d}k} \tag{6.34}$$

and the weighted spectrum

$$I = \int (k-k_c)^2\, E(k)\mathrm{d}k = \underbrace{\int k^2 E(k)\mathrm{d}k}_{Z} - k_c^2 \underbrace{\int E(k)\mathrm{d}k}_{K} \,. \tag{6.35}$$

Assuming that $\frac{dI}{dt} > 0$ (i.e., the energy spreads in spectral space which is a natural assumption for this nonlinear system) and using enstrophy conservation leads to

$$\frac{dk_c^2}{dt} < 0 \,. \tag{6.36}$$

The energy spectrum centroid is decreasing with time, energy is on average transferred toward large scales, this is the so-called inverse cascade regime first postulated by Kraichnan (1967). Using the same dimensional analysis as in 3D, the inverse cascade satisfies

$$E(k) \sim \epsilon^{2/3} k^{-5/3}, \tag{6.37}$$

where ϵ is now the energy flux carried by the inverse cascade. This prediction is only valid for scales larger than the energy injection scale.

Experimental and Numerical Evidences

The theoretical concept of inverse cascade predicted by Kraichnan (1967) has been verified both experimentally and numerically. Some experiments used a thin layer of light conducting fluid on top of a heavier fluid forced electromagnetically at small scales. They observed an accumulation of energy at the scale of the container (Paret and Tabeling 1997; Chen et al. 2006). Similar conclusions were achieved using soap films (Couder et al. 1989; Gharib and Derango 1989). 2D turbulence is ideally studied numerically (see Fig. 6.3 and Smith and Yakhot (1993) for example), and the double cascade scenario where energy is flowing toward large scales while enstrophy cascades toward small scales were recently achieved in one unique simulation with very large-scale separation (Boffetta and Musacchio 2010).

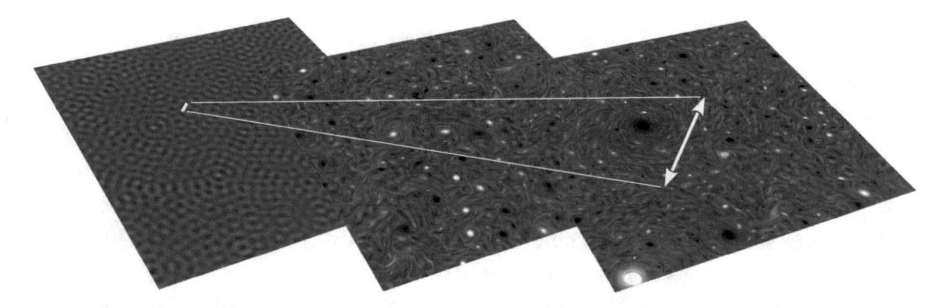

Fig. 6.3 Time evolution of the vertical vorticity in a standard doubly periodic direct numerical simulation of the forced 2D Navier–Stokes equations. The forcing is imposed at small scales only but large-scale coherent structures dynamically emerge. Bright and dark colors correspond to negative and positive values of the vorticity, respectively

Fate of the Energy at Large Scales

In the direct cascade regime, energy is transferred toward small scales, far from the nonuniversal behavior associated with the energy input at large scales. This gives some hope for the forward cascade mechanism characteristic of 3D turbulence to be universal. In the inverse cascade regime, however, energy piles up at the largest available scale so that the equilibrium state is not expected to be universal, but will depend on the details of the model considered. In the following, we give some examples of mechanisms responsible for the saturation of the inverse cascade.

In a doubly periodic domain, energy piles up at the box size generating a so-called condensate (Chertkov et al. 2007): a large-scale dipolar structure that interacts nonlocally with the cascade and modifies the $k^{-5/3}$ spectrum (Xia et al. 2008; Laurie et al. 2014).

In a closed container, no-slip boundary conditions can continuously reinject enstrophy in the system preventing the formation of large scales (Clercx et al. 2001).

If there is a linear friction term $-\lambda\zeta$ in the vorticity equation (to model the possible viscous coupling with a bottom boundary for example), an infrared cutoff is introduced at the transitional wave number $k_0 \sim \epsilon^{-1/2}\lambda^{3/2}$, where friction balances the upscale energy flux.

With additional physics (β-plane, finite-Rossby radius, etc.), the inverse cascade can compete with other mechanisms (see section Geostrophic Turbulence).

Thin-Layer Turbulence

While 2D turbulence is a necessary step to better understand energy transfers in more realistic flows, it is highly idealized and never exactly realized in nature. There have been several studies attempting to bridge the gap between 3D and 2D turbulence dynamics by considering the case of thin-layer turbulence (Smith et al. 1996; Celani et al. 2010).

Consider a thin layer of fluid contained between two stress-free horizontal plates a distance h apart. 3D turbulence is expected when the energy injection scale L_I is much smaller than the depth of the layer, $L_I \ll h$, while 2D turbulence is expected for $L_I \gg h$. In the intermediary regime, both inverse and direct energy cascades can exist, connected by a direct enstrophy cascade. Vortex stretching is negligible at large scales but feeds the direct energy cascade at small scales. This interesting split cascade scenario has been studied experimentally (Xia et al. 2009, 2011) and numerically (Benavides and Alexakis 2017).

There are other examples of inverse cascade of energy in three-dimensional systems. Rapidly rotating Rayleigh–Bénard convection is one of them: although the instability is intrinsically 3D, a spontaneous inverse cascade can develop in the so-called geostrophic turbulence regime (Julien et al. 2012; Favier et al. 2014, 2019).

Geostrophic Turbulence

The concept of homogeneous isotropic turbulence, both 2D and 3D, is very appealing from a theoretical point of view, but lacks several physical phenomena ubiquitous in geophysical and astrophysical applications. Fortunately, control parameters are often extreme in applications, leading to the derivation of reduced 2D or quasi-2D models. The purpose of this section is to give a brief overview of some of these models, and a more detailed description of the so-called β-plane turbulence.

Fundamental Concepts

There are several fundamental concepts very useful to study large-scale geophysical and astrophysical flows. Note that these are particularly useful to describe mid to high latitudes atmospheric or oceanic rotating flows in thin layers and are not necessarily relevant to other geophysical or astrophysical flows.

β-plane approximation: the lowest order effect of a planet sphericity appears only through the projection of the rotation vector on the local vertical axis

$$f = 2\boldsymbol{\Omega} \cdot \boldsymbol{e}_z \approx \underbrace{2\Omega \sin\theta_0}_{f_0} + \underbrace{\frac{2\Omega \cos\theta_0}{R_e}}_{\beta} y, \tag{6.38}$$

where Ω is the rotation rate, $\boldsymbol{e}_z$ is the local vertical axis, θ_0 is the latitude, R_e is the radius of the planet, and y is the local meridional coordinate (see Fig. 6.4). This effectively corresponds to a local linear variation of the Coriolis force along the meridional direction. The Coriolis parameter f is maximum at the pole while the β effect is negligible there.

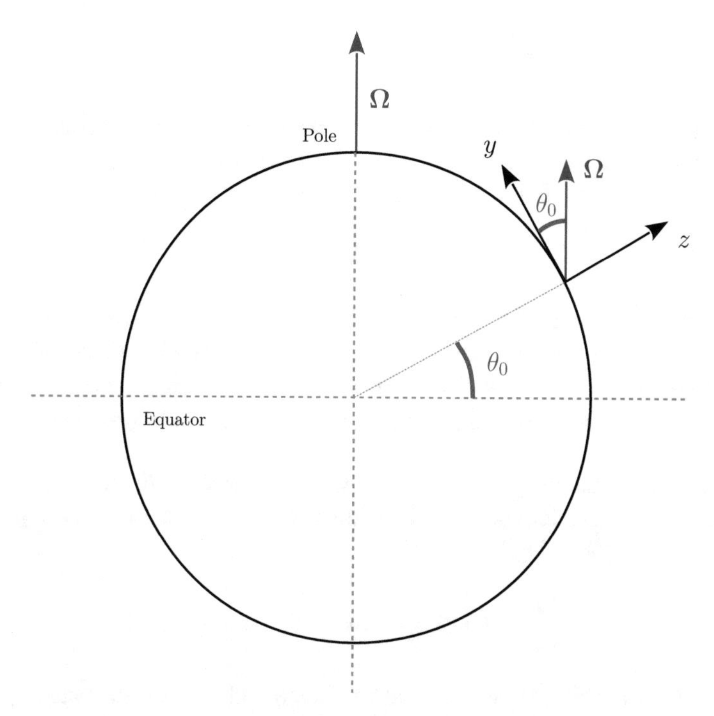

Fig. 6.4 Local β plane approximation. Only the local vertical projection of the rotation vector $\boldsymbol{\Omega}$ matters. θ_0 is the latitude

Hydrostatic balance: the vertical momentum balance reduces to

$$\partial_z P = -\rho g \; , \tag{6.39}$$

and inertial terms can be neglected. P is the pressure, ρ the density, and g is the gravitational acceleration.

Geostrophic balance: the horizontal momentum balance reduces to

$$f \boldsymbol{e}_z \times \boldsymbol{u} = -\frac{1}{\rho} \nabla_h P \; , \tag{6.40}$$

where ∇_h is the horizontal gradient operator.

Rossby radius of deformation: this is the fundamental scale at which rotation effects become as important as gravitational effects:

$$L_D = \frac{\sqrt{gh}}{f}, \tag{6.41}$$

where h is the fluid depth.

Quasi-geostrophy (QG)

In essence, quasi-geostrophic models consider perturbations from an exact geostrophic balance. Equations are obtained by a formal expansion in Rossby number $Ro = U/(fL)$, where U is a typical velocity and L its horizontal length scale:

$$\boldsymbol{u} = \boldsymbol{u}_0 + Ro\boldsymbol{u}_1 + \cdots \tag{6.42}$$

The first order gives the diagnostic geostrophic balance (6.40) while the next order gives the prognostic vorticity equation. In addition to the Rossby number being small, there are additional assumptions: the Ekman number $E = \frac{\nu}{fH^2}$, where H is a vertical length scale, must be small and $T \gg f^{-1}$, where T is a typical timescale (Pedlosky 1992; Vallis 2006).

There exists a hierarchy of QG models with an increasing degree of complexity. All of the models are derived from the vertical vorticity equation and can generally be written in the following form:

$$\frac{Dq}{Dt} = \text{Forcing} - \text{Dissipation} + O(Ro), \tag{6.43}$$

where we have introduced the potential vorticity q and D/Dt is the material derivative. In the absence of forcing or dissipative effects, q is a conserved quantity. Here is a list of some classical models, whose complete derivations can be found in classical textbooks (Pedlosky 1992; Vallis 2006):

- 2D Euler $\quad q = \nabla^2\psi,$
- β-plane (rigid lid approximation) $\quad q = \nabla^2\psi + \beta y,$
- Shallow water QG (Charney–Hasegawa–Mima) $\quad q = \nabla^2\psi - \frac{1}{L_D^2}\psi + \beta y,$
- Full shallow water QG $\quad q = \frac{\nabla^2\psi + f}{h}$, and
- Boussinesq continuous QG $\quad q = \nabla^2\psi + \frac{\partial}{\partial z}\left(\frac{f^2}{N^2}\frac{\partial\psi}{\partial z}\right).$

2D Turbulence on a β Plane

We focus in the rest of this section on the simplest QG model, the β-plane, and consider the dynamics of a 2D turbulent flow on such a plane.

Consider first a 2D horizontal flow rotating around the vertical. The vorticity equation is

$$\frac{\partial\zeta}{\partial t} + \boldsymbol{u}\cdot\nabla(\zeta + f) = \frac{D(\zeta + f)}{Dt} = \nu\nabla^2\zeta\,. \tag{6.44}$$

If $f = \text{cste}$, rotation does not affect the 2D flow. On the β-plane, however, $f = f_0 + \beta y$ as we saw previously in Eq. (6.38), and the vorticity equation is

$$\frac{\partial \zeta}{\partial t} + \boldsymbol{u} \cdot \nabla \zeta + \beta u_y = \frac{D(\zeta + \beta y)}{Dt} = \nu \nabla^2 \zeta \,, \tag{6.45}$$

where we see that the potential vorticity $q = \zeta + \beta y$ is materially conserved in the absence of viscosity. By analogy to 2D flows, the two inviscid invariants are also

$$K = -\frac{1}{2}\int_S \psi \zeta \mathrm{d}S \quad \text{and} \quad Z = \frac{1}{2}\int_S \zeta^2 \mathrm{d}S \,. \tag{6.46}$$

So far, the properties of β-plane turbulence seem very similar to standard 2D turbulence. A key difference is the existence of linear waves which were nonexistent in 2D isotropic turbulence. The vorticity equation in its linearized and inviscid form is indeed simply

$$\frac{\partial \zeta}{\partial t} = -\beta u_y \,, \tag{6.47}$$

where u_y is the velocity component in the meridional direction. Looking for plane wave solution of the form $\psi \sim \exp\left[i\left(\boldsymbol{k}\cdot\boldsymbol{x} - \omega t\right)\right]$ leads to the dispersion relation

$$\omega = -\beta \frac{k_x}{k^2} \,. \tag{6.48}$$

These Rossby waves propagate westward with a zonal phase velocity

$$c_{p,x} = \frac{\omega}{k_x} = -\frac{\beta}{k^2} \,. \tag{6.49}$$

The question is now to understand the importance of these Rossby waves in the dynamics of fully developed β-plane turbulence. We compare two simulations of β-plane turbulence in a channel in Fig. 6.5, one with $\beta = 0$ (standard 2D turbulence) and one with $\beta \neq 0$. Both are forced at small spatial scales. While we observe large-scale flows in both cases, vortices are favored in the standard 2D case while jets are sustained in the direction perpendicular to the background potential vorticity gradient for the β-plane turbulence. These large-scale jets are reminiscent of zonal jets observed on the atmosphere of Jupiter.

A balance between inertial and β terms in Eq. (6.45) gives

$$\frac{|\boldsymbol{u}\cdot\nabla\zeta|}{|\beta u_y|} \sim \frac{\zeta}{\beta L} \sim \frac{U}{\beta L^2} \sim 1 \quad \Rightarrow \quad L_{Rh} \sim \sqrt{U/\beta}, \tag{6.50}$$

where we introduced the length scale L_{Rh} called the Rhines scale (Rhines 1975). For scales much smaller than L_{Rh}, inertia dominates and the inverse cascade mechanism characteristic of classical 2D turbulence persists. At these small spatial scales, the frequency of Rossby waves is much lower than the typical turnover time of turbulent eddies and no significant interaction is expected. For scales comparable with L_{Rh}, however, β-plane dynamics and associated Rossby waves come into play. Note that

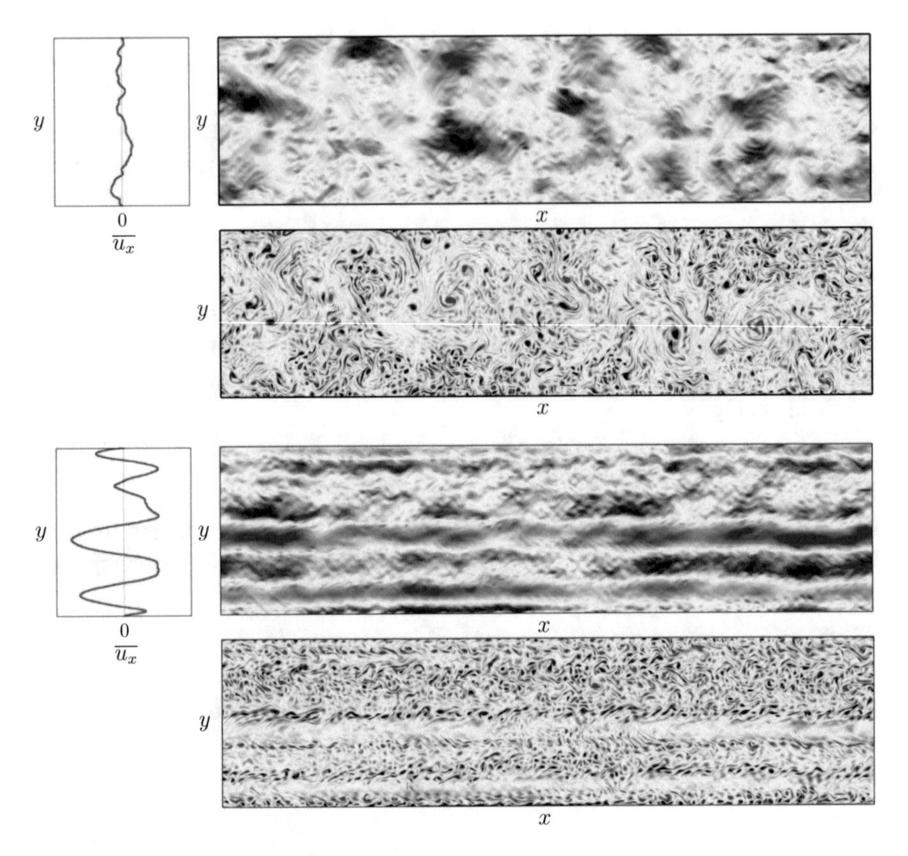

Fig. 6.5 2D turbulence forced at small scales in a periodic channel (boundaries are periodic in the horizontal direction and no-slip at the top and bottom of the channel). There is no β-effect at the top while there is one at the bottom (the gradient of potential vorticity is vertical). We show the horizontal velocity on the upper panel, the vorticity on the lower panel and the horizontally averaged flow $\overline{u_x} = 1/L_x \int u_x \mathrm{d}x$ on the left plot

we assumed that velocity and vorticity are evolving on the same length scale (see Eq. (6.50)), which is an arbitrary choice. One could equally choose another measure for the vorticity, such as $Z = \left\langle \zeta^2 \right\rangle^{1/2}$, where $<>$ denotes some averaging process, leading to another transition length scale

$$L_Z = \frac{Z}{\beta}\,. \tag{6.51}$$

Discussions about the relevant length scales for Rossby waves excitation and jet formation can be found in Sukoriansky et al. (2007) and references therein.

The formation of zonal jets in β-plane turbulence can be explained as follows. Turbulence is expected to excite Rossby waves when the wave period and eddy turnover time are matched, leading to a transition between an isotropic inverse cascade (generating isotropic large-scale vortices) to an anisotropic inverse cascade favoring zonal jets. This explanation was, for example, put forward by Vallis and Maltrud (1993). Equating the (anisotropic) frequency of a Rossby wave to the (isotropic) frequency of a turbulent eddy at wave number k leads to

$$\beta\frac{k_x}{k^2} \sim \epsilon^{1/3}k^{2/3} \quad \Rightarrow \quad k_x = k_\beta \cos^{8/5}\theta \quad \text{and} \quad k_y = k_\beta \sin\theta \cos^{3/5}\theta\,, \tag{6.52}$$

where we have introduced the polar angle $\theta = \tan^{-1}(k_y/k_x)$ and the transitional wave number

$$k_\beta = \left(\frac{\beta^3}{\epsilon}\right)^{1/5}\,. \tag{6.53}$$

Note that the Rhines scale is recovered by neglecting anisotropy and by replacing the turbulence eddy frequency $\omega \sim \epsilon^{1/3}k^{2/3}$ by the so-called sweeping frequency $\omega \sim kU$ where U is the same undetermined velocity scale as in Eq. (6.50). At scales smaller than k_β^{-1}, the usual 2D double cascade mechanism is expected to hold from the energy injection scale, with a direct cascade of enstrophy with rate η and an inverse energy cascade with rate ϵ:

$$E(k) \sim \eta^{2/3}k^{-3} \quad \text{and} \quad E(k) \sim \epsilon^{2/3}k^{-5/3}\,. \tag{6.54}$$

At scales comparable with k_β^{-1}, an anisotropic competition between Rossby waves and turbulent eddies appear as predicted by Eq. (6.52). The initially isotropic energy flux from the inverse cascade is channeled toward the plane $k_x \approx 0$ (structures invariant along the zonal axis, i.e., jets). Rhines (1975) postulated a critical balance between waves and turbulent eddies to predict the zonal flow spectrum (see also Huang et al. (2001))

$$|\boldsymbol{u}\cdot\nabla\zeta| \sim |\beta u_y| \quad \Rightarrow \quad \hat{u}(k) \sim \beta/k^2 \quad \Rightarrow \quad E_Z(k) \sim \beta^2 k^{-5}, \tag{6.55}$$

where $E_Z(k)$ is the energy spectrum of the zonally averaged flow. This scaling was observed in Jupiter atmosphere (Galperin et al. 2014) and in experiments (Cabanes et al. 2017). For more details about jets formation and dynamics, the interested reader is referred to the recent book by Galperin and Read (2019).

3D Homogeneous Turbulence in Rotating Fluids

The two previous sections discussed some properties of two-dimensional turbulence, with or without β effect. While these models are probably relevant for the largest scales of planetary flows, for which the Rossby number is very small, going down in scales gradually increases the local value of the Rossby number so that three-dimensional effects come back into play. The purpose of this section is to discuss some properties of three-dimensional rotating turbulence, especially focusing on the similarities and differences with two-dimensional turbulence.

Governing Equations

We start with the Navier–Stokes equations for an incompressible fluid in a frame rotating with an angular frequency $\boldsymbol{\Omega}$:

$$\frac{\partial \boldsymbol{u}}{\partial t} + \boldsymbol{u} \cdot \nabla \boldsymbol{u} + \underbrace{2\boldsymbol{\Omega} \times \boldsymbol{u}}_{\text{Coriolis}} = -\nabla \Pi + \nu \nabla^2 \boldsymbol{u} + \boldsymbol{F} \tag{6.56}$$

$$\nabla \cdot \boldsymbol{u} = 0, \tag{6.57}$$

where Π is the modified pressure taking into account centrifugal effects. In addition to the Reynolds number, we have introduced a second important dimensionless number, the Rossby number Ro defined as a balance between inertial and Coriolis terms

$$\frac{|\boldsymbol{u} \cdot \nabla \boldsymbol{u}|}{|\boldsymbol{\Omega} \times \boldsymbol{u}|} \sim \frac{U}{\Omega L} \equiv Ro\,. \tag{6.58}$$

Three important regimes can be discussed as follows:

- When $Ro \gg 1$, the turbulence ignores rotation and behaves like 3D homogeneous isotropic turbulence.
- When $Ro \sim 1$, the turbulence becomes anisotropic and is dynamically affected by rotation. A gradual transition to quasi-2D dynamics can be observed.
- When $Ro \ll 1$, inertial waves and quasi-geostrophic motions dominate the dynamics.

Very importantly, the length scale L introduced to define the Rossby number is left to be defined for now. In a turbulent flow with a wide range of spatial scales, we naturally expect the local Rossby number (based on an eddy size l for example) to go from very small values at large scales to very large values at small scales. Note also that quasi-geostrophic models discussed previously are effectively filtering out inertial waves. They can nevertheless have an important impact on the dynamics at low Rossby numbers.

Inertial Waves

Similarly to the case of β-plane turbulence discussed previously, linearizing the equations leads to nontrivial dynamics. Considering the linear inviscid limit of the

Navier–Stokes equations in a rotating frame, one gets the Poincaré equation (Poincaré 1885):

$$\frac{\partial^2 \nabla^2 \boldsymbol{u}}{\partial t^2} + 4\Omega^2 \frac{\partial^2 \boldsymbol{u}}{\partial z^2} = 0, \tag{6.59}$$

which, looking for plane wave solutions, leads to the dispersion relation of inertial waves:

$$\omega = \pm 2\Omega \frac{k_z}{k} = \pm 2\Omega \cos\theta, \tag{6.60}$$

where we have introduced the polar angle θ between the wave vector $\boldsymbol{k}$ and the vertical axis, k_z being the vertical component of the wave vector. Inertial waves frequency is bounded by 2Ω so that they are expected to dominate the low-frequency part of the spectrum. The dispersion relation is anisotropic and the frequency tends to zero for $k_z \to 0$. A visualization of inertial waves excited by an oscillating object in a rotating fluid can be seen in Fig. 6.6.

The role of inertial waves in the dynamics of rotating turbulence is still an active area of research. On the one hand, they are widely observed both in experiments (Yarom et al. 2013) and in numerical simulations (Favier et al. 2010). On the other hand, their role in the dynamics is still unclear. Are they passively generated by turbulent eddies or do they significantly contribute to energy transfers through resonant interactions? Note finally that it is also possible to consider the so-called wave turbulence limit, where low-amplitude inertial waves weakly interact (Galtier 2003; Bellet et al. 2006; Le Reun et al. 2017) without leading to the formation of turbulent eddies. This limit will not be discussed in the following and only finite-Rossby rotating turbulence will be considered.

Phenomenological Observations

There are several empirical observations about the behavior of homogeneous turbulence submitted to background rotation.

First, as the rotation rate is increased, the dissipation rate ϵ is reduced. This has been observed experimentally by rotating grid-generated turbulence (Jacquin et al. 1990) or by spinning a propeller in a rotating tank filled with water (Campagne et al. 2016). This reduction of dissipation can be understood by a reduction of the forward energy flux through the direct cascade.

The second empirical property of rotating turbulence is the systematic formation of columnar structures invariant along the rotation axis. This is related to the Taylor–Proudman theorem, which is valid in the limit of steady, vanishing Rossby number, and inviscid motions. In that case, the curl of the geostrophic balance shows that motions are invariant along the rotation axis. Note however that the hypotheses required to justify this result are all violated in rotating turbulence, so that explaining the dynamical formation of columnar structures in rotating turbulence solely based on the Taylor–Proudman theorem is an approximation at best.

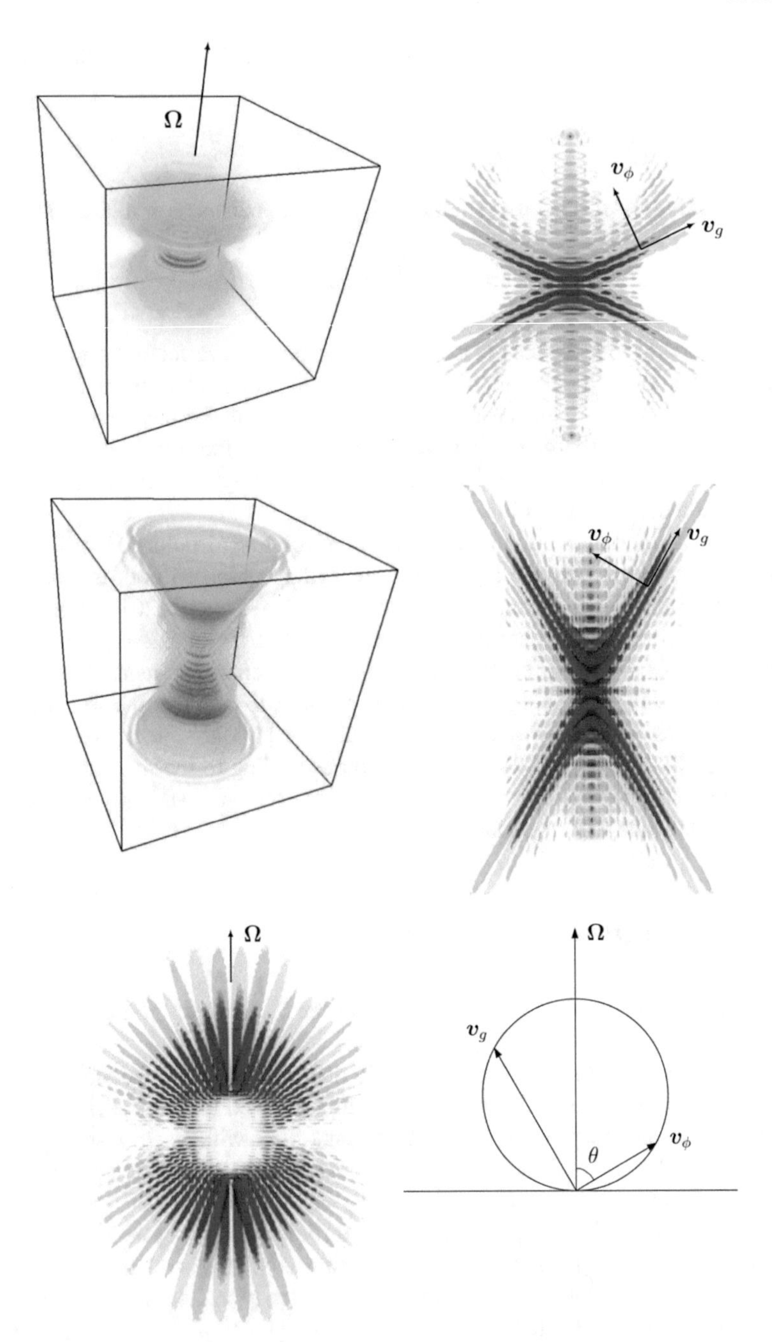

Fig. 6.6 Visualizations of the vertical velocity in a numerical simulation of inertial waves excited by an oscillating force localized at the center of the domain. The frequency is decreasing from the top panel to the middle one. At the bottom, we show the response to an impulse exciting a large band of frequencies. $\boldsymbol{v}_g$ is the group velocity while $\boldsymbol{v}_\phi$ is the phase velocity. Taken from Favier (2009)

Finally, the last important empirical observation concerns the cyclone–anticyclone asymmetry. During the process of anisotropy growth and columnar structure formation, the distribution of elongated vortices becomes asymmetric, such that cyclonic vorticity is favored. Cyclone–anticyclone asymmetry is indeed a generic feature of rotating flows, which originates from the modification of stretching and tilting of the vorticity by the Coriolis force, suggesting a more pronounced asymmetry at $Ro \sim 1$ and a restoration of symmetry for vanishing Rossby numbers. From an initially isotropic velocity field, it can be shown that, for short times (Gence and Frick 2001)

$$\frac{d}{dt}\left\langle\omega_z^3\right\rangle = \frac{4}{5}\Omega\left\langle\omega_i\omega_j S_{ij}\right\rangle, \tag{6.61}$$

which explains the asymmetric nature of the probability density function of the vertical vorticity observed in many rotating turbulent flows. From an initially symmetric distribution function (i.e., $\left\langle\omega_z^3\right\rangle \approx 0$), rotation breaks the cyclone–anticyclone symmetry. In addition, it is known that cyclonic vortices are more robust against centrifugal and elliptic instabilities than anti-cyclonic vortices (Sipp et al. 1999).

Anisotropic Energy Transfers

A very important observation is that the Coriolis force does no work ($\boldsymbol{u} \cdot (\boldsymbol{\Omega} \times \boldsymbol{u}) = 0$) so that the kinetic energy equation (which is a second order quantity) is unchanged in the rotating frame. Rotation does affect the kinetic energy transfers (which is a third-order quantity) by reducing the forward cascade, reducing dissipation, and generating anisotropic energy transfers (Cambon et al. 1997). The Coriolis force being anisotropic, energy transfers along the rotation axis are inevitably different from those perpendicular to it. The concept of isotropization of the turbulence where all statistical quantities can be reduced to isotropic tensors is not applicable anymore. In Fourier space, this means that statistical quantities do not only depend on the amplitude k of the wave vector $\boldsymbol{k}$, but also on its orientation. The angular energy spectrum $E(k,\theta)$ can be defined by

$$K = \frac{1}{2}\int \hat{R}_{ii}(\boldsymbol{k})\mathrm{d}\boldsymbol{k} = \int e(\boldsymbol{k})\mathrm{d}\boldsymbol{k} = \int_0^\infty \int_0^\pi \underbrace{\int_0^{2\pi} e(\boldsymbol{k})k^2 \sin\theta \mathrm{d}\phi \, \mathrm{d}\theta \mathrm{d}k}_{E(k,\theta)}, \tag{6.62}$$

where the dependence on the polar angle θ is conserved, contrary to standard isotropic spectral statistics. Rotating turbulence is typically characterized by

$$E(k, \theta \approx 0) \ll E(k, \theta \approx \pi/2)$$

where $\theta = 0$ corresponds to purely vertical wave vectors (associated with spatial structures nearly invariant along the horizontal directions), whereas $\theta = \pi/2$ corresponds to horizontal wave vectors (associated with spatial structures invariant along the vertical direction). Energy is not only transferred across wave numbers but also across angles from the pole to the equator. Note that as $\theta \to \pi/2$, $k_z \to 0$, and $\partial_z \to 0$, which is of course consistent with the formation of columnar structures discussed earlier. The question remains at which spatial scales are these columnar anisotropic structures expected?

Using the usual scaling $\epsilon \sim u^3/l$, one can build a typical length scale L_Z where the local Rossby number is around unity (Zeman 1994):

$$Ro(L_Z) = \frac{(L_Z \epsilon)^{1/3}}{\Omega L_Z} \sim 1 \quad \Rightarrow \quad L_Z = \left(\frac{\epsilon}{\Omega^3}\right)^{1/2} . \tag{6.63}$$

Scales larger than L_Z have a low Rossby number and are dynamically constrained by rotation. Scales smaller than L_Z have a large Rossby number and should recover an isotropic behavior. This transitional length scale separating anisotropic and isotropic dynamics has been observed in simulations (Delache et al. 2014).

Inverse Cascade

Since rotating turbulence naturally tends to transfer energy toward the $k_z = 0$ plane, it is then natural to focus on the dynamics of the $k_z = 0$ modes. Remember that modes with $k_z = 0$ are steady non-propagative motions when considering the dispersion relation of inertial waves (6.60). Starting from the Navier–Stokes equations without rotation but assuming that $\partial/\partial_z = 0$ leads to

$$\frac{\partial \boldsymbol{u}_h}{\partial t} + \boldsymbol{u}_h \cdot \nabla \boldsymbol{u}_h = -\nabla_h p + \nu \nabla_h^2 \boldsymbol{u}_h \tag{6.64}$$

$$\frac{\partial w}{\partial t} + \boldsymbol{u}_h \cdot \nabla_h w = \nu \nabla_h^2 w, \tag{6.65}$$

where $\boldsymbol{u}_h$ and w are the horizontal and vertical velocity components, respectively, and ∇_h is the horizontal gradient operator. The vertical component is passively advected by the horizontal flow while the horizontal flow satisfies the standard 2D Navier–Stokes equation. This state is often referred to as 2D-3C turbulence for two-dimensional three-components turbulence. We therefore expect that, in the limit where the flow becomes vertically invariant, an inverse cascade can develop in analogy with truly 2D turbulence. Of course, rotating turbulence is not exactly invariant along the rotation axis (see Gallet (2015) though), so that a transition is expected as the Rossby number is varied. An inverse cascade of the depth-invariant geostrophic modes is indeed observed in numerical simulations (Pouquet et al. 2013; Yokoyama and Takaoka 2017) and experiments (Campagne et al. 2014) when the Rossby number is low enough.

3D Homogeneous Turbulence in Stratified Fluids

This section is devoted to the case of fully three-dimensional turbulence in a linearly stratified fluid. As usual, we consider a homogeneous situation without boundaries and there is an external force injecting energy in the system.

Equations

Using the Boussinesq approximation, we consider the equations of motions

$$\frac{\partial \boldsymbol{u}}{\partial t} + \boldsymbol{u} \cdot \nabla \boldsymbol{u} = -\frac{1}{\rho_0} \nabla p + \nu \nabla^2 \boldsymbol{u} - \frac{\rho}{\rho_0} g \boldsymbol{e}_z \tag{6.66}$$

$$\nabla \cdot \boldsymbol{u} = 0 \tag{6.67}$$

$$\frac{\partial \rho}{\partial t} + \boldsymbol{u} \cdot \nabla \rho = \kappa \nabla^2 \rho + u_z \frac{\rho_0 N^2}{g}, \tag{6.68}$$

where ρ_0 is mean density, κ is the diffusivity of the stratifying agent, ν is the viscosity, ρ is the density perturbation around the equilibrium profile $\overline{\rho}(z) = \rho_0 \left(1 - N^2 z/g\right)$, and $N = \sqrt{-\frac{g}{\rho_0} \frac{\partial \overline{\rho}}{\partial z}}$ is the Brunt–Väisälä frequency.

There are several important dimensionless numbers in this problem:

- The Reynolds number $Re = \frac{UL}{\nu}$,
- The Froude number $Fr = \frac{U}{NL}$,
- The Richardson number $Ri = \frac{N^2}{|\partial \boldsymbol{u}_h / \partial z|^2}$,
- The Schmidt number $Sc = \frac{\nu}{\kappa}$,

where we have introduce a typical velocity scale U and a typical length scale L.

These equations can be made dimensionless using various approaches. Scaling velocity with U, length with L, time with N^{-1}, and buoyancy $b \equiv \rho g/\rho_0$ with UN gives the following set of dimensionless equations (Lilly 1983):

$$\frac{\partial \boldsymbol{u}}{\partial t} + Fr\, \boldsymbol{u} \cdot \nabla \boldsymbol{u} = -\nabla p + \frac{Fr}{Re} \nabla^2 \boldsymbol{u} + b \boldsymbol{e}_z \tag{6.69}$$

$$\frac{\partial b}{\partial t} + Fr\, \boldsymbol{u} \cdot \nabla b = -u_z + \frac{Fr}{Re\, Sc} \nabla^2 b \tag{6.70}$$

$$\nabla \cdot \boldsymbol{u} = 0. \tag{6.71}$$

In the low Froude and ideal (i.e., neglecting viscous and diffusive terms) limit, one recovers the internal wave equation

$$\frac{\partial}{\partial t} \nabla^2 \phi + N^2 \nabla_h^2 \phi = 0, \tag{6.72}$$

where ∇_h is the horizontal gradient operator and ϕ is a velocity potential function. Internal waves correspond to the linear oscillatory response of a linearly stratified fluid and share several properties with inertial waves (dispersive, orthogonality between group and phase velocities, bounded frequencies). Similarly to rotating turbulence, stratified turbulence is characterized by a superposition and interaction between turbulent eddies and linear or weakly nonlinear waves.

Let us consider another way of scaling the equations. Scaling horizontal velocities with U, length with L, time with L/U, b with U^2/L, and vertical velocity with $U^3/(N^2L^2)$ (this particular scaling is obtained by a balance between $\partial_t b$ and $N^2 u_z$ in the buoyancy equation) gives

$$\frac{\partial \boldsymbol{u}_h}{\partial t} + \boldsymbol{u}_h \cdot \nabla_h \boldsymbol{u}_h + Fr^2 u_z \partial_z \boldsymbol{u}_h = -\nabla_h p + \frac{1}{Re}\nabla^2 \boldsymbol{u}_h \tag{6.73}$$

$$Fr^2 \left(\frac{\partial u_z}{\partial t} + \boldsymbol{u} \cdot \nabla u_z \right) = -\partial_z p + b + \frac{1}{Re}\nabla^2 u_z \tag{6.74}$$

$$\frac{\partial b}{\partial t} + \boldsymbol{u}_h \cdot \nabla b + Fr^2 u_z \partial_z b = -u_z + \frac{1}{Re\ Sc}\nabla^2 b \tag{6.75}$$

$$\nabla_h \cdot \boldsymbol{u}_h + Fr^2 \partial_z u_z = 0. \tag{6.76}$$

In the low Froude inviscid limit, one gets quasi-2D layered motions

$$\frac{\partial \nabla^2 \psi}{\partial t} + J(\psi, \nabla_h^2 \psi) = 0, \tag{6.77}$$

where $\psi(x, y, z, t)$ is the streamfunction. This equation is similar to the standard 2D Euler equation discussed previously in the context of 2D turbulence, except that the stream function depends on all three spatial coordinates.

These two particular scalings illustrate that strongly stratified flows can support two types of motions: three-dimensional internal waves and quasi-2D layered motions. While this could suggest that stratified turbulence and 2D turbulence share some similarities, there is an obvious limitation to the previous scaling. There is no information about the vertical correlation length between each layer. Are they decoupled or is there some vertical correlation in a turbulent stratified flow? Note also that this is reminiscent of rotating turbulence and the duality between inertial waves and geostrophic vortices.

The Zig-Zag Instability and the Buoyancy Scale

The Zig-Zag instability is characteristic of an initially vertically invariant vortex pair moving in a linearly stratified fluid (Billant and Chomaz 2000a). The vortex pair breaks into several layers with a well-defined vertical thickness. The most unstable vertical wavelength of the Zig-Zag instability is found to be Billant and Chomaz (2000b):

$$\lambda_z \sim U/N, \tag{6.78}$$

where U is the vortex pair traveling velocity and N is the Brunt–Väisälä frequency. Billant and Chomaz (2001) also showed that the inviscid governing equations in the limit $Fr \to 0$ are self-similar with respect to the variable zN/U. Vortex structures with a vertical size scaling like U/N have been also reported in stratified Taylor–Couette flows in the strongly stratified regime (Boubnov et al. 1995). It was therefore suggested that the intrinsic vertical scale, when no vertical length scales are imposed by initial or boundary conditions and when the fluid is strongly stratified, is given by

$$L_B = U/N\ , \tag{6.79}$$

the so-called buoyancy scale.

It is natural to focus on the dominant horizontal flow $\boldsymbol{u}_h = \left(u_x, u_y, 0\right)$ and on its correlation length scales

$$l_h = \frac{1}{\overline{\boldsymbol{u}_h^2}} \int \overline{\boldsymbol{u}_h(\boldsymbol{x})\cdot\boldsymbol{u}_h(\boldsymbol{x}+r\boldsymbol{e}_h)}\mathrm{d}r \quad \text{and} \quad l_v = \frac{1}{\overline{\boldsymbol{u}_h^2}} \int \overline{\boldsymbol{u}_h(\boldsymbol{x})\cdot\boldsymbol{u}_h(\boldsymbol{x}+r\boldsymbol{e}_z)}\mathrm{d}r\ , \tag{6.80}$$

where the overbar represents an averaging process, $\boldsymbol{e}_h$ and $\boldsymbol{e}_z$ are unit vectors in the horizontal and vertical directions, respectively. One can then define horizontal and vertical Froude numbers:

$$Fr_h = \frac{u_h}{Nl_h} \quad \text{and} \quad Fr_v = \frac{u_h}{Nl_v} \quad \text{where} \quad u_h = \sqrt{\overline{\boldsymbol{u}_h^2}}\ . \tag{6.81}$$

The incompressibility constraint leads to $u_v/l_v \sim u_h/l_h$ where u_v is a typical vertical velocity. In the strong stratification regime, one might expect $u_v \ll u_h$ so that $l_v \ll l_h$. The empirical observation that $l_v \sim L_B = U/N$ leads to $Fr_v \sim 1$ and not $Fr_v \to 0$ as assumed in the 2D layered turbulence characterized by Eq. (6.77).

The Buoyancy Reynolds Number

Strongly stratified turbulence is characterized by $Fr_h \ll 1$ and it organizes itself such that $Fr_v \sim 1$ or equivalently $l_v \sim U/N$ (Billant and Chomaz 2001; Riley and deBruynKops 2003). Using this scaling, the ratio between the horizontal inertial forces and the vertical viscous forces is

$$\frac{u_h^2/l_h}{\nu u_h/l_v^2} = \frac{u_h l_h}{\nu}\left(\frac{u_h}{Nl_h}\right)^2 = Re_h Fr_h^2 \equiv \mathcal{R}\ , \tag{6.82}$$

where we have introduced the buoyancy Reynolds number $\mathcal{R}$. For the horizontal motions to be viscously decoupled in the vertical direction, one requires $\mathcal{R} \gg 1$ or equivalently $Re_h \gg Fr_h^{-2} \gg 1$. This double limit is very difficult to achieve numerically and a transition between viscously locked layers when $\mathcal{R} \ll 1$ and strong stratified turbulence when $\mathcal{R} \gg 1$ has only been achieved recently (Brethouwer et al. 2007; Waite 2011).

From numerical simulations, it is observed that $\epsilon \sim u_h^3/l_h$. Following Kolmogorov's approach this leads to (Lindborg 2006):

$$E_h \sim \epsilon^{2/3} k_h^{-5/3} ,$$

where E_h is the horizontal energy spectrum and k_h is the horizontal wave number. It is still not clear if this scaling is indeed observed in the regime of large buoyancy Reynolds number.

Using the usual scaling $\epsilon \sim u^3/l$, one can build a typical length scale L_0, the so-called Ozmidov scale (Ozmidov 1965), where the local Froude number is around unity

$$Fr(L_0) = \frac{(L_0\epsilon)^{1/3}}{NL_0} \sim 1 \quad \Rightarrow \quad L_0 = \left(\frac{\epsilon}{N^3}\right)^{1/2} . \tag{6.83}$$

Scales larger than L_0 have a low Froude number and are dynamically constrained by stratification. Scales smaller than L_0 have a large Froude number and should recover an isotropic behavior. This is similar to the Zeman scale of rotating turbulence.

- In the ocean, $L_0 \sim 1$ m.
- In strongly stable atmosphere boundary layers $L_0 \sim 1$ m.
- In the upper troposphere or lower stratosphere, $L_0 \sim 10$ m.

Conclusion

These notes are at best a very broad introduction to the many subtleties of turbulence in rapidly rotating and stratified fluids. Many fundamental aspects have been neglected to focus on simpler physical processes or qualitative descriptions. The interested reader can read more complete classical textbooks cited throughout this review, in particular Davidson (2013).

To bridge the gap between this introduction and more applied issues relevant to the fluid dynamics of planets and stars, several additional effects have to be taken into account. Compressibility is one of them, since density can change by many orders of magnitude in stellar interiors or planetary atmospheres. Some properties of compressible turbulence, which are also relevant to engineering applications such as turbo-machinery, are, for example, discussed in Sagaut and Cambon (2008). Astrophysical fluids are often electrically conducting and therefore interact dynamically with the magnetic field. An introduction to magnetohydrodynamics is given in the chapter by Deguen and Lasbleis of the same book. The topic of magnetohydrodynamical turbulence is vast since it applies to many geophysical and astrophysical flows from the Earth's liquid metal core to the ionized gases of stellar interiors. Finally, coupling all these effects together remain a tremendous task involving many dimensionless numbers and many different regimes depending on the spatial scales considered (Alexakis and Biferale 2018).

References

Alexakis, A. (2011). Two-dimensional behavior of three-dimensional magnetohydrodynamic flow with a strong guiding field. *Physical Review E, 84*, 056330.

Alexakis, A., & Biferale, L. (2018). Cascades and transitions in turbulent flows. *Physics Reports*,.

Batchelor, G. K. (1953). *The Theory of Homogeneous Turbulence*. Cambridge Science Classics: Cambridge University Press. ISBN 9780521041171.

Bellet, F., Godeferd, F. S., Scott, J. F., & Cambon, C. (2006). Wave turbulence in rapidly rotating flows. *Journal of Fluid Mechanics, 562*, 83–121.

Benavides, S. J., & Alexakis, A. (2017). Critical transitions in thin layer turbulence. *Journal of Fluid Mechanics, 822*, 364–385.

Billant, P., & Chomaz, J.-M. (2000a). Experimental evidence for a new instability of a vertical columnar vortex pair in a strongly stratified fluid. *Journal of Fluid Mechanics, 418*, 167–188.

Billant, P., & Chomaz, J.-M. (2000b). Three-dimensional stability of a vertical columnar vortex pair in a stratified fluid. *Journal of Fluid Mechanics, 419*, 65–91.

Billant, P., & Chomaz, J.-M. (2001). Self-similarity of strongly stratified inviscid flows. *Physics of Fluids, 13*(6), 1645–1651.

Boffetta, G., & Musacchio, S. (2010). Evidence for the double cascade scenario in two-dimensional turbulence. *Physical Review E, 82*, 016307.

Boubnov, B. M., Gledzer, E. B., & Hopfinger, E. J. (1995). Stratified circular couette flow: Instability and flow regimes. *Journal of Fluid Mechanics, 292*, 333–358.

Brethouwer, G., Billant, P., Lindborg, E., & Chomaz, J.-M. (2007). Scaling analysis and simulation of strongly stratified turbulent flows. *Journal of Fluid Mechanics, 585*, 343–368.

Cabanes, S., Aurnou, J., Favier, B., & Le Bars, M. (2017). A laboratory model for deep-seated jets on the gas giants. *Nature Physics, 13*(4), 387.

Cambon, C., Mansour, N. N., & Godeferd, F. S. (1997). Energy transfer in rotating turbulence. *Journal of Fluid Mechanics, 337*, 303–332.

Campagne, A., Gallet, B., Moisy, F., & Cortet, P.-P. (2014). Direct and inverse energy cascades in a forced rotating turbulence experiment. *Physics of Fluids, 26*(12), 125112.

Campagne, A., Machicoane, N., Gallet, B., Cortet, P.-P., & Moisy, F. (2016). Turbulent drag in a rotating frame. *Journal of Fluid Mechanics, 794*, R5.

Celani, A., Musacchio, S., & Vincenzi, D. (2010). Turbulence in more than two and less than three dimensions. *Physical Review Letters, 104*, 184506.

Chen, S., Ecke, R. E., Eyink, G. L., Rivera, M., Wan, M., & Xiao, Z. (2006). Physical mechanism of the two-dimensional inverse energy cascade. *Physical Review Letters, 96*, 084502.

Chertkov, M., Connaughton, C., Kolokolov, I., & Lebedev, V. (2007). Dynamics of energy condensation in two-dimensional turbulence. *Physical Review Letters, 99*, 084501.

Clarke, C., Carswell, B., & Carswell, R. F. (2007). *Principles of astrophysical fluid dynamics*. Cambridge: Cambridge University Press.

Clercx, H. J. H., Nielsen, A. H., Torres, D. J., & Coutsias, E. A. (2001). Two-dimensional turbulence in square and circular domains with no-slip walls. *European Journal of Mechanics - B/Fluids, 20*(4), 557–576.

Couder, Y., Chomaz, J. M., & Rabaud, M. (1989). On the hydrodynamics of soap films. *Physica D: Nonlinear Phenomena, 37*(1), 384–405.

Davidson, P. A. (2013). *Turbulence in rotating, stratified and electrically conducting fluids*. Cambridge: Cambridge University Press.
Delache, A., Cambon, C., & Godeferd, F. S. (2014). Scale by scale anisotropy in freely decaying rotating turbulence. *Physics of Fluids*, *26*(2), 025104.
Favier, B. (2009). *Modélisation et simulations en turbulence homogène anisotrope: effets de rotation et magnétohydrodynamique*. Ph.D. thesis, Ecole Centrale de Lyon.
Favier, B., Godeferd, F. S., & Cambon, C. (2010). On space and time correlations of isotropic and rotating turbulence. *Physics of Fluids*, *22*(1), 015101.
Favier, B., Godeferd, F. S., Cambon, C., Delache, A., & Bos, W. J. T. (2011). Quasi-static magnetohydrodynamic turbulence at high reynolds number. *Journal of Fluid Mechanics*, *681*, 434–461.
Favier, B., Silvers, L. J., & Proctor, M. R. E. (2014). Inverse cascade and symmetry breaking in rapidly rotating Boussinesq convection. *Physics of Fluids*, *26*, 096605.
Favier, B., Guervilly, C., & Knobloch, E. (2019). Subcritical turbulent condensate in rapidly rotating Rayleigh-Bénard convection. *Journal of Fluid Mechanics*, *864*, R1.
Frisch, U. (1995). *Turbulence: the legacy of AN Kolmogorov*. Cambridge: Cambridge University Press.
Gallet, B. (2015). Exact two-dimensionalization of rapidly rotating large-reynolds-number flows. *Journal of Fluid Mechanics*, *783*, 412–447.
Galperin, B., & Read, P. L. (2019). *Zonal jets: Phenomenology, genesis, and physics*. Cambridge: Cambridge University Press.
Galperin, R. M. B., Young, R. M., Sukoriansky, S., Dikovskaya, N., Read, P. L., Lancaster, A. J., et al. (2014). Cassini observations reveal a regime of zonostrophic macroturbulence on jupiter. *Icarus*, *229*, 295–320.
Galtier, S. (2003). Weak inertial-wave turbulence theory. *Physical Review E*, *68*, 015301.
Gence, J. N., & Frick, C. (2001). Birth of the triple correlations of vorticity in an homogenous turbulence submitted to a solid body rotation. *Comptes Rendus de l'Académie des Sciences*, *329*(5), 351–356.
Gharib, M., & Derango, P. (1989). A liquid film (soap film) tunnel to study two-dimensional laminar and turbulent shear flows. *Physica D: Nonlinear Phenomena*, *37*(1), 406–416.
Huang, H.-P., Galperin, B., & Sukoriansky, S. (2001). Anisotropic spectra in two-dimensional turbulence on the surface of a rotating sphere. *Physics of Fluids*, *13*(1), 225–240.
Jacquin, L., Leuchter, O., Cambon, C., & Mathieu, J. (1990). Homogeneous turbulence in the presence of rotation. *Journal of Fluid Mechanics*, *220*, 1–52.
Julien, K., Rubio, A. M., Grooms, I., & Knobloch, E. (2012). Statistical and physical balances in low Rossby number Rayleigh-Bénard convection. *Geophysical and Astrophysical Fluid Dynamics*, *106*, 392–428.
Kolmogorov, A. N. (1941). The local structure of turbulence in incompressible viscous fluid for very large reynolds numbers. *Comptes rendus de l'Académie des sciences de l'URSS*, *30*, 301–305.
Kraichnan, R. H. (1967). Inertial ranges in 2D turbulence. *Physics of Fluids*, *10*, 1417–1423.
Laurie, J., Boffetta, G., Falkovich, G., Kolokolov, I., & Lebedev, V. (2014). Universal profile of the vortex condensate in two-dimensional turbulence. *Physical Review Letters*, *113*, 254503.
Le Reun, T., Favier, B., Barker, A. J., & Le Bars, M. (2017). Inertial wave turbulence driven by elliptical instability. *Physical Review Letters*, *119*, 034502.
Lilly, D. K. (1983). Stratified turbulence and the mesoscale variability of the atmosphere. *Journal of the Atmospheric Sciences*, *40*(3), 749–761.
Lindborg, E. (2006). The energy cascade in a strongly stratified fluid. *Journal of Fluid Mechanics*, *550*, 207–242.
Ozmidov, R. V. (1965). On the turbulent exchange in a stably stratified ocean. *Izvestiya - Academy of Sciences USSR, Atmospheric and oceanic physics*, *1*, 493–497.
Paret, J., & Tabeling, P. (1997). Experimental observation of the two-dimensional inverse energy cascade. *Physical Review Letters*, *79*, 4162–4165.
Pedlosky, J. (1992). *Geophysical fluid dynamics. Springer study edition*. New York: Springer.

Poincaré, H. (1885). Sur l'équilibre d'une masse fluide animée d'un mouvement de rotation. *Acta mathematica*, *7*(1), 259–380.
Pouquet, A., Sen, A., Rosenberg, D., Mininni, P. D., & Baerenzung, J. (2013). Inverse cascades in turbulence and the case of rotating flows. *Physica Scripta*, *T155*, 014032.
Rhines, P. B. (1975). Waves and turbulence on a beta-plane. *Journal of Fluid Mechanics*, *69*(3), 417–443.
Richardson, L. F. (1922). *Weather prediction by numerical process*.
Riley, J. J., & deBruynKops, S. M. (2003). Dynamics of turbulence strongly influenced by buoyancy. *Physics of Fluids*, *15*(7), 2047–2059.
Sagaut, P., & Cambon, C. (2008). *Homogeneous turbulence dynamics*. Cambridge: Cambridge University Press.
Sipp, D., Lauga, E., & Jacquin, L. (1999). Vortices in rotating systems: Centrifugal, elliptic and hyperbolic type instabilities. *Physics of Fluids*, *11*(12), 3716–3728.
Smith, L. M., Chasnov, J. R., & Waleffe, F. (1996). Crossover from two- to three-dimensional turbulence. *Physical Review Letters*, *77*, 2467–2470.
Smith, L. M., & Yakhot, V. (1993). Bose condensation and small-scale structure generation in a random force driven 2d turbulence. *Physical Review Letters*, *71*, 352–355.
Sukoriansky, S., Dikovskaya, N., & Galperin, B. (2007). On the arrest of inverse energy cascade and the rhines scale. *Journal of the Atmospheric Sciences*, *64*(9), 3312–3327.
Vallis, G. K. (2006). *Atmospheric and oceanic fluid dynamics*. UK: Cambridge University Press.
Vallis, G. K., & Maltrud, M. E. (1993). Generation of mean flows and jets on a beta plane and over topography. *Journal of Physical Oceanography*, *23*(7), 1346–1362.
Waite, M. L. (2011). Stratified turbulence at the buoyancy scale. *Physics of Fluids*, *23*(6), 066602.
Xia, H., Punzmann, H., Falkovich, G., & Shats, M. G. (2008). Turbulence-condensate interaction in two dimensions. *Physical Review Letters*, *101*, 194504.
Xia, H., Shats, M., & Falkovich, G. (2009). Spectrally condensed turbulence in thin layers. *Physics of Fluids*, *21*, 125101.
Xia, H., Byrne, D., Falkovich, G., & Shats, M. (2011). Upscale energy transfer in thick turbulent fluid layers. *Nature Physics*, *7*, 321–324.
Yarom, E., Vardi, Y., & Sharon, E. (2013). Experimental quantification of inverse energy cascade in deep rotating turbulence. *Physics of Fluids*, *25*(8), 085105.
Yokoyama, N., & Takaoka, M. (2017). Hysteretic transitions between quasi-two-dimensional flow and three-dimensional flow in forced rotating turbulence. *Phys. Rev. Fluids*, *2*, 092602.
Zeman, O. (1994). A note on the spectra and decay of rotating homogeneous turbulence. *Physics of Fluids*, *6*(10), 3221–3223.

Printed in the United States
By Bookmasters